M. A. Ali · M. Anctil

Retinas of Fishes

An Atlas

With 364 Figures in 125 Plates

Springer-Verlag
Berlin Heidelberg New York 1976

Professor M.-Ather Ali
Dr. Michel Anctil
Département de Biologie, Université de Montréal, Canada

ISBN 3-540-07840-1 Springer-Verlag Berlin Heidelberg New York
ISBN 0-387-07840-1 Springer-Verlag New York Heidelberg Berlin

Library of Congress Cataloging in Publication Data. Ali, Mohamed Ather, 1932– . Retinas of fishes. Bibliography: p. Includes index. 1. Retina-Atlases. 2. Fishes-Anatomy-Atlases. I. Anctil, Michel, 1945– joint author. II. Title. QL639.A427 597′.04′8 76-22204

Printed in Germany.

Typesetting, printing, and bookbinding by Universitätsdruckerei H. Stürtz AG, Würzburg

Preface

A considerable amount of information on the retinal morphology in fishes has been accumulating during the past century. Among the vertebrates, fishes are a highly successful group, both in number of species and in the adaptive radiation of forms. For instance, 415 teleost families are now recognised (Greenwood, Rosen, Weitzmann and Myers, 1966), and the 20,000 odd fish species mentioned in text-books have been by far out numbered. The fish retina also shows considerable variations, in conformity with the extreme morphological diversification reached by piscine forms, in colonising all conceivable aquatic habitats and developing a wide spectrum of life habits.

We intend to illustrate this in the present Atlas, a collection of short texts and photomicrographs of the retina from about one hundred fish families. This Atlas is intended also to fulfil other purposes. One of them is to present in a phylogenetic order the rather scattered data on fish retinal structure, with appropriate illustrative material; another is to assist the visual physiologist or biochemist in his search for a retina with particular morphological features compatible with his specific requirements. In other words, what we aim at is a ready pool of information for laymen, students, and specialists of varied interests.

The material used for this Atlas comes from various sources. Fishes from about 70% of the families represented herein were collected and their eyes processed by us in the last twenty-two years. Materials from fish made specially for the Atlas more recently, and from continuing projects, have been incorporated. The remaining fish families which compose the Atlas come from materials (fish specimens, histological slides, photomicrographs) kindly made available to us by many colleagues around the world (see Acknowledgements).

The atlas is organised in three principal parts. The first part introduces the classical morphology of the fish retina and defines the terminology used in the characterisation of the retinas of each fish family. It also includes a brief list of the main histological techniques applied to the study of retinal morphology and a chapter on the preparation and examination of the retina. The second part constitutes the bulk of the Atlas. It is conceived as a catalogue of fish retinal structure, arranged in families according to the modern phylogenetic classification (Bigelow and Schroeder, 1948; Greenwood et al., 1966; McAllister, 1968; Rosen and Patterson, 1969). The mode of presentation is as follows. A textual page faces the corresponding photographic plate on the right. The text includes (1) the phylogenetic position of the fish group (Class or Subclass, Superorder, Series, Order, Suborder, Family), (2) a line diagram and a paragraph briefly describing the fish family (taxonomic position, habitat and habits), (3) a paragraph describing the most striking features of the retina of one or more species of that family, (4) a collection of references concerning the retinal structure of fishes belonging to that family, (5) when available, a list of the categories (vitamins A_1, A_2) and maximal spectral absorptions (λ_{max}) of the visual pigments investigated in the members of that family, and (7) the legend for the facing photographic plate (s). A short text arranged according to the above scheme (except for the legend) is also included for fish families whose retinal structure has been described, but for which no photographic material is available. As a whole, the Atlas contains about 150 fish families, some 100 with photographic material. The third part is an exhaustive, though not complete, bibliography on fish retinal structure; it includes, of course, all the references mentioned in the text. Since many readers may not be familiar with the modern classification of fishes (especially the teleosts), a synopsis of families is also given and the index includes all the taxonomic terms used, including the generic, specific and common names.

Acknowledgements. We are grateful to the following colleagues for providing material in the form of fixed eyes, paraffin or plastic blocks, histological slides, or photomicrographs:

Dr. D.D. Beatty (Hiodontidae)
Dr. Geneviève Chèze (Anguillulidae)
Dr. P.W. Gilbert (Carcharinidae)
Drs. D.E. Hoss and D. Peters (Batrachoididae)
Dr. H.-J. Lang (Poecilidae)
Mr. V. Legendre and the Abitibi Division of the Ministère de la Chasse, du Tourisme et de la Pêche, Québec (Hiodontidae)
Dr. N.A. Locket (Alepocephalidae, Latimeridae, Melamphaeidae, Scopelarchidae, Sternoptychidae)
Dr. O. Munk (Bathylagidae, Bathypteroidae, Ceratiidae, Diretmidae, Gobiidae, Idiacanthidae, Ipnopidae, Omosudidae)
Prof. J.A.C. Nicol (Ophidiidae, Priacanthidae)
Mr. V. Parthe (Centropomidae, Gerridae)
Dr. H.W. Reno (Notopteridae)
Miss K. Schoebitz (Galaxiidae)
Prof. H.O. Schwassmann (Anablepidae, Serranidae)
Prof. F. Vrabec (Callionymidae)
Dr. H.-J. Wagner (Cichlidae, Lebiasinidae, Mochokidae)

Dr. H.-J. Wagner was also kind enough to help with the preparation of a number of plates and the chapter on the preparation and examination of the retina during his stay in Montréal in the summer of 1975.

The suggestion that an Atlas based on the material in our possession be assembled was made by Professor A.M. Monnier of l'Université de Paris VI and we thank him for it. We are grateful to Dr. Donn E. Rosen who kindly allowed us to reproduce some of the outlines of representatives of fish families from a publication he co-authored (Greenwood et al., 1966).

We also thank Miss Margaret Pertwee and Professor P.K. Menon for editorial help. Miss Pertwee also assisted a great deal in the organisation of the material and bibliography. Monsieur V. Legendre gave valuable advice with reference to some taxonomical questions.

Mademoiselle Nicole Valiquette did a conscientious job of the typing and Mrs. Hoda Farid was ever ready with her cooperation in the preparation of the plates. Monsieur Jean-Luc Verville was always willing to help us with the developing and printing of the photomicrographs during the last few hectic months. We are thankful to Dr. K.F. Springer, Mrs. Th. Deigmöller and Mr. E. Kirchner of Springer-Verlag, Heidelberg for their cooperation, patience, and understanding.

Montréal, Autumn 1976

M.A. Ali
Michel Anctil

Contents

List of Plates

1. INTRODUCTION

The aim of this chapter is to give a brief description of a generalised fish retina to enable easy understanding of the photomicrographs presented in the following sections. However, as mentioned earlier, the great morphological diversity displayed by fishes should deter us from using this scheme for purposes other than introducing the material and terminology.

The fish eye is oval (Plate 1) with a spherical lens. Due to the peculiar accommodatory process wherein the lens is moved back and forth, unlike in the higher vertebrates, the lens muscles are usually well developed. The lens plays a predominant role in image formation in fishes, whereas the cornea has optical characteristics similar to those of water and therefore is not involved in dioptric processes. Other ocular structures resemble more or less those in the rest of the vertebrates; WALLS (1942) and ROCHON-DUVIGNEAUD (1943) should be consulted for details on various fish eye structures and their relationships with those of higher vertebrates.

Many fishes, especially the elasmobranchs, have pigment in the choroid. In the elasmobranch eye this pigment is capable of movement in response to light. In light it expands and occludes the reflecting material in the choroid, while in the dark it retracts, thus leaving the reflecting material exposed (NICOL, 1963). This results in the maximal utilisation of the available light due to reflexion within the eye cup.

1.1. Retinal Circulatory System

Besides the choroidal blood supply and its rete mirabile involved in active secretion of oxygen into the eye (WITTENBERG and WITTENBERG, 1962), the fish retina is nourished by other intraocular vascularisations. These include several types of falciform process, with or without branches on the inner (vitreal) surface of the retina, and vitreal vessels, all of which originate from the hyaloid artery and join a peripheral (ora terminalis) annular vein (HANYU, 1959, 1962; ALI et al., 1968; ANCTIL, 1968). The eel families constitute a notable exception in that the choroid blood supply is absent and is replaced by a rich supply of retinal vessels running on and, in the retina, converging into a hyaloid vein (HANYU, 1959; ALI et al., 1968). In some cases (e.g. sharks and skates), both the vitreous chamber and the retina are avascular (HANYU, 1962).

1.2. The Retina

The retina is a projection of the brain and consists of various cell types which are traditionally arranged in eight "layers" and two "membranes". Although this layer arrangement does not reflect the actual situation (see Plate 2), it will be retained in the following description for convenience and ease of comprehension. From sclerad to vitread these layers are as numbered below (Plates 2 and 3):

1. Epithelial layer
2. Visual cell layer (rods and cones)
3. External limiting membrane
4. External nuclear layer
5. External plexiform layer
6. Internal nuclear layer
7. Internal plexiform layer
8. Ganglion cell layer
9. Nerve fibre layer
10. Internal limiting membrane

The *epithelial layer* is made up of cells which have a basal part, the cell body (usually hexagonal in shape) containing the nucleus, and one or more finger-like processes. The latter are interspersed among the outer segments of the visual cells. Except in hagfishes and the elasmobranchs (sharks, skates and rays) the epithelial cells contain melanin granules and/or reflecting material. In fishes which are nocturnal, or which inhabit deep or murky waters these granules are usually concentrated in the cell body. In most teleosts (bony fishes) the pigment is considerable in quantity and composed of granules and needle-shaped crystals. In many cases this pigment, like the rods and cones, is capable of responding to light. It migrates vitread within the cell processes. In the dark it vacates the cell processes and concentrates in the cell body, thus forming a dense strip (Plate 3).

Visual cell layer. A visual cell or photoreceptor is made up of an outer segment, with lamellar pilings (saccules) containing the visual pigment. The outer segment is attached to the inner segment by a connecting cilium. There is also a ciliary structure, the accessory outer segment, alongside the outer segment, whose origin lies in a centriole near that of the connecting cilium. The inner segment is made up of an ellipsoid containing mostly mitochondria and a myoid containing ribosomes, vesicles, rough and smooth endoplasmic reticulum, Golgi bodies, and microtubules. The nucleus and perikaryon are followed by the inner fibre (axon) containing microtubules and by the terminal pedicle containing synaptic vesicles and ribbons. In a few fishes the outer tip of the ellipsoid may contain an "oil" droplet of mitochondrial origin,

---

Plate 1. Transverse section through the eye of a juvenile chum salmon (*Oncorhynchus keta*).
CH choroid, *CO* cornea, *FP* falciform process, *I* iris, *L* lens, *ON* optic nerve, *RE* retina, about ×100

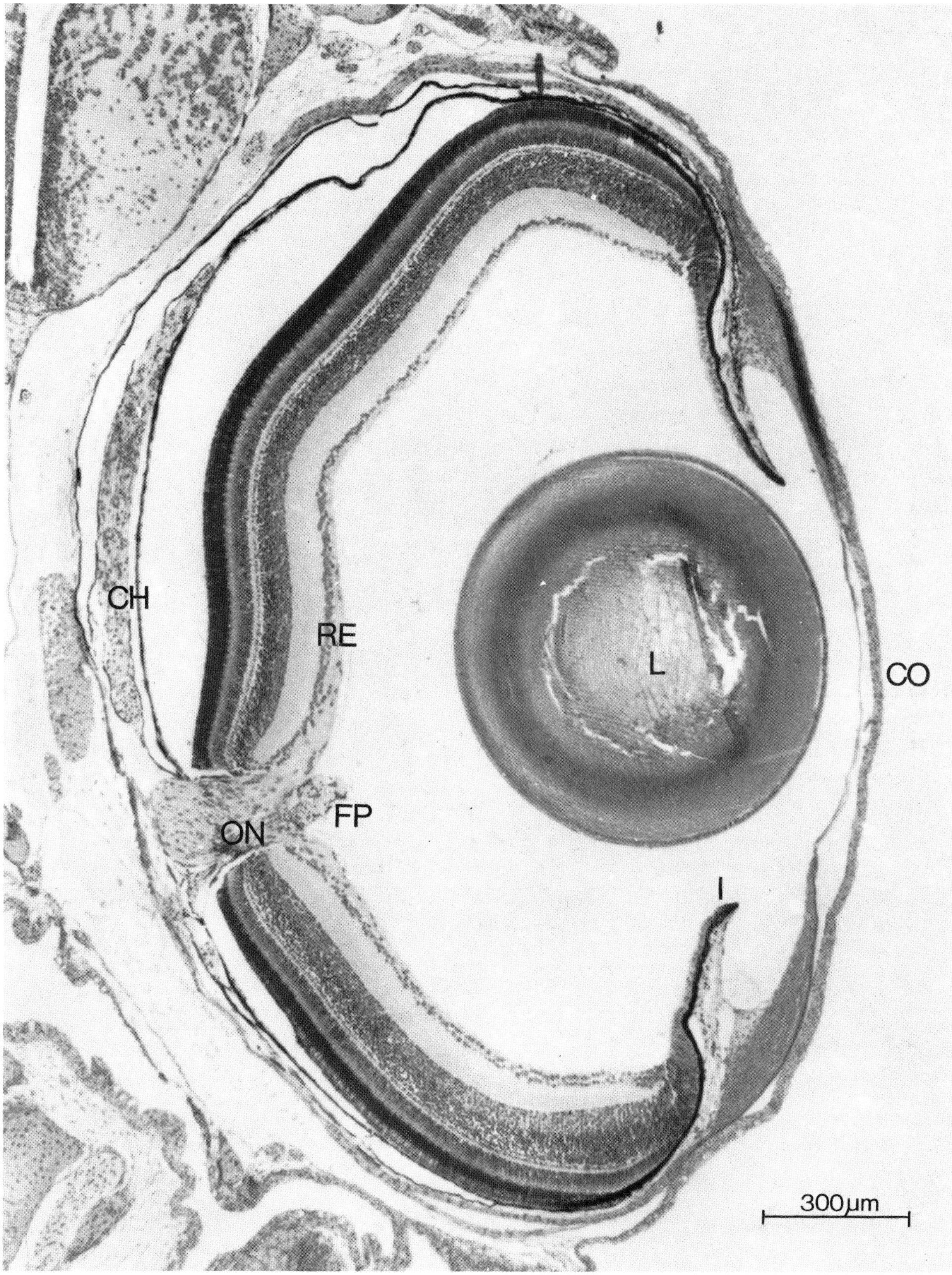
CH
RE
L
CO
FP
ON
I
300µm

while in some a paraboloid (glycogen) is present in the myoid region (Plate 4).

In general, two types of photoreceptors may be distinguished: the rods (scotopic) and the cones (photopic). The former are generally more numerous and slender with a cylindrical outer segment, while the latter are fewer and bulkier with a shorter and tapering outer segment (Plate 5). The outer segments of the rods and cones differ also in their ultrastructure (COHEN, 1969, 1972). Autoradiographic studies coupled with electron microscopy have shown that while the discs of the rod outer segments are continually regenerated to replace the older ones when phagocytosed by the retinal epithelial cells (in frogs and rats), the cone outer segment discs are not subject to such a process (YOUNG, 1969). The cones may be of different kinds. Usually, only two types are present, the single and double (or twin) cones. Sometimes triple or even quadruple cones are encountered. The double cones may be either equal or unequal in size and shape. When they are unequal, one member is referred to as the principal and the other, presumably of rod origin, is labelled accessory (Plate 5). The shape and density of the mitochondria vary between the rods and the cones. It is also probable that even within one type, electron microscopy will demonstrate differences in the density of the mitochondria between light- and dark-adapted states.

In many teleosts the rods and cones respond to light and dark by undergoing changes in their positions. These changes go hand-in-hand with the response of the epithelial pigment mentioned earlier. In light, the rods are expanded and their outer segments are generally masked by the pigment that has migrated into the cell processes surrounding them. Meanwhile the cone ellipsoids are situated adjacent to the external limiting membrane (Plate 3, right). In the dark, the cones are expanded and their outer segments

---

▷

Plate 2. Routine histological transverse section of the retina of *Nannacara anomala* on the left and a number of silver-impregnated receptors and neurones of the same retina arranged according to the normal layering on the right. The classical description of the retina in terms of nuclear and plexiform layers *(left part)* does not take into account the range of the dendritic or terminal arborisations of the different classes of nerve or interneurone cells, which can only be seen in the right part. It would appear that this latter approach is more suited to give an idea of the functional morphology of the neural network in the vertebrate retina
PEL pigmented epithelial layer, *ENL* external nuclear layer, *EPL* external plexiform layer, *INL* internal nuclear layer, *IPL* internal plexiform layer, *GCL* ganglion cell layer, *NFL* optic nerve fibre layer, *og* oligodendrocyte, *r* rod, *rm* rod myoid, *c* cone, *dc* double cone, *sc* single cone, *cp* cone pedicle, *eho* external horizontal cell, *iho* internal horizontal cell, *rbi* rod bipolar cell, *cbi* cone bipolar cell, *sam* stratified amacrine cell, *bam* bistratified amacrine cell, *bgc* bistratified ganglion cell, *sgc* stratified ganglion cell, *1–7* subdivisions of the inner plexiform layer. ×1,000. (Courtesy of Dr. H.-J. WAGNER)

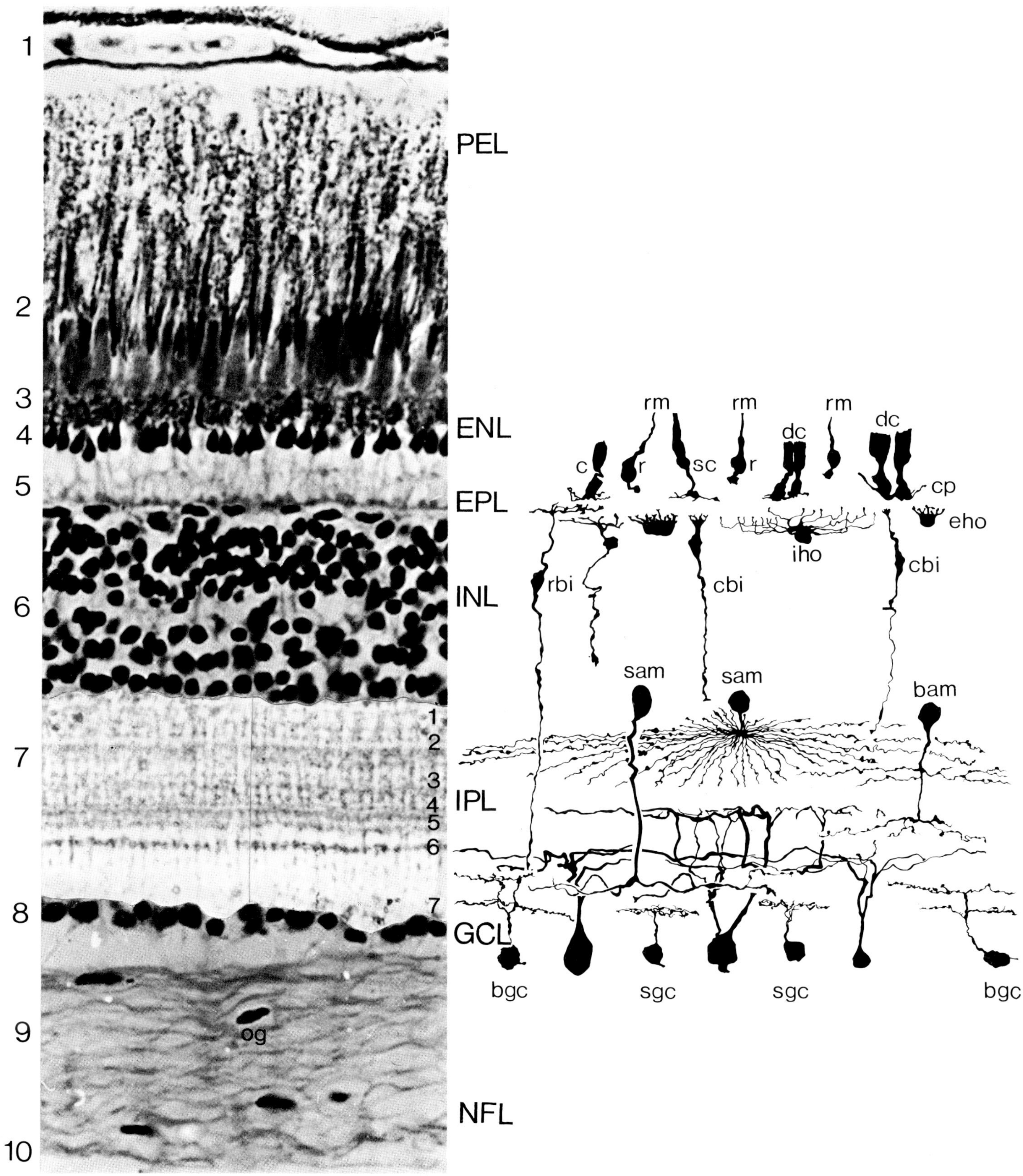
1
PEL
2
3
ENL
4
5
EPL
6
INL
7
1
2
3
IPL
4
5
6
7
8
GCL
9
og
NFL
10
rm
rm
rm
dc
dc
c
r
sc
r
cp
EPL
eho
iho
rbi
cbi
cbi
sam
sam
bam
bgc
sgc
sgc
bgc

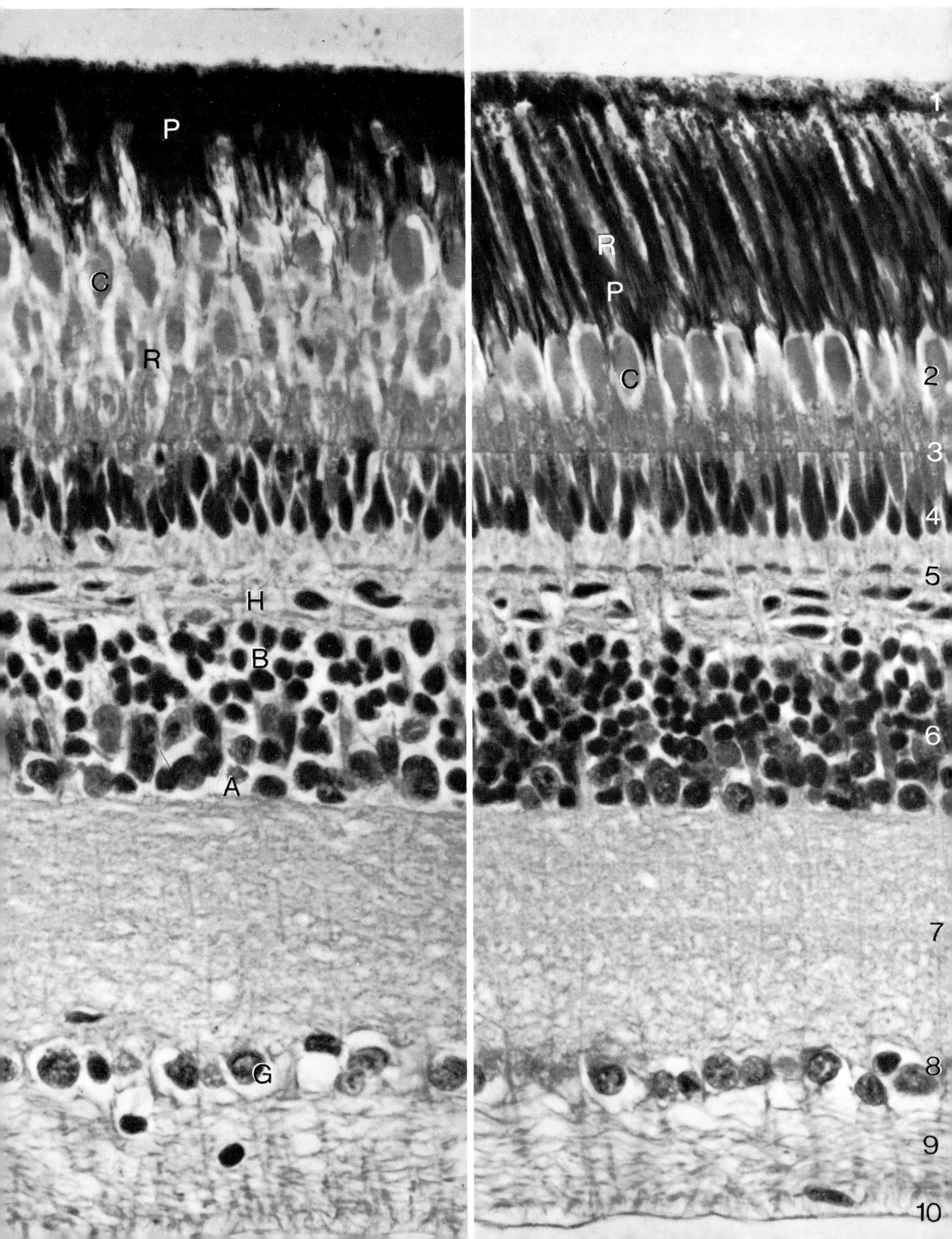
P
C
R
H
B
A
G
R
P
C
1
2
3
4
5
6
7
8
9
10

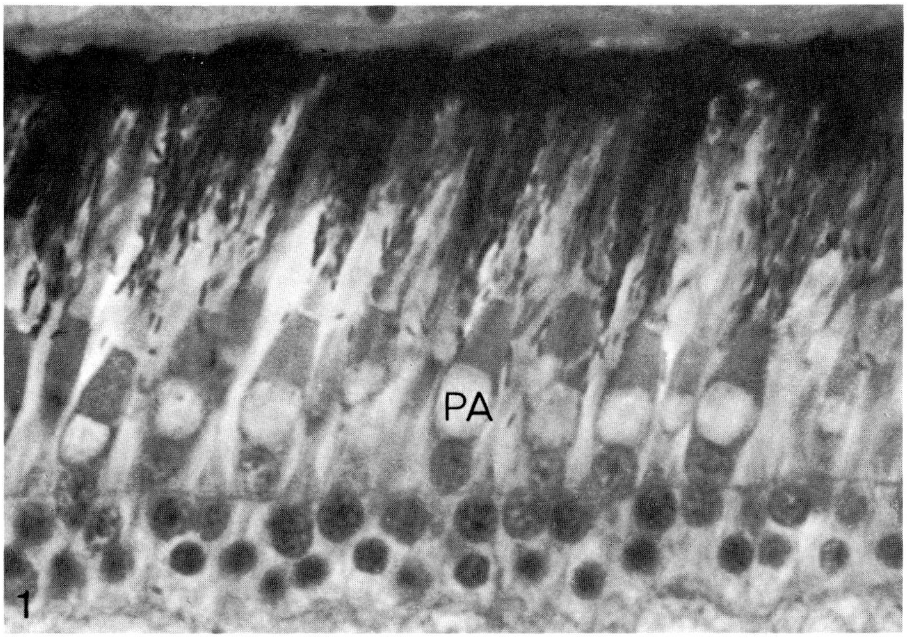

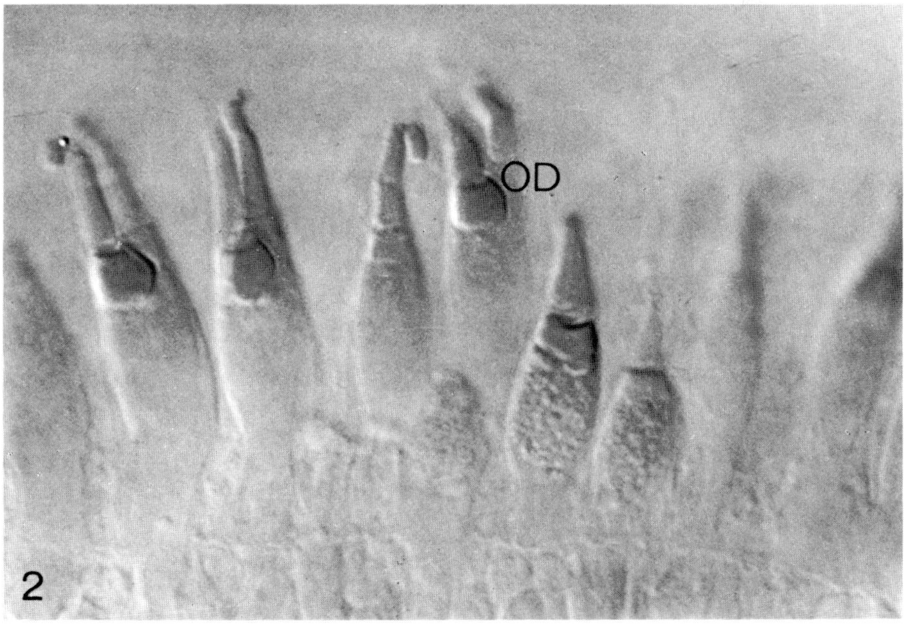

Plate 4. 1, 2. Transverse sections of the retinas of *Amia calva*. (1) Amiidae, × 1,000, showing paraboloid *PA* in the myoid region and of *Fundulus heteroclitus*. (2) Cyprinodontidae, × 1,900, showing cone droplets *OD*

◁ Plate 3. Transverse sections through the dark-adapted *(left)* and light-adapted *(right)* retinas of the Arctic char (*Salvelinus alpinus*)
1 pigmented epithelial layer, *2–10* see Plate 2 and Introduction
A amacrine cell, *B* bipolar cell, *C* cone ellipsoid, *G* ganglion cell, *H* horizontal cell, *P* epithelial pigment, *R* rod ellipsoids. About × 1,000

and ellipsoids are approximately where the rods were in light, while the rods have moved vitread, closer to the external limiting membrane and consequently closer to the incoming light quanta (Plate 3, left). There is considerable variation from this generalised pattern among the various fish families depending on their retinal structure, habitat, and behaviour. These movements of the epithelial pigment and visual cells are collectively known as the *retinomotor responses* and have been known for about a century. Their characteristics, the influence of environmental factors on them, and their relationship to the fish's habitat and behaviour have been studied by various workers and the literature has been reviewed from time to time (AREY, 1915; VON STUDNITZ, 1952; ALI, 1971, 1975). It is probable that the myoid and the accessory outer segment of the cone play a role in the retinomotor responses. The myoids might pull the photoreceptors vitread by contracting while the accessory outer segment of the cone may act as a spring, pulling it sclerad in the dark when the myoid is relaxed.

The *external limiting membrane* usually straddles most of the cone nuclei, which are more sclerad. It is usually very clearly seen and is often the dividing line between the myoids and the nuclei of the visual cells. This membrane is made up of the sclerad extremities of the Müllerian cell branches. These are the main supporting cells of the retina, and are also called radial fibres.

The *external nuclear layer* is composed of the visual cell nuclei. In rare cases (some elasmobranchs, holosteans, sturgeons, and the pike) one encounters some displaced bipolar cells also in the vitread part of this layer. It is usually possible to distinguish the rod and cone nuclei. In most teleosts, the former are located more vitread and are smaller while the latter are larger and located very close to the external limiting membrane, some of them even protruding into it. In the elasmobranchs, these respective positions are inverted.

The *external plexiform layer* is composed of the foot-pieces (inner fibres and pedicles) of the visual cells, and the sclerad processes and dendrites of the horizontal and bipolar cells, which make synaptic contacts with the visual cells.

The *internal nuclear layer* includes three distinct sublayers of neuronal cell bodies and nuclei. The sclerad-most sublayer is composed of horizontal cells, the intermediate one of bipolar cells, and the vitread-most of the amacrine cells. All these sublayers, especially the horizontal and bipolar cells, may themselves be multilayered and interdigitate with other sublayers. The horizontal cells are usually flat or cuboidal with large nuclei and smooth fibrillar cytoplasm. They send out lateral processes which connect with various visual cell pedicles over a certain receptive field. The bipolar cells are second order neurones with slender cell bodies, round nuclei, and more or less elaborate dendritic ramifications. The diverse categories of horizontal and bipolar cells are often connected to

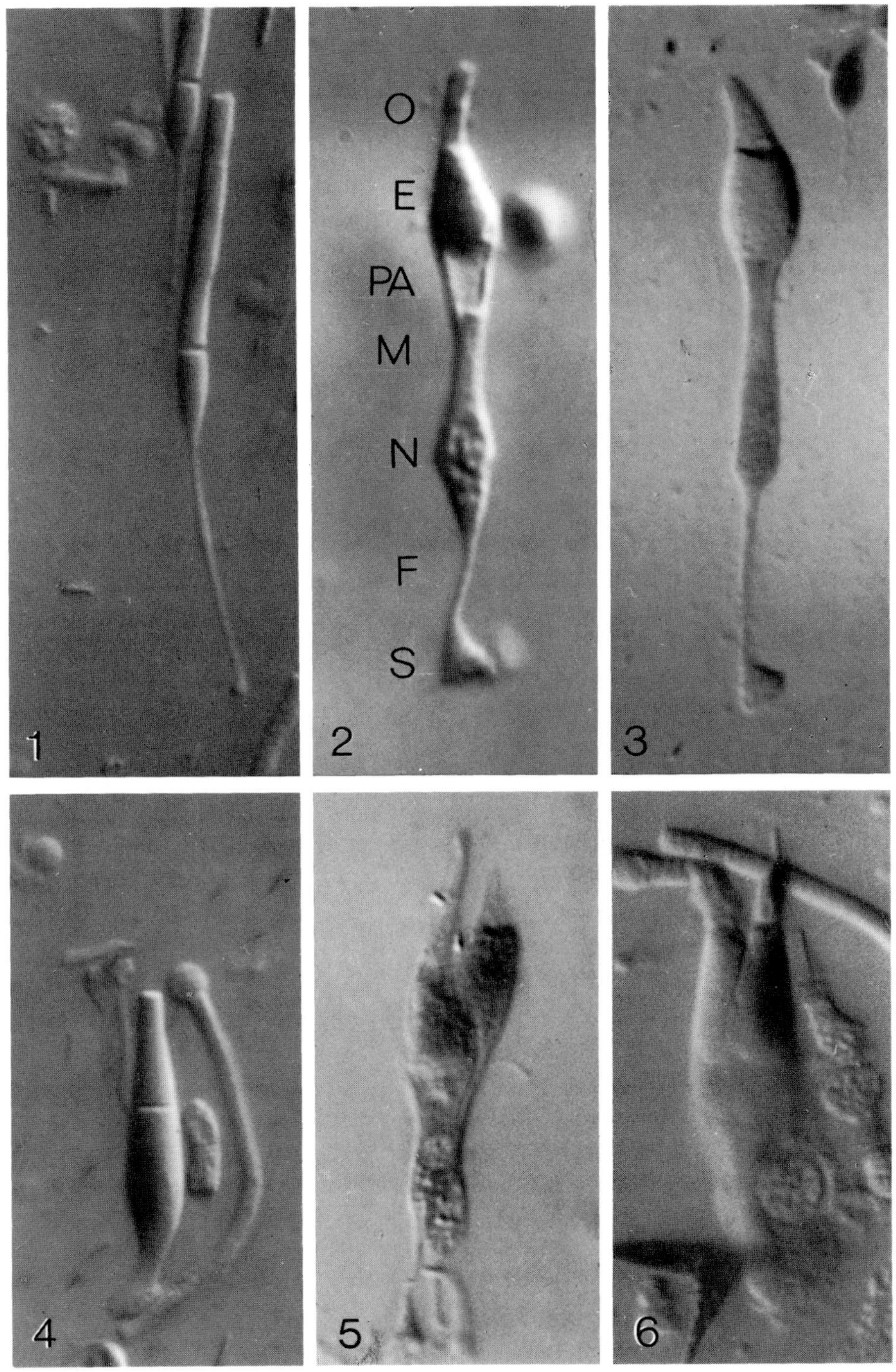

Plate 5.1-6. Photomicrographs of photoreceptors isolated by chemical or mechanical means (see ANCTIL et al., 1973). (1) brook trout (*S. fontinalis*) rod, (2) bowfin (*A. calva*) single cone, (3) goldfish (*Carassius auratus*) single cone, (4) yellowperch (*Perca flavescens*) single cone, (5) bowfin unequal double cone, (6) brook trout equal double cone

O outer segment, *E* ellipsoid, *PA* parabolid, *M* myoid, *N* nucleus, *F* inner fibre, *S* synaptic pedicle. All ×1,950

a corresponding type of visual cell (rod, single, or double cone). The amacrine cell bodies are larger than those of the bipolars and their nuclei stain less deeply with haematoxylin than the bipolar nuclei. Also part of this layer are the Müllerian cell or radial fibre nuclei.

The internal plexiform layer comprises the axonal processes of the bipolar cells, the amacrine cell processes, and the dendritic ramifications of the ganglion cells which make synaptic contacts with the two former. The result is a dense meshlike network of fibres running both horizontally and vertically. In some cases (agnathes, elasmobranchs), ganglion cell bodies are scattered throughout this layer.

The retinal *ganglion cell* bodies of all except the jawless and cartilaginous fishes are confined to one layer, immediately vitread to the internal plexiform layer in which two or more types of cells may be distinguished according to size and ramification patterns.

The axons of the ganglion cells form the *nerve fibre layer*. These fibres converge in the region of the optic disc to form the optic nerve (or tract). In the jawless fishes, this layer is located between the internal nuclear layer and internal plexiform layer. In several primitive fishes (bowfin, gar, deep-sea salmoniform teleosts, lungfish) the optic nerve fibres are located as fascicles between nuclei in the ganglion cell layer.

The *internal limiting membrane* separates the nerve fibre layer from the adjoining vitreous humour and vitreal vessels. This apparent membrane, analogous to the external limiting membrane, is the result of the expansion of the vitreal ends of contiguous Müller's or radial fibres.

2. TECHNIQUES

The methods mentioned here are those used both by the authors and other workers whose material has been incorporated. Most of the eyes and retinas used in the preparation of this Atlas were fixed in Bouin's fixative. Some were fixed in formalin, Kolmer's fluid, Pereny's fluid, etc. Some eyes, whose cornea and lens had been removed, were also fixed in buffered glutaraldehyde and retinal pieces were postfixed in buffered osmium tetroxide. Whole eyecups or retinal pieces were embedded in paraffin and, in a few cases, in celloidin; paraffin or celloidin sections were cut at three to ten microns. The small retinal pieces fixed in glutaraldehyde and osmic acid were embedded in Epon 812 and cut at one to two microns for light microscopy. In some fishes, the retinas were treated using the rapid Golgi-impregnation technique (see PARTHE, 1972), from which celloidin sections were prepared.

Stained and unstained histological sections were used. The main staining techniques employed were (1) Harris' hematoxylin and eosin, (2) Mallory's one-step triple stain, (3) Heidenhein's Azan, and (4) fuchsin-paraldehyde and one-step trichrome (GABE, 1968). Some Epon sections were unstained; others were stained with the Azure II-Methylene Blue technique. The unstained preparations were examined with phase contrast and Nomarski differential interference-contrast optics. The stained preparations were examined in bright field microscopy and, in some cases, with the interference-contrast optics.

For further details, the reader is referred to the pertinent literature on retinal morphology listed in the bibliography and to the following chapter.

3. PREPARATION AND EXAMINATION OF THE RETINA

1. Measure body length.
2. Inspect intact eyes: eye shine, colour of cornea, colour of lens, shape of iris etc.
3. Decapitate fish and enucleate eyes (not necessary in specimens with less than 3 cm body length destined for light microscopy). Smaller fish to be fixed in toto.
4. Measure horizontal eye diameter (or temporo-nasal).

5.1. Puncture the eyes at the sclero-corneal junction with a fine needle at least twice (sufficient for light microscopy)—or

5.2. Remove cornea and lens with a sharp razor blade or a De-Wecker type of scissors (essential for electron microscopy).

5.3. Careful inspection of retinal fundus: intraretinal vessels, falciform process, ventral eye fissure, optic papilla and fovea. In larger eyes, separate ventral, dorsal, nasal, and temporal parts as well as the fundus (Plate 6→).

6. Fixation in one of the following media:

6.1. Light microscopy:

   1. Bouin's fixative (75 parts saturated picric acid; 25 parts formaldehyde; 5 parts glacial acetic acid).
   2. Kolmer's fixative (5% aqueous potassium bichromate solution, 20 cc; 10% formalin, 20 cc; glacial acetic acid, 5 cc; 50% trichloracetic acid, 5 cc; saturated aqueous (10%) uranyl acetate, 5 cc).

6.2.1. Electron microscopy (see 15 for recipes): 2.5% glutaraldehyde and 1% paraformaldehyde diluted in 0.06 M sodium phosphate buffer containing 3% sucrose for freshwater species and 5% for marine fishes. Fix for 30 min at room temperature and 90 more min at 4° C.

6.2.2. Wash twice in buffer and carefully remove sclera in a way that does not separate retina, pigment epithelium and choroidal gland (binocular microscope). Trim pieces of tissue to convenient size (3×3 mm). If not done earlier, separate ventral, dorsal, nasal, temporal, and fundus regions now (Plate 6).

6.2.3. Post-fixation in 1% osmic acid (2 h at 4° C). Wash thoroughly in distilled water after postfixation.

6.3. Material fixed for light microscopy destined for semithin (1 μm) sectioning may be treated according to 6.2 and processed further as for electron microscopy.

7. Dehydratation in graded series of alcohol (i.e. methanol) at 4° C starting at 50% (5 min), 70% (10 min), 80% (15 min), 90% (20 min), 95% (20 min) and 2×100% (15 min each).

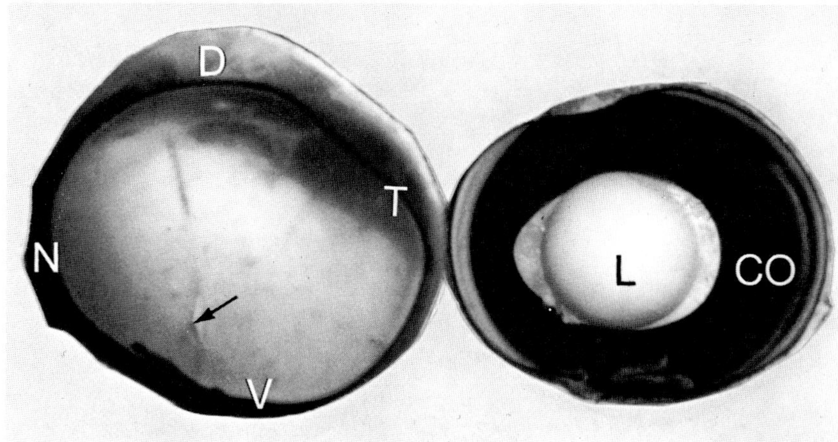

Plate 6. Photograph of the dissected left eye of *Labrisomus nuchipinnis* (Clinidae) showing the eye-cup on the left and the cornea *CO* and lens *L* on the right. The ventral fissure *(arrow)* is clearly seen. Dorsal *D*, ventral *V*, temporal *T* and, nasal *N* regions are indicated. (×11.7)

8. Infiltration.
8.1. Light microscopy.
8.2. In acetone (or propylenoxide) (30 min, 4° C) and in three stages of graded mixtures of acetone-resin (3:1, 15 min; 1:1, 30 min; 1:3, 45 min and pure resin, 60 min).
9. Embedding: make sure that the orientation of the specimens will allow both transversal and tangential sectioning.
9.1. Light microscopy: paraplast; if there is no need to examine lens, cornea and sclera, you may gently peel off these structures and remove the lens with a hot needle after a first embedding stage and then re-embed the remaining retinal tissue.
9.2. Electron microscopy: Epon or Araldite.
10. Sectioning: transversal and tangential sections of the retina at different levels must be obtained.
10.1. Light microscopy: 5 µm sections for paraffin-embedded material; 1 µm sections for plastic-embedded material.
10.2. Electron microscopy: silber-gray sections.
11. Staining.
11.1. Light microscopy: Haematoxylin-Eosin or Masson's trichromatic stain for paraffin-embedded material. Methylene blue according to Richardson et al. (1960) for semithin plastic sections.
11.2. Electron microscopy: double "staining" in uranyl acetate and lead citrate.
12. Suggestions for examination of retinal structure (light microscopy; see Plate 7).
12.1. Pigment epithelium
    1. Extent of development

2. Distribution of melanin granules
3. Relation of processes to outer segments of rods and cones
4. Capacity for migration

12.2. Tapetum lucidum.

12.3. Photoreceptor cells

a) Rods
1. Shape and size
2. Distribution: single, grouped, in banks
3. Density (tangential sections or counts of nuclei in the external nuclear layer (ENL)
4. Capacity for movement

b) Cones
1. Types: equal or unequal double cones, different types of single cones
2. Size and shape of cone types
3. Structural peculiarities (e.g. oil droplets, paraboloids etc.)
4. Distribution: mosaic, row or square pattern with central and/or additional single cones (tangential sections)
5. Density (tangential sections or counts of nuclei in the ENL).

12.4. External plexiform layer
1. Thickness.

12.5. Horizontal cells
1. Number of rows
2. Number of cell types (amounts of cytoplasm stained)
3. Distribution (mosaic?)
4. Presumably horizontal axons.

12.6. Bipolar cells
1. Density.

12.7. Glial cells.

12.8. Amacrine cells
1. Density.

Plate 7.1-8. Transverse section of the dorsal retina (area) of *Blennius cristatus* surrounded by the tangential sections of the parts of retina indicated by the brackets. (1) Pigmented epithelial layer *PEL* and rods *R*. (2) Cone *C* ellipsoids forming a square mosaic of double cones with additional single cone, as seen within the circle. Single cones, if present, when situated at the corners of a square mosaic are called "additional", and when situated in the centre of the mosaic are referred to as "central". (3) Section through the external limiting membrane *ELM* showing tangental sections of cone myoids. Note the square mosaics. (4) External nuclear layer *ENL* darker nuclei are those of rods. (5) Tangential section across the external plexiform layer *EPL* showing square mosaics formed by the cone pedicles. Dark spots within these correspond to the ribbon synaptic complexes. (6) Horizontal cells *H*. Note their regular arrangement in rows. (7(Rod horizontal cells. (8) Bipolar cells *B* form the major part of the inner nuclear layer *INL*. Semithin (1 μm) sections stained with methylene blue (×700) ▷

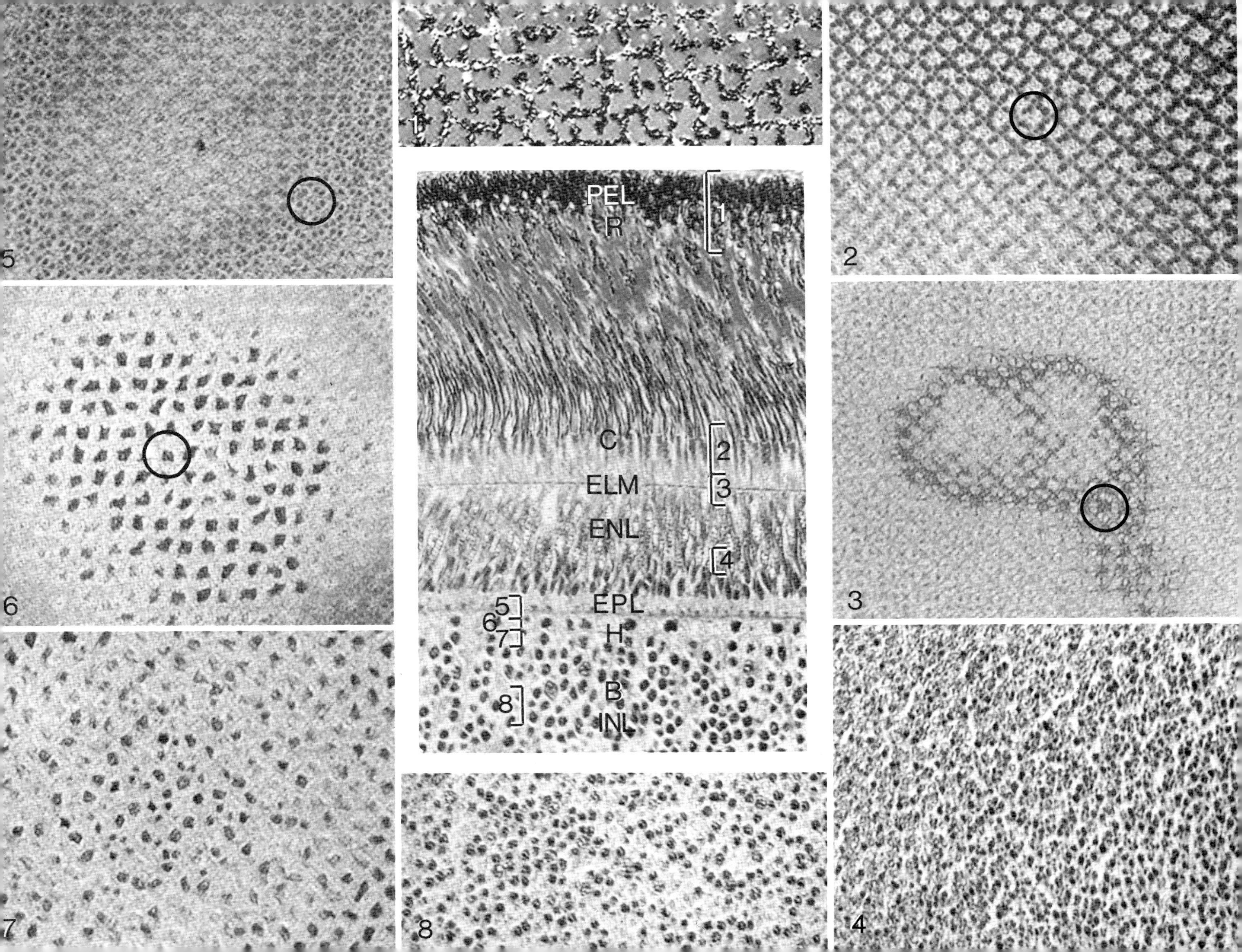
1
2
3
4
5
6
7
8
PEL
R
C
ELM
ENL
EPL
H
B
INL
1
2
3
4
5
6
7
8

12.9. Inner plexiform layer
   1. Number of horizontal sublayers
   2. Regular pattern in tangential sections
   3. Displaced ganglion cell nuclei.

12.10. Ganglion cell layer
   1. Density of nuclei.

13. Quantitative assessment of observations.

13.1. Summation rates are obtained by determining the ratio of receptor cell nuclei to bipolar cell nuclei to one ganglion cell.

13.2. The degree of integration is indicated by the number of horizontal and amacrine cell nuclei.

13.3. The items mentioned should ideally be examined and compared in the dorsal, ventral, nasal, temporal, and central region of the eye.

14. Special conditions: sometimes specimens cannot be processed without major interruptions in this schedule. During an expedition, for example, it may be possible only to fix the tissue and then store it. Storage is most convenient both for light and electron microscopy at the 70% alcohol stage. If it has to occur earlier, tissue may be kept in buffer after the end of the fixation period. For electron microscopy, it is essential to carry out the postfixation in $Os\ O_4$ because the aldehydes are not capable of fixing all ultrastructural details.

15. Routine fixation for electron microscopic examination

15.1. Preparation of 100 ml 0.06 M phosphate buffer
   1. Dilute 0.828 g $Na\ H_2PO_4 \cdot H_2O$ in 100 ml H_2O (sol. A) and 1.608 g $Na_2H\ PO_4$ in 100 ml H_2O (sol. B)
   2. To reach pH 7.3 take 23 ml of sol. A and add 77 ml of sol. B;
   3. Adjust with concentrated Na OH or HCl, if necessary
   4. Add 3 g of sucrose for freshwater specimens and 5 g for marine fishes.

15.2. Preparation of 20 ml aldehyde-fixative
   1. Dilute 0.2 g of paraformaldehyde in above-mentioned buffer (20 ml); heat to about 70° C and stir for 30 min. Solution must be filtered after cooling.
   2. Add 2.5 ml of 25% glutaraldehyde.

15.3. Preparation of 10 ml of osmic acid (1%)
   1. Dilute 2.5 ml of stock solution (4%) in 7.5 ml of the above buffer.

15.4. Use the remaining buffer for thorough rinsing.
If need be: stain tissue blocks in 2% uranyl acetate in sodium maleate buffer at pH 5 for 90 min.

4. DESCRIPTIONS

4.1. Agnatha (*Jawless Fishes*)

Myxinidae (*Hagfishes*)

Agnatha, Myxiniformes

Hagfishes belong to a very primitive and well-adapted group referred to as the cyclostomes, because of the circular suctorial mouth equipped with rasping teeth. The jawless fishes (hagfishes and lampreys) lack bones, jaws, paired limbs and girdles, scales, and their bodies are supported by an unsegmented notochord. They possess a single nostril and respiration takes place through gill pouches. The hagfish is a blind, elongate, wormlike fish with several whiskers extending from around the mouth, and is a voracious scavenger. It inhabits soft mud bottoms where the food is located entirely by smell.

The retina of the hagfish (*Myxine glutinosa, M. garmani, Polistotrema stouti, Eptatretus*) is degenerate to some extent and its infolding fills the space of the vitreous chamber (Plate 8.1, *M. glutinosa*). There are distinct and well-developed external and internal nuclear layers (Plate 8.1), but the ganglion cells cluster near the optic tract. The visual cells are cone-like in appearance in *M. garmani,* with long, tapering outer segments, whereas they are rod-like in *P. stouti,* with cylindrical outer segments. The visual cells of *M. glutinosa* cannot be classified into rods or cones; the outer segments are short, whorl-like, and somewhat disorganised (Plate 8.2), and there is no differentiation of the cell organelles into inner segment fibre and synaptic body. In *M. glutinosa,* there are no synaptic ribbons in the visual cell foot-piece, but only synaptic agranular vesicles adjacent to membrane densities are found (Plate 10). Synaptic ribbons are present in the better-developed retina of *P. stouti.*

References: Kohl (1892); Allen (1905); Eigenmann (1909); Dücker (1924); Franz (1934); Walls (1942); Kobayashi (1964); Holmberg (1969, 1970, 1971); Kleerekoper (1972); Fernholm and Holmberg (1975)

Plate 8.1, 2. (1) Median section through the eye, *1* Epithelial cell layer, *2* Receptor cell layer, *3* Inner layer of cell bodies, *4* Inner fibrous layer, *5* Connective tissue strand, *6* Optic tract, *ES* extracellular space. An outer fibrous layer *x* often separates the receptor cell layer from the inner layer of cell bodies. Azan (Heidenhain). ×300. (2) Median section through the eye. The receptor cell layer *RL* as well as follicle cells *Fo* show a PAS-positive reaction. *IL* inner fibrous layer, *OT* optic tract. Periodic-acid/Schiff, ×220

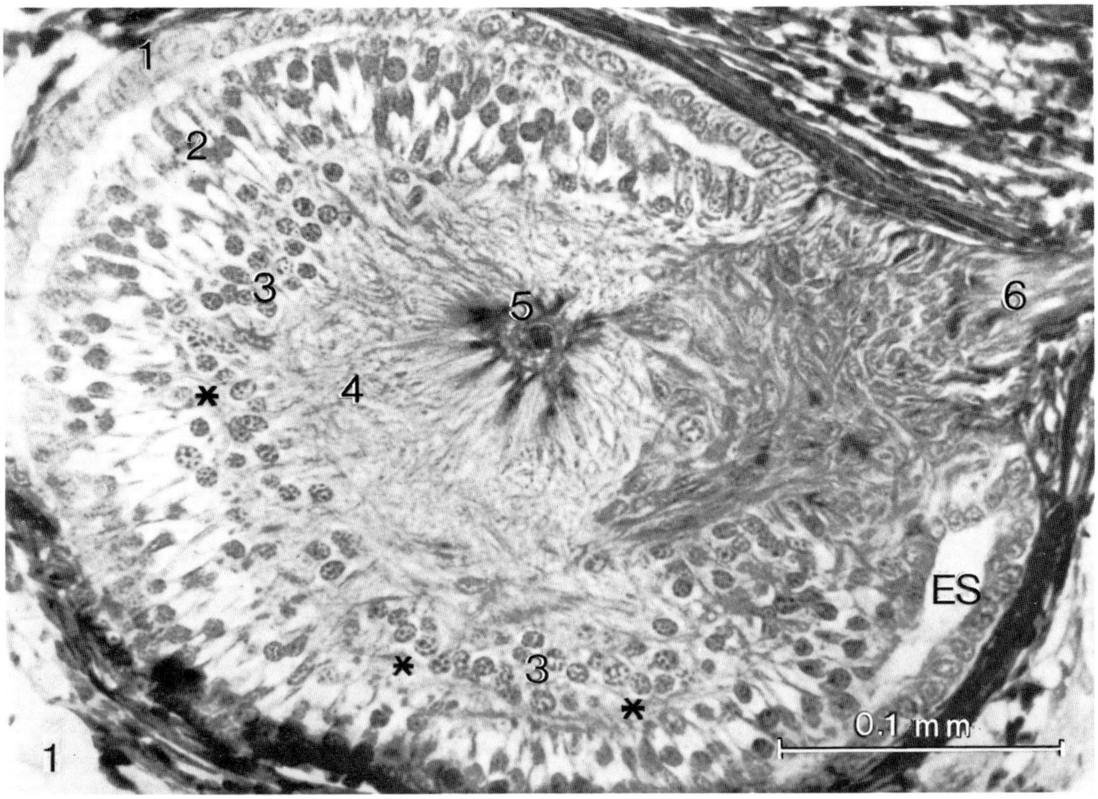
1
2
3
5
6
4
3
ES
0.1 mm
1

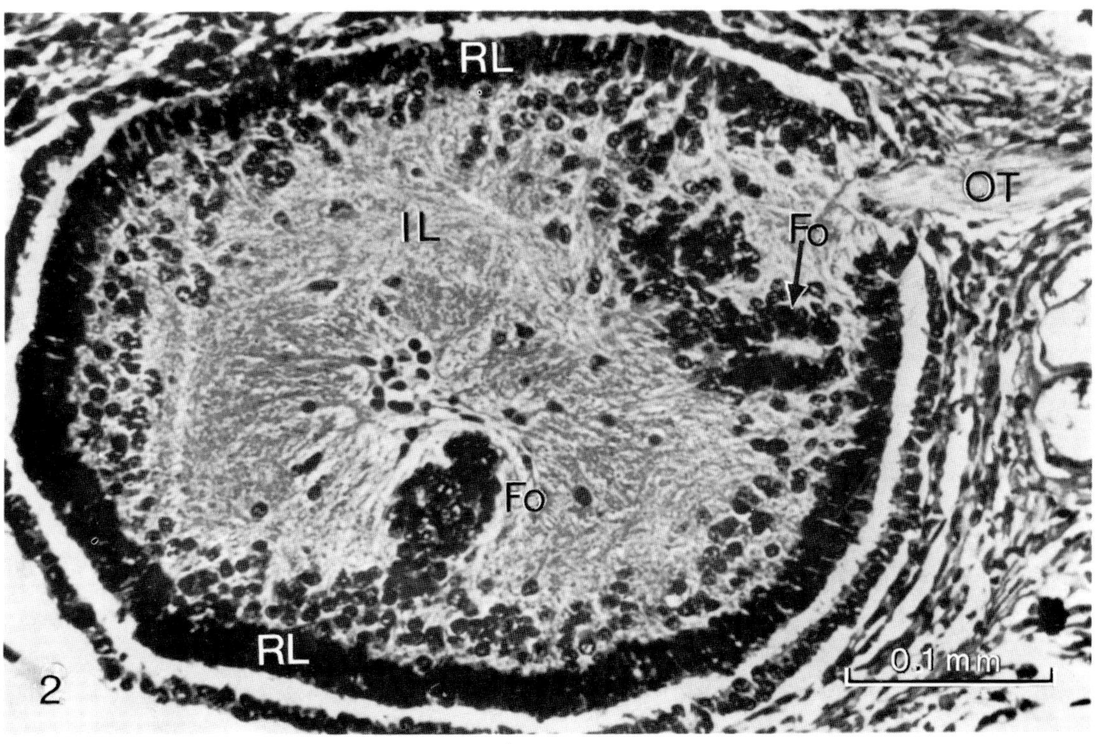
RL
OT
IL
Fo
Fo
RL
0.1 mm
2

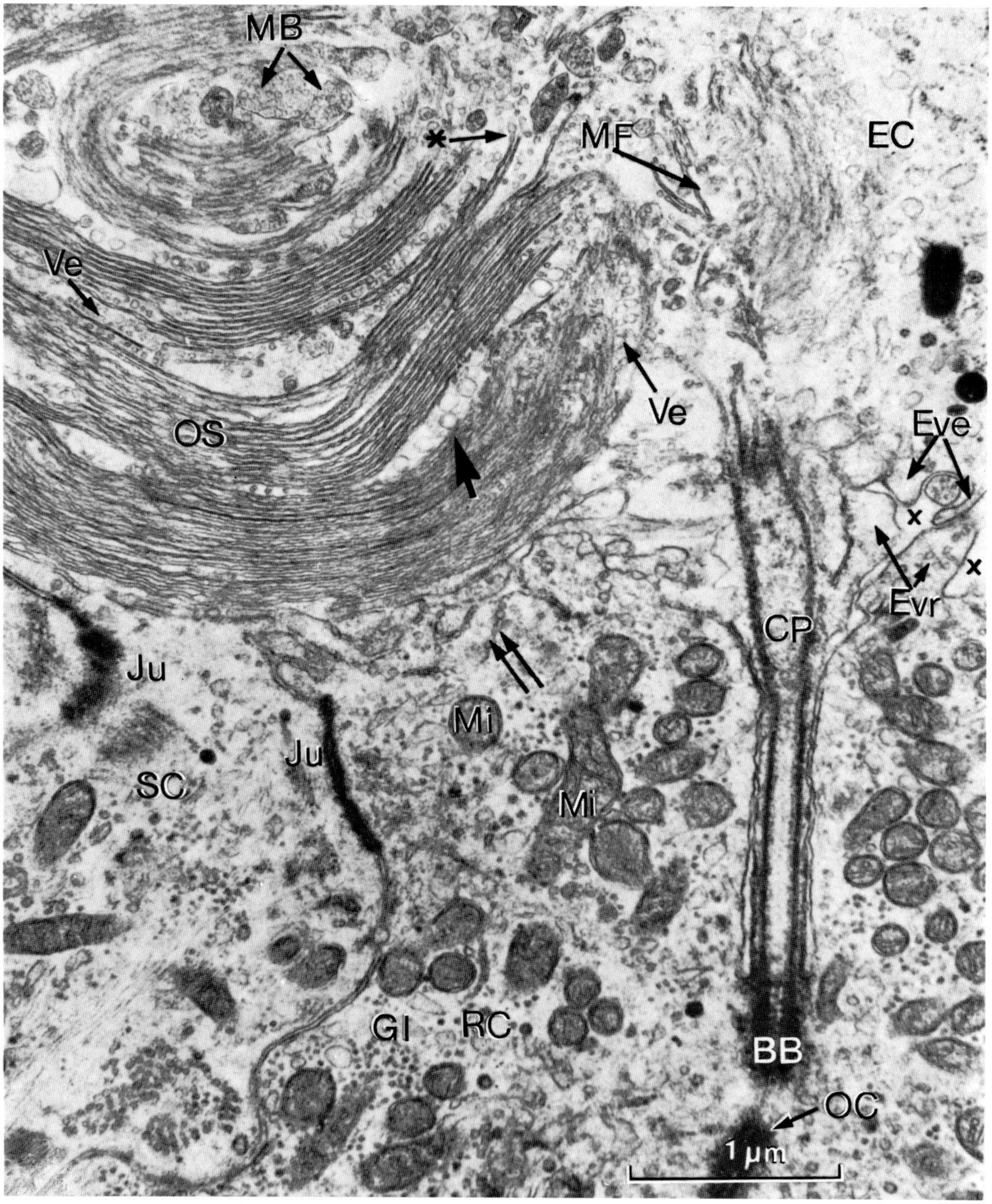

Plate 9. Longitudinal section through the apical part of a receptor cell *RC*. *BB* basal body, *CP* connecting piece, *EC* epithelial cell, *Eve* evagination from the apical surface of epithelial cell, *Evr* evagination from the apical surface of receptor cell, *Gl* glycogen granules, *Ju* cell junction, *Mi* mitochondria, *OC* oblique centriole, *SC* "Supporting" cell, *x* extracellular space. Multivesicular bodies *MB* and vesicles *Ve* are seen in spaces between the arrays of membranes. Larger vesicles *(Broad arrow)*, membrane fragments *MF* and discs which pinch off small vesicles *x* are found in material fixed in osmium tetroxide alone. Notice rows of small vesicles *(Double arrows)* near the plasma membrane of the receptor cell. Osmium tetroxide, ×26,000

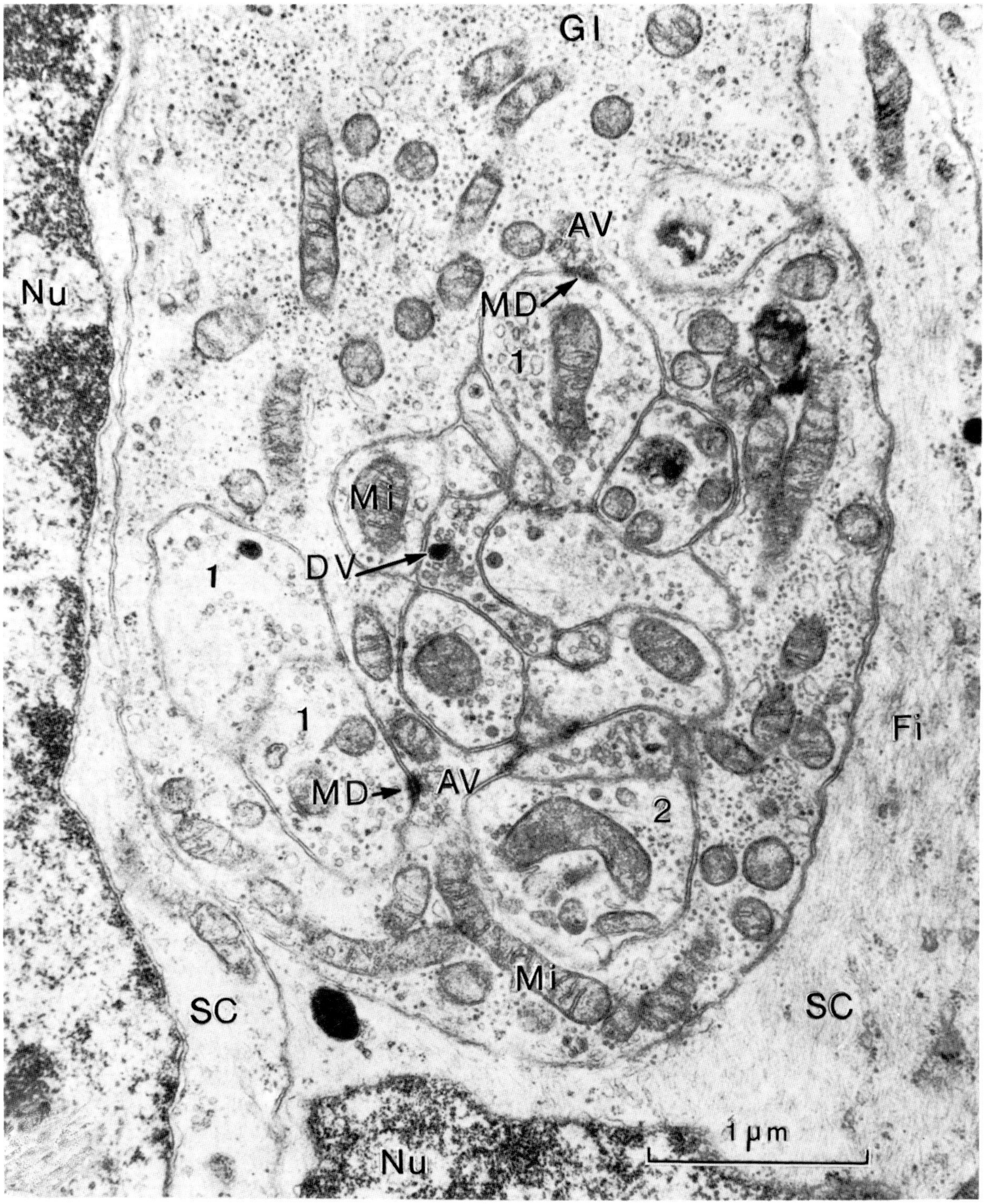

Plate 10. Receptor base in which agranular vesicles *AV* are aggregated near membrane densities *MD*. *DV* dense-core vesicle, *Fl* filaments, *Gl* glycogen granules, *Mi* mitochondria, *Nu* nuclei of "supporting" cells *SC*, 2 type 2 process without agranular vesicles. Osmium tetroxide, ×26,000. Reproduced with modifications from HOLMBERG (1970)

Petromyzontidae (*Lampreys*)

Agnatha, Petromyzontiformes

Lampreys have freshwater as well as marine representatives. Although they are capable of rapid, undulating swimming, the sea lampreys (*Petromyzon marinus*) travel mostly attached to host fishes while sucking out their blood. In general, the parasitic forms of lampreys remain for a shorter time on the host than do the hagfishes, the latter being internal parasites from deeper water.

The retinas of *Geotria australis, Ichthyomyzon, Petromyzon marinus, Entosphenus* and *Lampreta fluviatilis* (listed in a primitive-to-advanced evolutionary gradient) have been studied. The retina of the lamprey differs from all others in that its ganglion cells are not separated from the internal nuclear layer (Plate 11.2). Dissimilar to hagfishes, large horizontal cells are present in the adult *Lampreta* retina (Plate 11.1, 3). The visual cells are all conelike *Geotria*; they belong to two types according to length in *Ichthyomyzon,* but both have tapering outer segments; there are rodlike (cylindrical outer segment) and conelike visual cells which are longer in *Petromyzon*; in *Entosphenus* the conelike cells are still longer; in *Lampreta,* there are two visual cell types which differ in the length of the myoid segment and in the fluorescence of the slightly tapered, truncated outer segment with Procion Yellow. Synaptic ribbons are present in the foot-piece of *Lampreta* visual cells.

References: DÜCKER (1924); WALLS (1928, 1935, 1942); DUKE-ELDER (1958); YAMADA and ISHIKAWA (1967); GOVARDOVSKY and TSIRULIS (1969); ÖHMAN (1970, 1971); KLEEREKOPER (1972)

Visual Pigments: A_1 max 497, A_2 max 518 (LYTHGOE, 1972)

Plate 11.1-3. (1) Transverse section through the retina of *P. marinus* showing the pigment epithelium with its short processes *(upper inset),* the photoreceptor layer made of rodlike *R* and conelike *C* cells, the two layers of large horizontal cells and the small number of bipolar cells. Note the prominent Müller's fibres *M.* ×560. (2) Another transverse section through the retina of *P. marinus* showing displaced bipolar cells *B* and optic nerve fibres *ON* running between the bipolar cell layer and the inner plexiform layer. ×560. (3) Enlarged view of the photoreceptor synaptic layer and the horizontal cells. Note the displaced bipolar cells *B*, the rodlike *R* and conelike *C* synaptic pedicles. ×1,400

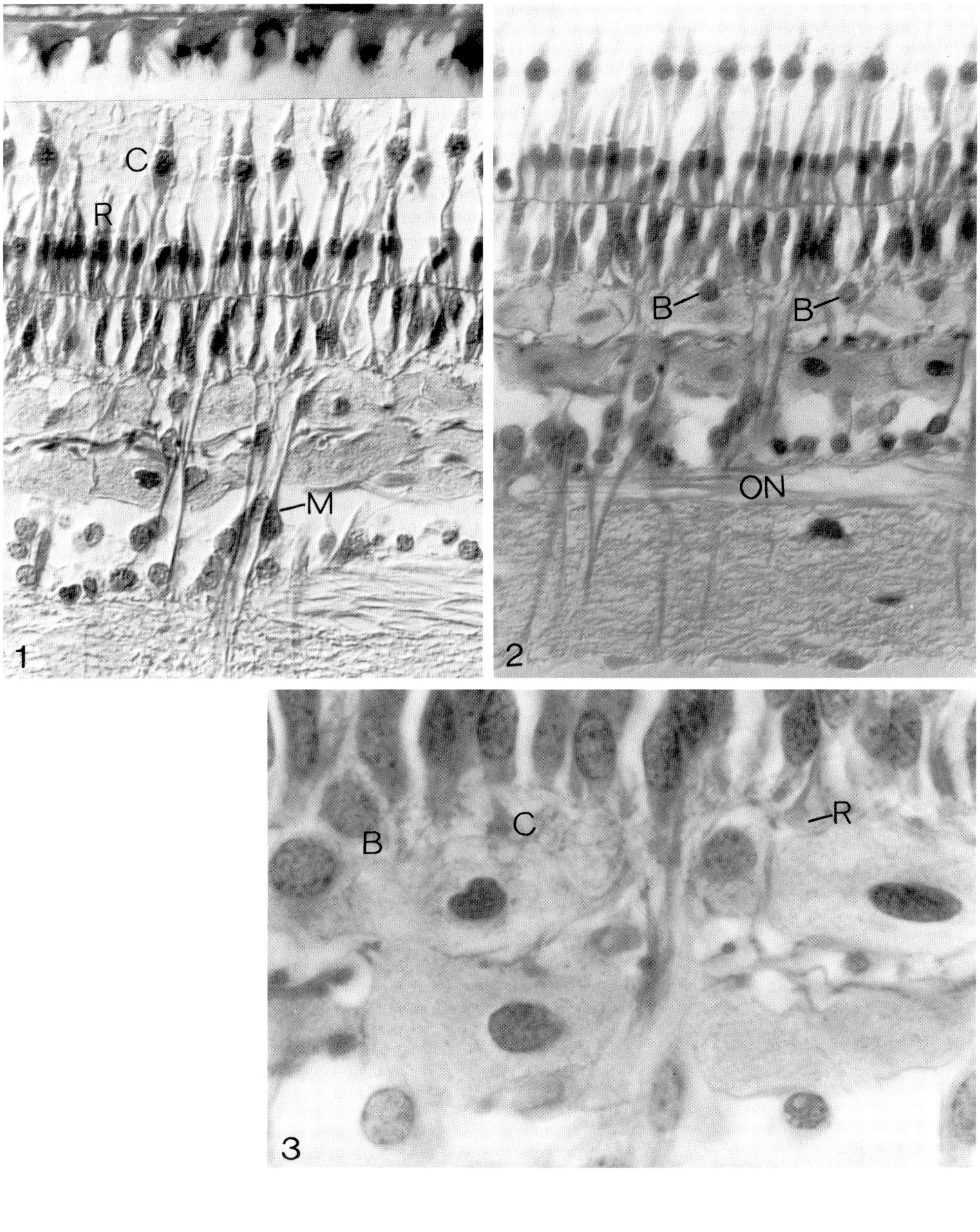

Plate 11.1-3. (Legend see opposite page)

4.2. Chondrichthyes (*Cartilaginous Fishes*)

Isuridae (*Mackerel Sharks*)

Chondrichthyes, Elasmobranchii, Selachii, Galeoidea

Chondrichthyes are fish with a cartilaginous skeleton and toothlike scales. The elasmobranch subclass includes the sharks (Selachii), skates and rays (Batoidei). In the selachians, the gill openings are at least partly lateral and the upper margin of the orbit is free from the eyeball (free eyelid), whereas in the batoids the gill openings are confined to the ventral surface and the upper margin of the orbit is not free from the eyeball.

Mackerel sharks are diurnally active, and swim strongly in pursuit of their prey (schools of mackerels and herrings). They are distributed from the surface to the bottom of the continental shelf waters.

The elasmobranch retina shows a number of characteristic features: the epithelial cells are devoid of pigment and the processes are very short or absent; as a rule the rods are the preponderant photoreceptor elements; the retinal cells, especially in the inner neuronal layers, are relatively large. Cones have been reported in the mixed retina of *Lamna cornubica*.

Reference: Rochon-Duvigneaud (1943, 1958)

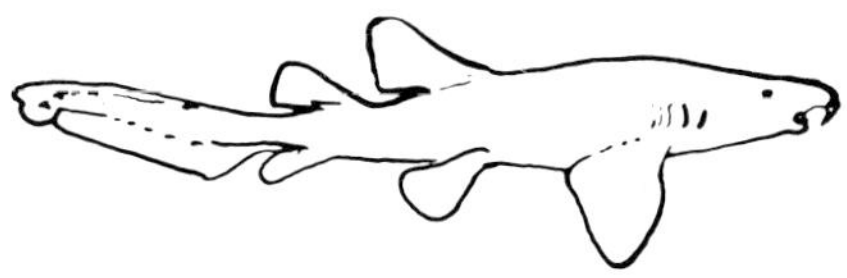

Orectolobidae (*Carpet and Nurse Sharks*)

Chondrichthyes, Elasmobranchii, Selachii, Galeoidea

The nurse shark, *Ginglymostoma cirratum,* is a sluggish fish which is found chiefly inshore, in relatively shallow waters. It forms small schools and feeds mostly on invertebrates.

Single cones have been reported in this species. The rod: cone ratios are 7:1 in the dorsal retina and 13:1 in the ventral retina. Cone myoids, which are found below the external limiting membrane, have been reported to elongate in the vitread direction in the dark-adapted state.

References: HAMASAKI and GRUBER (1965); WANG (1968); GALLEGO (1972); SCHROEDER and EBBESSON (1975)

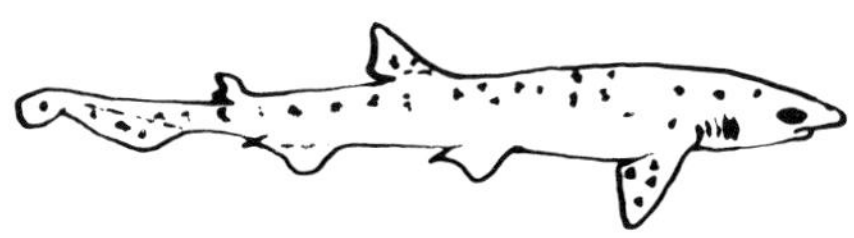

Scyliorhinidae (*Cat Sharks*)

Chondrichthyes, Elasmobranchii, Selachii, Galeoidea

These small sharks appear to be bottom dwellers from moderate depths of the continental shelf. The spotted dogfish, *Scyliorhinus canicula,* may inhabit shallower waters.

S. canicula has a simple pure-rod retina according to most authors with the exception of PROTASOV (1960) who reported cones. The rods are numerous (4–5 layers of rod nuclei), normal in shape, with a short myoid. There is one layer of horizontal cells. The bipolar and amacrine cells are fairly numerous, but the ganglion cells are few. Pure-rod retinas were also reported in *S. stellare, S. catulus, Galeus canis* and *Pristiurus* sp.

References: VERRIER (1930a); KOLMER (1936); ROCHON-DUVIGNEAUD (1943, 1958); PROTASOV (1960); NICOL (1961a)

Triakidae (*Whitetip Smoothhounds*)

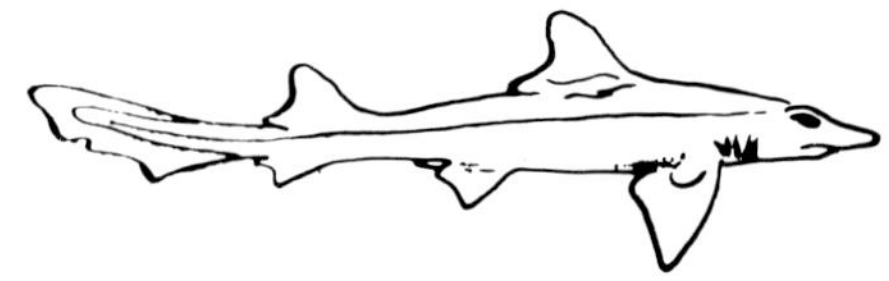

Chondrichthyes, Elasmobranchii, Selachii, Galeoidea

Triakid sharks are closely related to the Carcharhinidae. They inhabit both shallow and relatively deep waters. They are most active during the day but generally sluggish in habit.

The whitetip smoothhound, *Triaenodon obesus,* has been reported to have a pure-rod retina (Plate 12.1); the rods are slender and possess long myoids. The internal nuclear layer is composed of two layers of large horizontal cells, the sclerad one being made of nearly cuboidal cells, and a few bipolar and amacrine cells. The ganglion cells are few. Plate 12.2 shows the retina of an embryonic smoothhound (44 cm). The short visual cell layer has filamentous rod outer segments, and the internal nuclear layer is thick.

The retina of the smooth dogfish, *Mustelus canis,* differs from that described above by the presence of a few cones and a third horizontal cell layer; this retina is also characterised by the presence of three types of giant ganglion cells which are found at three distinct levels in the amacrine cell, internal plexiform, and ganglion cell layers.

References: Greeff (1899); Franz (1913, 1931); Verrier (1930a); Sverdlick (1940, 1955); Kato (1962); Yamada and Ishikawa (1965); Stell and Witkovsky (1973a, b); Witkovsky and Stell (1973); Stell et al. (1975)

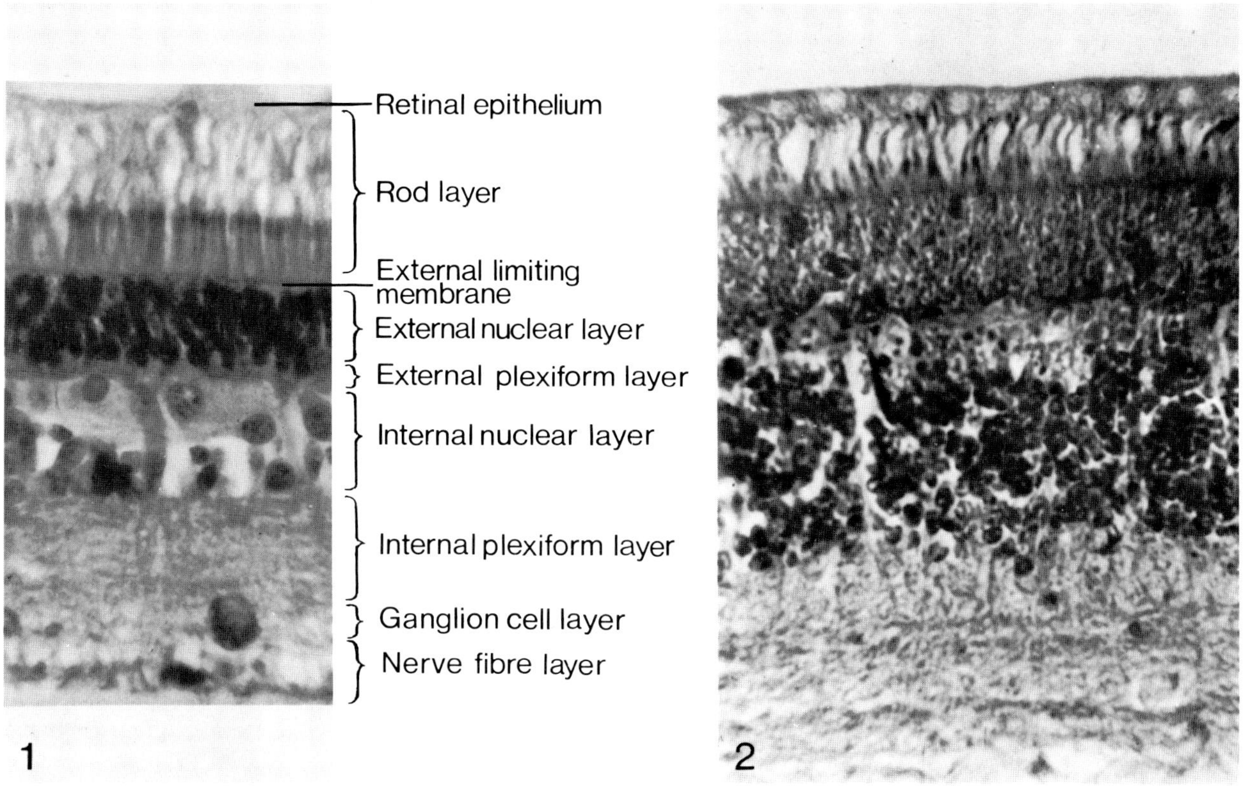

Plate 12.1, 2. Transverse section of the retina of the adult (1) and embryonic (2) whitetip shark, *T. obesus*. (1) × 560; (2) × 600. After KATO (1962)

Carcharhinidae (*Requiem Sharks*)

Chondrichthyes, Elasmobranchii, Selachii, Galeoidea

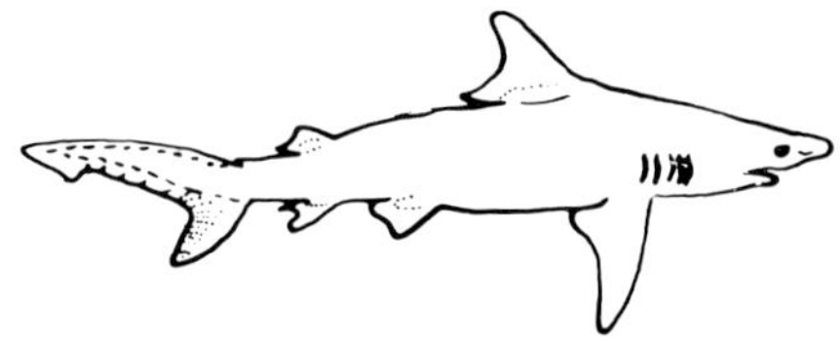

Requiem sharks constitute the largest family among the elasmobranch fishes. Their appearance resembles closely that of the shark species: two dorsal fins, paired pectoral and pelvic fins, an anal fin, and a long lopsided tail with the upper lobe much elongated. Carcharhinid sharks usually forage in relatively shallow waters and are diurnal. While feeding, vision apparently plays a major role during the final phase of search and attack.

The retinal structure in these sharks is strikingly similar to that of the hammerhead sharks (Sphyrnidae). In *Carcharhinus longimanus* the rods are largely predominant over the cones; they are slender cells with long inner segments (myoid and ellipsoid) and outer segments of about the same length (Plate 13.1). The cones are short and hard to distinguish (Plate 13.3). The three layers of horizontal cells appear to correspond to those present in the hammerheads (Plate 13.1, 4). There are few bipolar cells, but giant ganglion cells are present in this species (Plate 13.1, 4) as well as in the lemon shark (*Negaprion brevirostris*) and the tiger shark (*Galeocerdo cuvieri*). The retina of the latter species differs from that of other requiem sharks by the presence of large, blunt processes in the epithelial cell layer (Plate 13.2).

References: Gilbert (1961, 1963); Kato (1962); Gruber et al. (1963); Wang (1968); Gruber (1975)

Visual Pigments: A_1 max 497 (Lythgoe, 1972)

Plate 13.1-4. Transverse sections of the retinas of the whitetip shark, *C. longimanus* (1, 3 and 4), and the tiger shark, *G. cuvieri* (2). ×560. *G* giant ganglion cell, *H* large horizontal cell. *Arrows* indicate cones

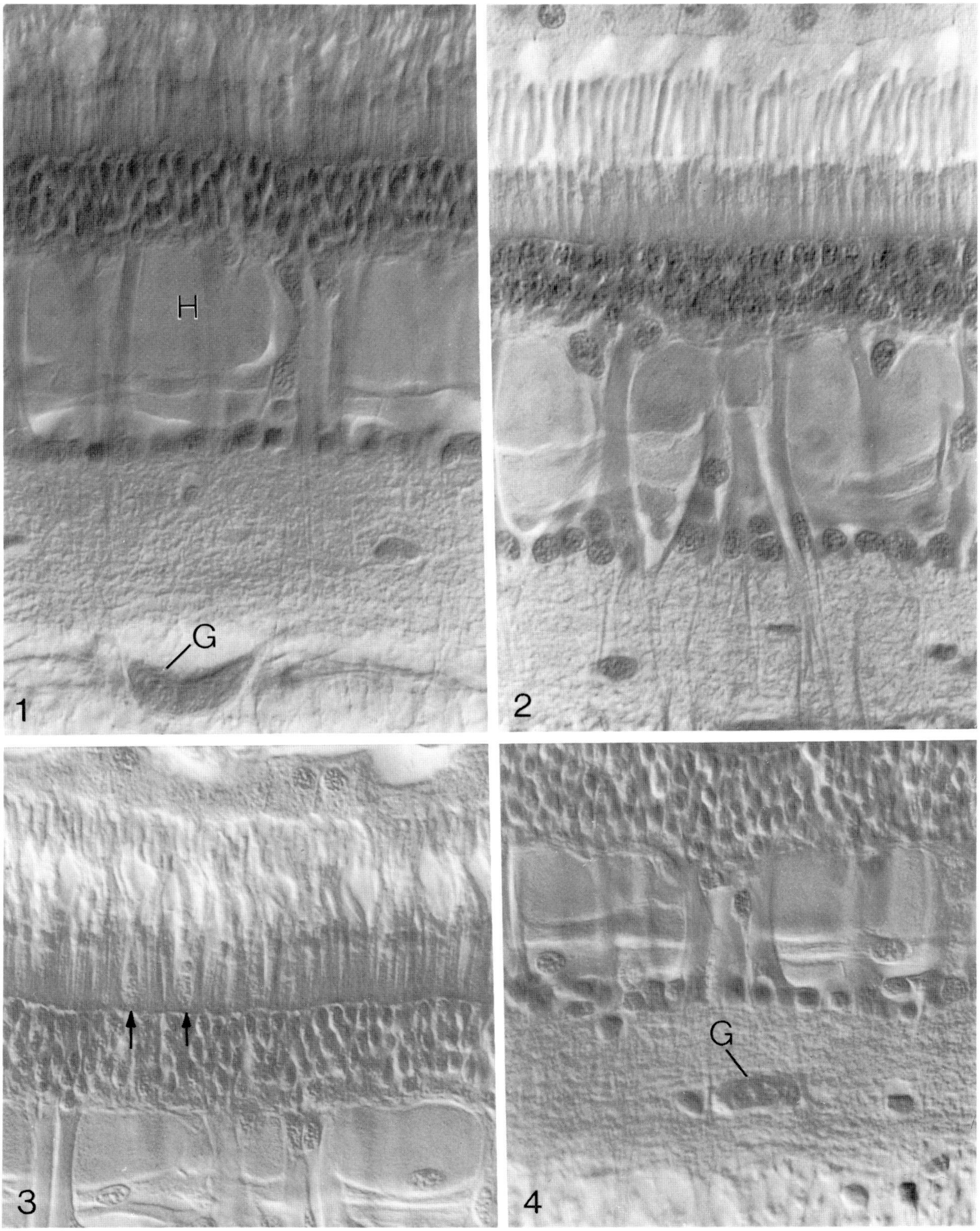

Plate 13.1-4. (Legend see opposite page)

Sphyrnidae (*Hammerhead Sharks*)

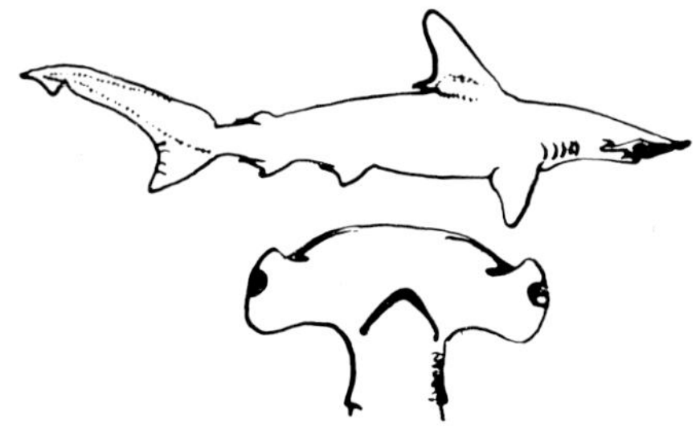

Chondrichthyes, Elasmobranchii, Selachii, Galeoidea

The characteristic features of the hammerhead sharks are those of the carcharhinids, except that the anterior portion of the head is much flattened dorso-ventrally and very widely expanded laterally, with the eyes at its outer edges. These sharks are found either very close to the shore or far offshore. Many are strong swimmers. The sluggish species feed on invertebrates in shallow water.

The retina of the young hammerhead shark is almost pure-rod and the single cones are difficult to distinguish (Plate 14.1). The rods are very slender and densely packed (ca. 160,000/sq mm). The photoreceptor synaptic pieces form a distinct row just above the horizontal cell layers (Plate 14.1). The latter form three layers of large cells: the external (Plate 14.1,2) made of cuboidal cells, the intermediate (Plate 14.1,3), and internal (Plate 14.1,4) layers made of flattened cells with long processes. Note the large Müller's fibres penetrating the intercellular spaces between the horizontal cells. The bipolar and amacrine cells are very numerous (Plate 14.1). Giant ganglion cells are found in the amacrine cell layer (Plate 15.1), in the internal plexiform layer (Plate 15.2), and in the ganglion cell layer proper (Plate 15.3).

References: GRUBER et al. (1963), report cones similar to those of *Negaprion* in *Sphyrna mokarran*. ANCTIL and ALI (1974)

Plate 14.1-4; Plate 15.1-3. Transverse (14.1; 15.1-3) and tangential (14.2-4) sections of the retina of the scalloped hammerhead shark, *Sphyrna lewini*. ×560. *G* giant gangliom cell ▷

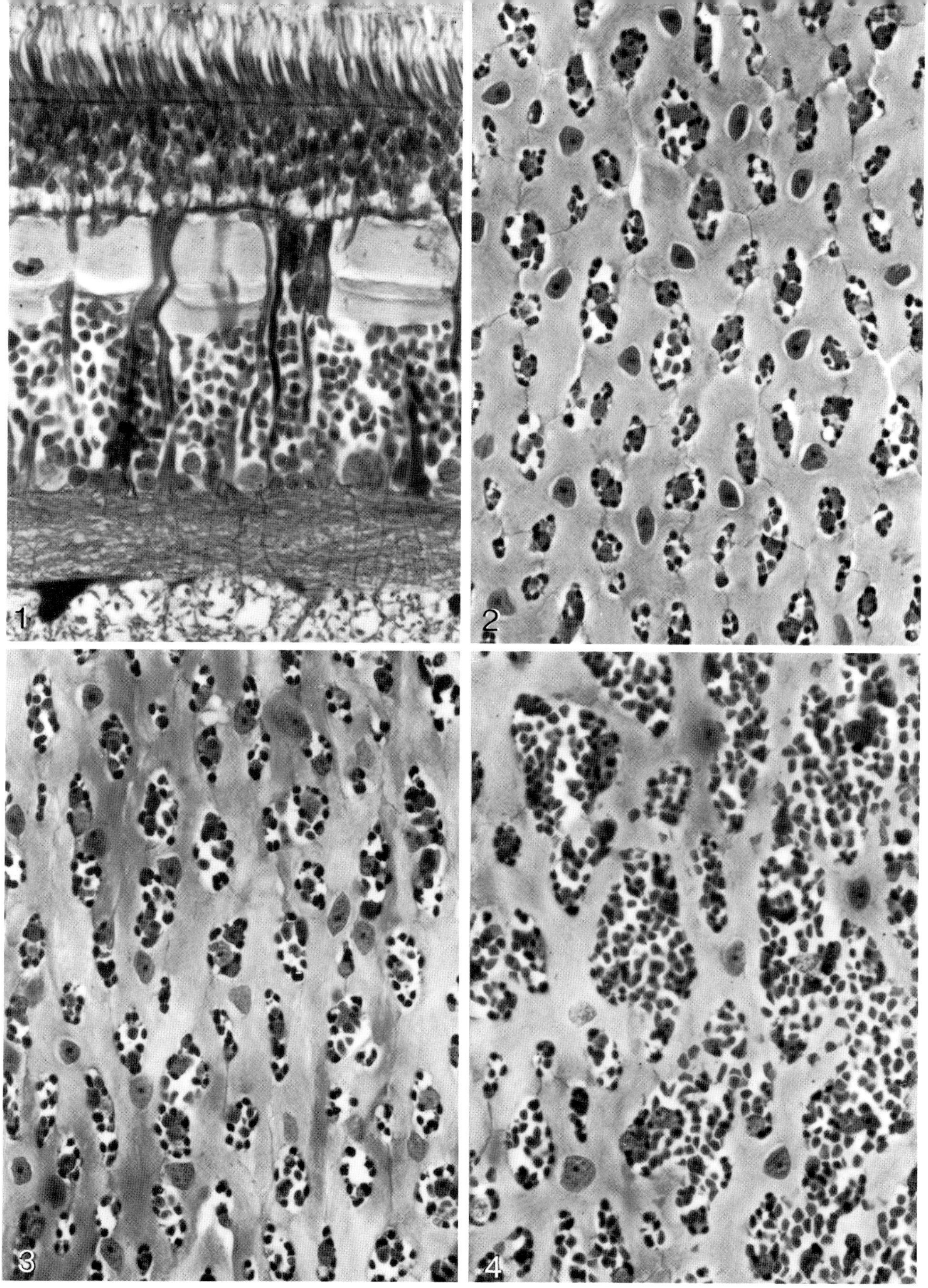
1
2
3
4

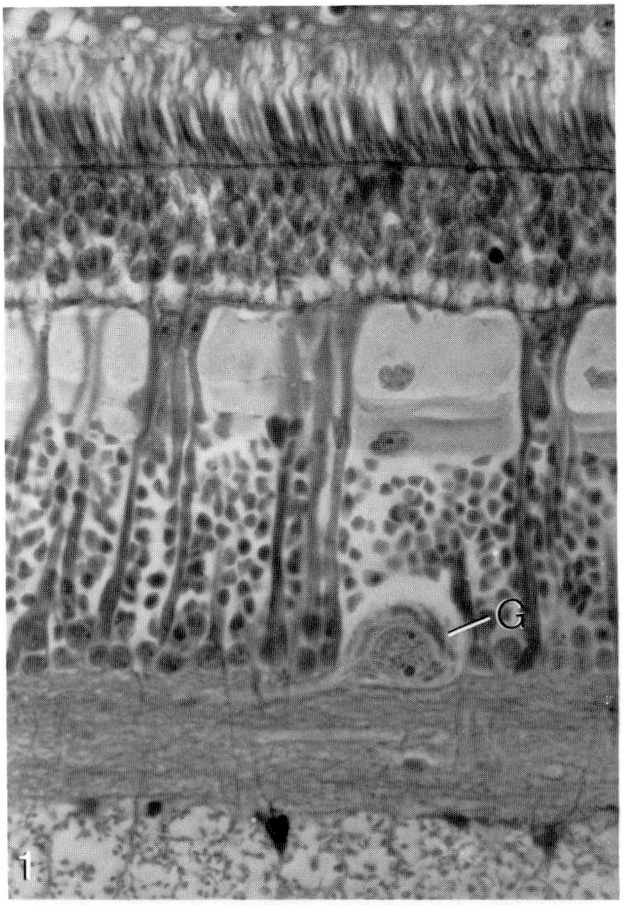

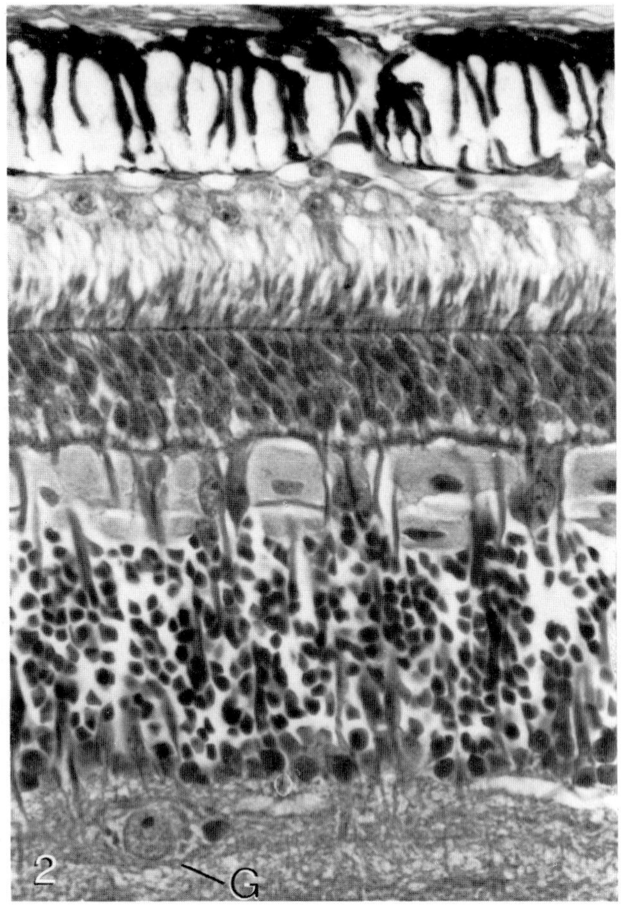

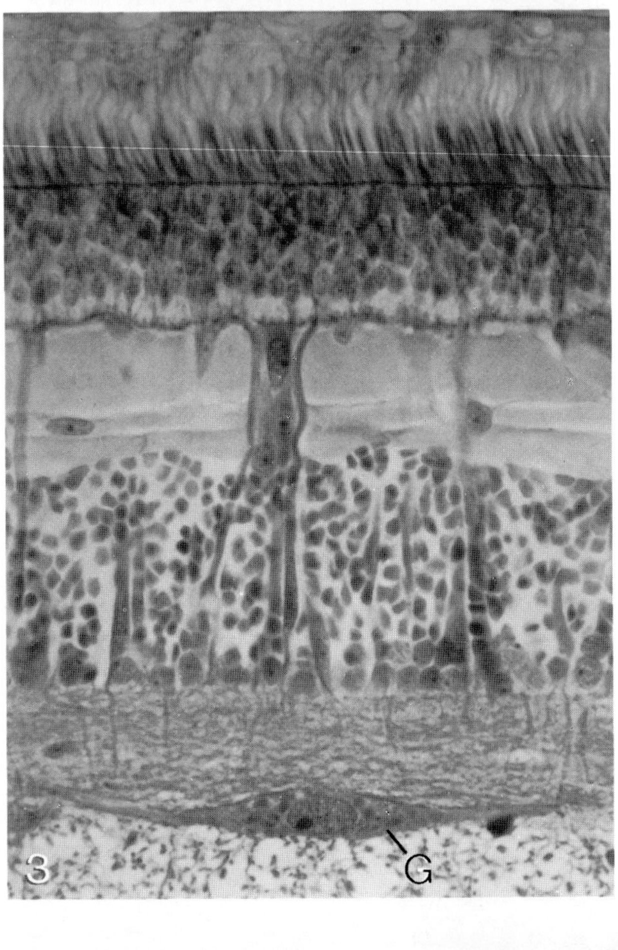

Plate 15.1-3. (Legend see page 32)

Squalidae (*Dogfish Sharks*)

Chondrichthyes, Elasmobranchii, Selachii, Squaloidea

Squalid sharks are characterised by the presence of a spine in front of each of the two dorsal fins. Squaloids have a fusiform and compressed body, lateral eyes, and nostrils without whiskers. *Squalus acanthias* is a sluggish fish and a facultative schooling species; it may live at various depths along the coast. Dogfishes of the genera *Centroscyllium* and *Etmopterus* are confined in distribution to the deep sea and possess light organs.

Both rods and single cones are present in the retina of *S. acanthias* and *S. suckleyi*. The rod: cone ratio is about 50:1. The rod and cone myoids are large, and the rod and cone pedicles are much alike. There are at least two layers of large horizontal cells. The bipolar, amacrine, and Müller's cell bodies form little more than a single layer. Pure-rod retinas were reported in *Centrophorus calceus, C. granulosus, Centrina salviani* and *Etmopterus* sp. (132,000 fundus-rods/sq mm).

References: Retzius (1896, 1905); Franz (1905, 1910, 1913, 1920, 1931); Verrier (1930a); Kolmer (1936); Rochon-Duvigneaud (1943); Stell (1972)

Visual Pigments: A_1 max 497.5 (Lythgoe, 1972)

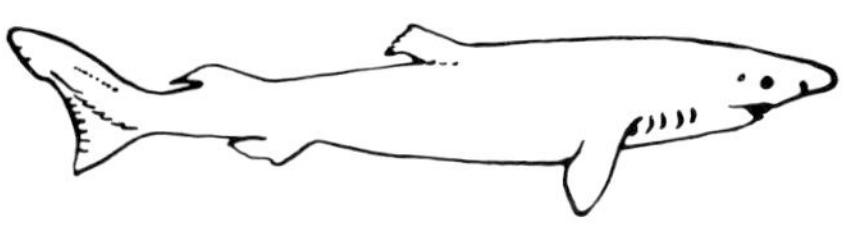

Dalatiidae (*Spineless Dogfishes*)

Chondrichthyes, Elasmobranchii, Selachii, Squaloidea

These sharks are very closely allied to the squalids, from which they differ by the loss of one or both dorsal spines. They are pelagic in habit.

Pure-rod retinas were reported in *Somniosus microcephalus* and the semi-abyssal *Scymnus lichia* (= *Dalatias*).

Reference: Kolmer (1936)

Squatinidae (*Angel Sharks*)

Chondrichthyes, Elasmobranchii, Selachii, Squatinoidea

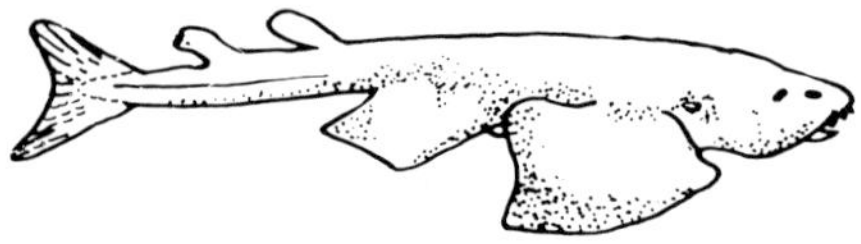

Although angel sharks resemble skates in general appearance and in some skeletal characters, they are classified with sharks because they share with them the selachian characters mentioned before (see Isuridae). Their eyes are placed dorsally and the nostrils possess branched barbels. Angel sharks live usually in very shallow water but also stray to deeper water. They are bottom fishes, often partially burying themselves in sand or mud. Both rods and cones have been reported in the retina of *Squatina* sp.

Reference: FRANZ (1913)

Torpedinidae (*Electric Rays*)

Chondrichthyes, Elasmobranchii, Batoidea, Torpedinoidea

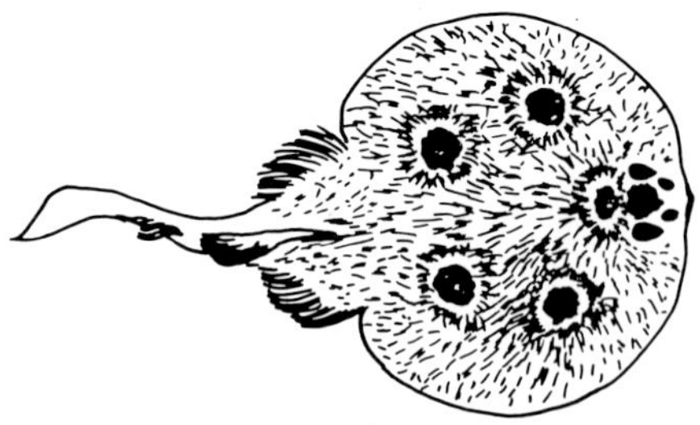

Skates and rays are highly specialised elasmobranchs that have gone far in developing adaptive features for a benthic life. The body is extremely flattened dorso-ventrally and the eyes are dorsal. The electric and torpedo rays have small eyes, which are functional in most species, but rudimentary in a few deep-sea forms. Electric rays, possessing two electric organs on the body, are sluggish in habit and lie on the bottom most of the time, partially buried in sand or mud. *Narcine brasiliensis* is a little electric ray found inshore in very shallow water.

The retina of *N. brasiliensis* is thin (Plate 16.1); rods and cones are present in the ratio 12:1. The rods resemble in shape those of the sharks, but are larger and fewer (Plate 16.1). The horizontal cells are of moderate size and form one or two layers. The bipolar and amacrine cells are few and sparsely distributed. Some ganglion cells are found in the internal plexiform layer (Plate 16.2). There is a difference in the retinal thickness and development between the posterior (Plate 16.1) and anterior part (Plate 16.2) of the eye.

References: BRAUER (1908); FRANZ (1905, 1913, 1931); VERRIER (1930a); ROCHON-DUVIGNEAUD (1958)

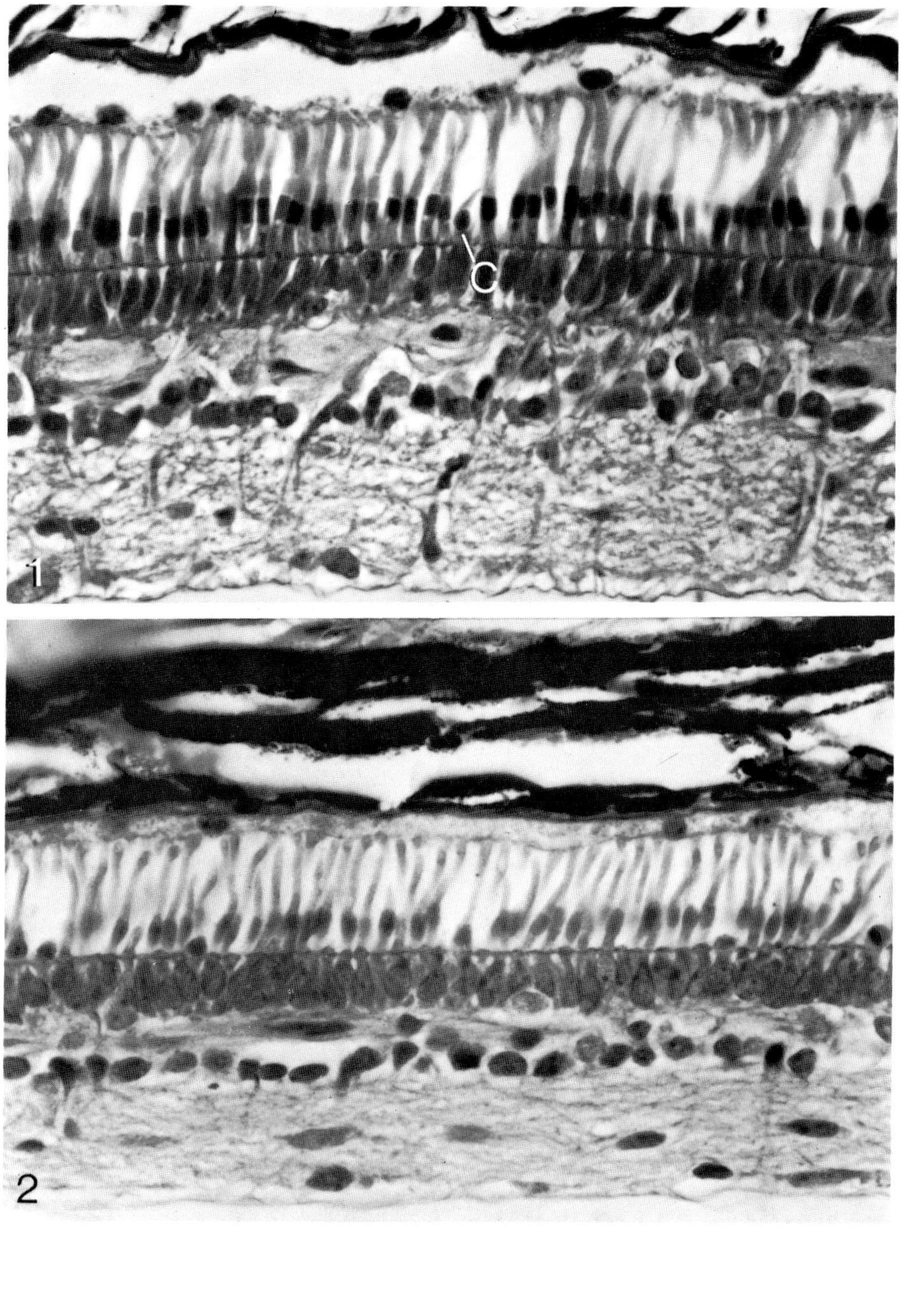

Plate 16.1, 2. Transverse sections of the posterior (1) and anterior (2) retina of the electric ray, *N. brasiliensis*. ×560. *C* cone

Rajidae (*Winter Skates*)

Chondrichthyes, Elasmobranchii, Batoidea, Rajoidea

Skates are benthic fishes inhabiting the continental shelf at various depths. They are sluggish and often live half-buried in the sand or mud. The skate eats mainly benthic invertebrates; it usually swims over its victim, then suddenly falls upon it and devours it.

The winter skate has a pure-rod retina (Plate 17); the rods are similar to those of the shark retina, with long myoids and small ellipsoids. The rod nuclei form about two layers. The horizontal cells are present and appear to form one or two layers. The bipolar and amacrine cells are moderately abundant in the internal nuclear layer. The ganglion cells are relatively few and tend to lie in rows parallel to the optic nerve fibre bundles.

References: PAPPENHEIM (1842); FRANZ (1905, 1913, 1931); VERRIER (1930a); ALI et al. (1968); DOWLING and RIPPS (1970)

Visual Pigments: A_1 max 497 (LYTHGOE, 1972)

Dasyatidae (*Sting or Whip Rays*)

Chondrichthyes, Elasmobranchii, Batoidea, Myliobatoidea

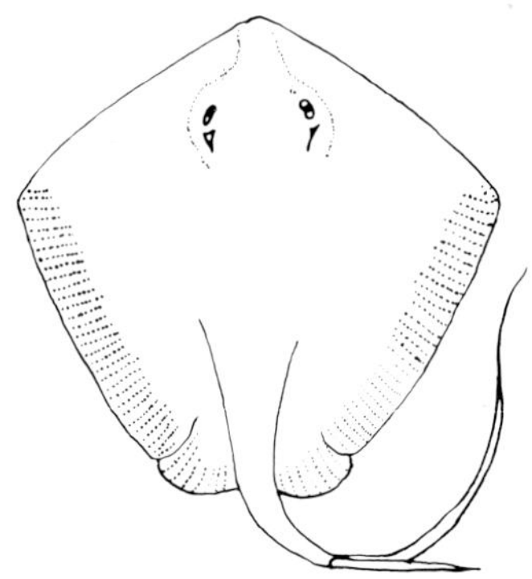

As in most batoids, stingrays have dorsal eyes and spiracles. They are characterised by a venomous spine associated with the dorsal fin or replacing it. Stingrays commonly lie on the bottom in shallow water, and often bury themselves in the sand or the mud with their tails, eyes, and spiracles exposed. They feed on benthic invertebrates by excavating the bottom.

Rods and cones are present in the retina of *Trygon pastinaca* (= *Dasyatis americana*). There are three layers of visual cell nuclei, and layers of large horizontal cells that strikingly resemble those of the smooth dogfish and hammerhead shark. Bipolar cells are sparse.

References: ROCHON-DUVIGNEAUD (1958); PROTASOV (1960); BERNSTEIN and DIETRICH (1960); GRUBER et al. (1963); HAMASAKI and GRUBER (1965); YAMADA and ISHIKAWA (1965)

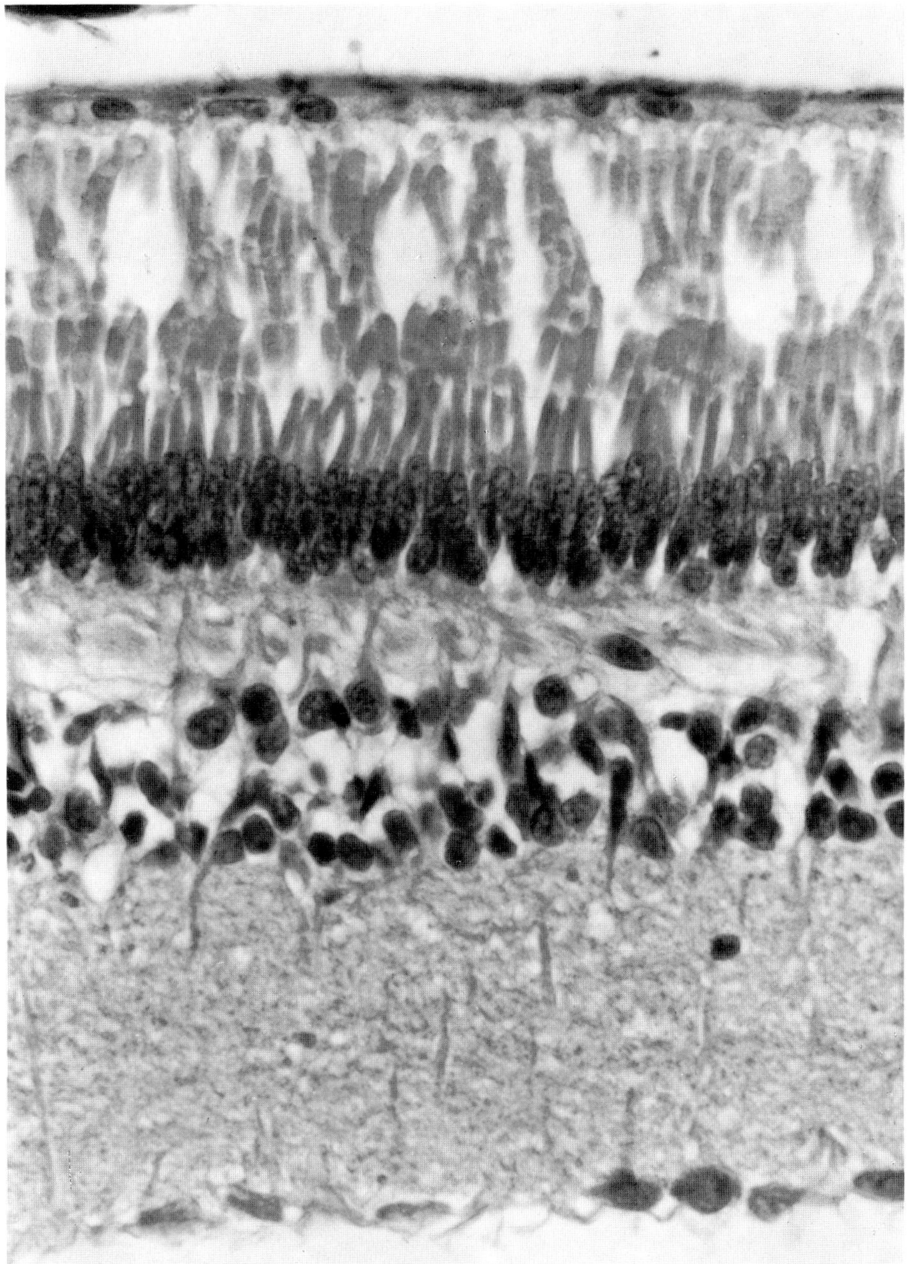

Plate 17. Transverse section of the ventral retina of the winter skate, *Raja ocellata*. ×785

Myliobatidae (*Eagle Rays*)

Chondrichthyes, Elasmobranchii, Batoidea, Myliobatoidea

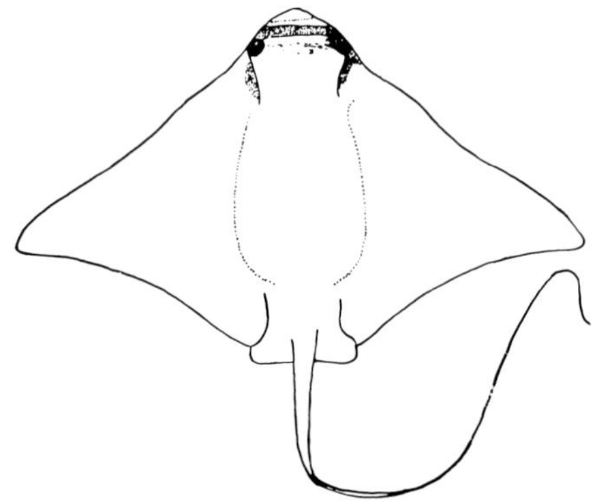

Eagle rays are characterised by the presence of a large, fleshy pad extending around the front end of the head. The eyes and spiracles, unlike in the other rays, are lateral. The eagle rays are pelagic and active swimmers, but they feed exclusively on the bottom.

Rods and cones are present in the retina of *Myliobatis aquila*. The rods are slender; the cones have large, oblong ellipsoids and short outer segments. The cone ellipsoids are found vitread to the rod ellipsoids. Although the cones have not been counted, they appear to be nearly as numerous as those of the freshwater stingray. At least one layer of large, cuboidal horizontal cells are present. The bipolar and amacrine cells form one or two layers.

Reference: VERRIER (1930a)

Potamotrygonidae (*Freshwater Stingrays*)

Chondrichthyes, Elasmobranchii, Batoidea, Myliobatoidea

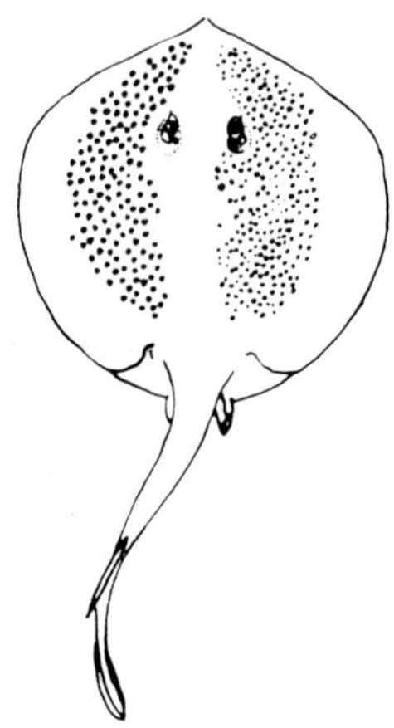

Freshwater stingrays are close relatives of the dasyatid-rays from which they are distinguished by the presence of a long, median, forward-directed process on the pelvis. Dasyatid-rays inhabit marine and brackish waters, whereas the potamotrygonids are exclusively freshwater rays from the tropics. The eyes of these rays are like longitudinal slits.

The retina of the freshwater stingray is very thin (Plate 18.1) when compared with the adjacent chorioidal tapetum lucidum. Both rods and cones are present, and the rod/cone ratios average around 7:1 and 6:1, which are the lowest ratios so far reported in elasmobranch retinas. Note in Plate 18 (3) (tangential section, fundus retina) the large number of cone ellipsoids. The horizontal cells are large and form one or two poorly defined rows (Plate 18.1,4). The bipolar and amacrine cells are large and moderately abundant. Large ganglion cells are present in the internal plexiform layer and in the nerve fibre layer (Plate 18.1,2). Note that the posterior part of the retina (Plate 18.2) is thicker than the lower (fundus) retina (Plate 18.1).

Reference: ALI and ANCTIL (1974a)

Visual Pigments: A_1 max 500 (MUNTZ et al., 1973)

Plate 18.1-4. Transverse (1, 2) and tangential (3, 4) sections of the retina of the freshwater stingray *Paratrygon motoro*. × 560. *CH* choriordal tapetum lucidum ▷

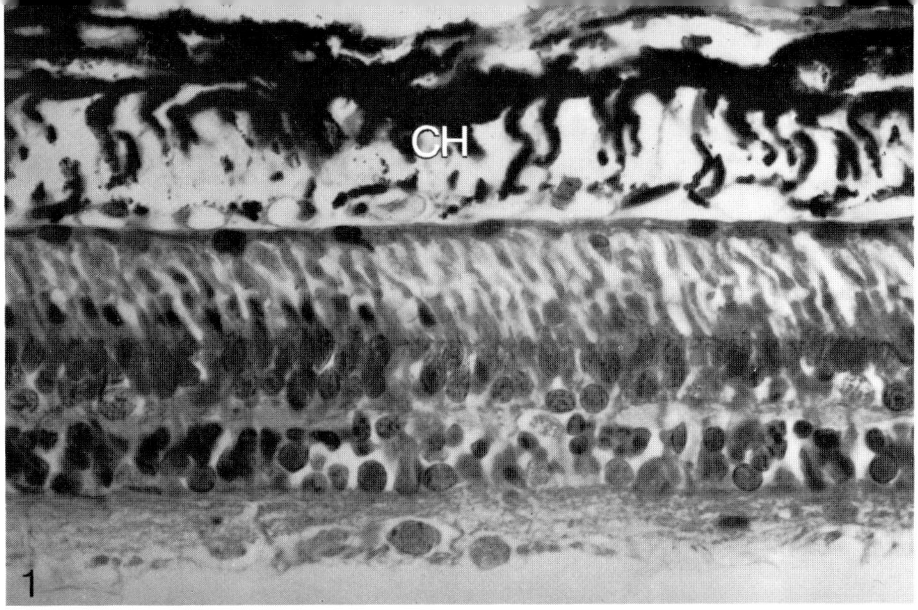
CH
1

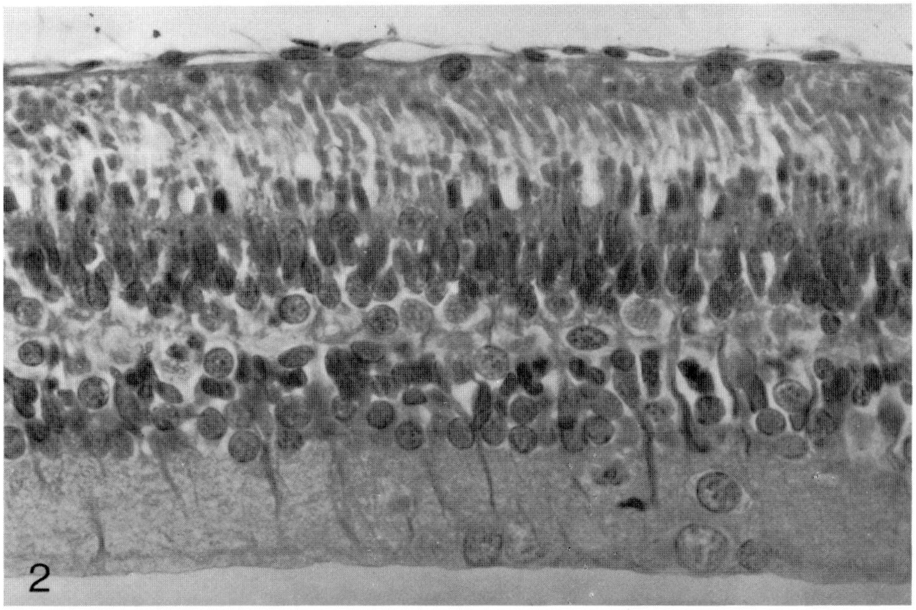
2

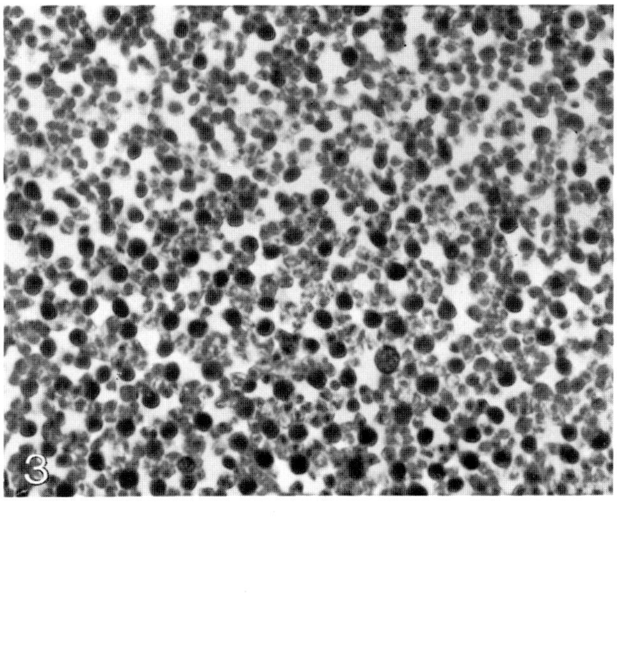
3

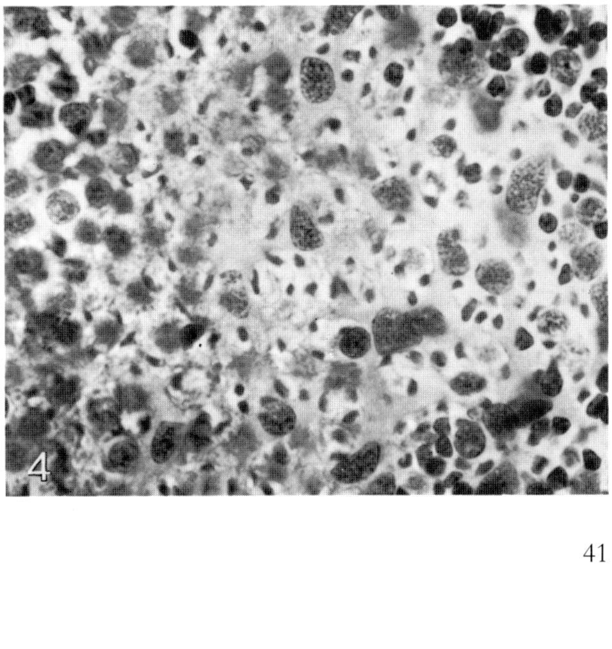
4

Chimaeridae (*Chimaerids and Ratfishes*)

Chondrichthyes, Holocephali, Chimaerae

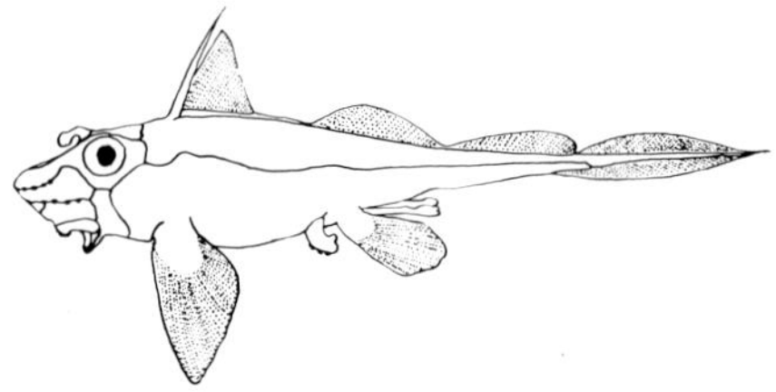

Chimaerids are aberrant forms of sharks thought to have evolved from a sharklike ancestor. As in sharks, they have a cartilagenous skeleton and paired claspers used for internal fertilisation. But they also have a dermal opercle covering the four pairs of gill openings, and an anal opening that is independant of the urogenital aperture, both characters shared by bony fishes. The short-nosed chimaerids are marine species living in deep shelf water and are benthic or bathybenthic in their habits. They have large eyes and are omnivorous.

The retina of *Chimaera* sp. is apparently a pure-rod one. The estimate of rod density in the fundus-retina is 100,000/sq mm.

Reference: Franz (1905, 1913, 1931)

Visual Pigments: A_1 max 484–499 (Lythgoe, 1972)

4.3. Osteichthyes (*Bony Fishes*)

Acipenseridae (*Sturgeons*)

Teleostomi, Actonopterygii, Chondrostei, Acipenseriformes

Sturgeons are regarded as primitive fishes, which have not changed significantly in general appearance and morphological features for millions of years. They have an elongate and fusiform body with five longitudinal rows of bony scutes. Their skeleton is cartilaginous for the most part. Their snout is greatly extended and they possess prominent chin barbels. Both freshwater and marine forms are known. The Atlantic sturgeon (*Acipenser oxyrhinchus*) is a bottom fish from the continental shelf and is anadromous. This is a relatively sluggish fish, which finds its food (worms, molluscs, sand lances) mainly by its tactile (barbels) and olfactory organs.

The retina of *A. oxyrhinchus* is poorly developed and made of relatively large cells (Plate 19.1). The short, blunt epithelial cell processes have little pigment. The visual cell layer is composed of large rods and single cones. The cones have each a large oil droplet in the centre of the ellipsoid, and a short, truncated outer segment. A row of displaced bipolar cells is seen vitread to the visual cell nuclei (Plate 19.1, 2). The horizontal cells appear to form two huge layers. The visual, horizontal, and bipolar cells are smaller and more abundant in the peripheral (Plate 19.2) than in the fundus-retina (Plate 19.1).

References: WALLS (1942); ALI et al. (1968); NICOL (1969); WAGNER (1972)

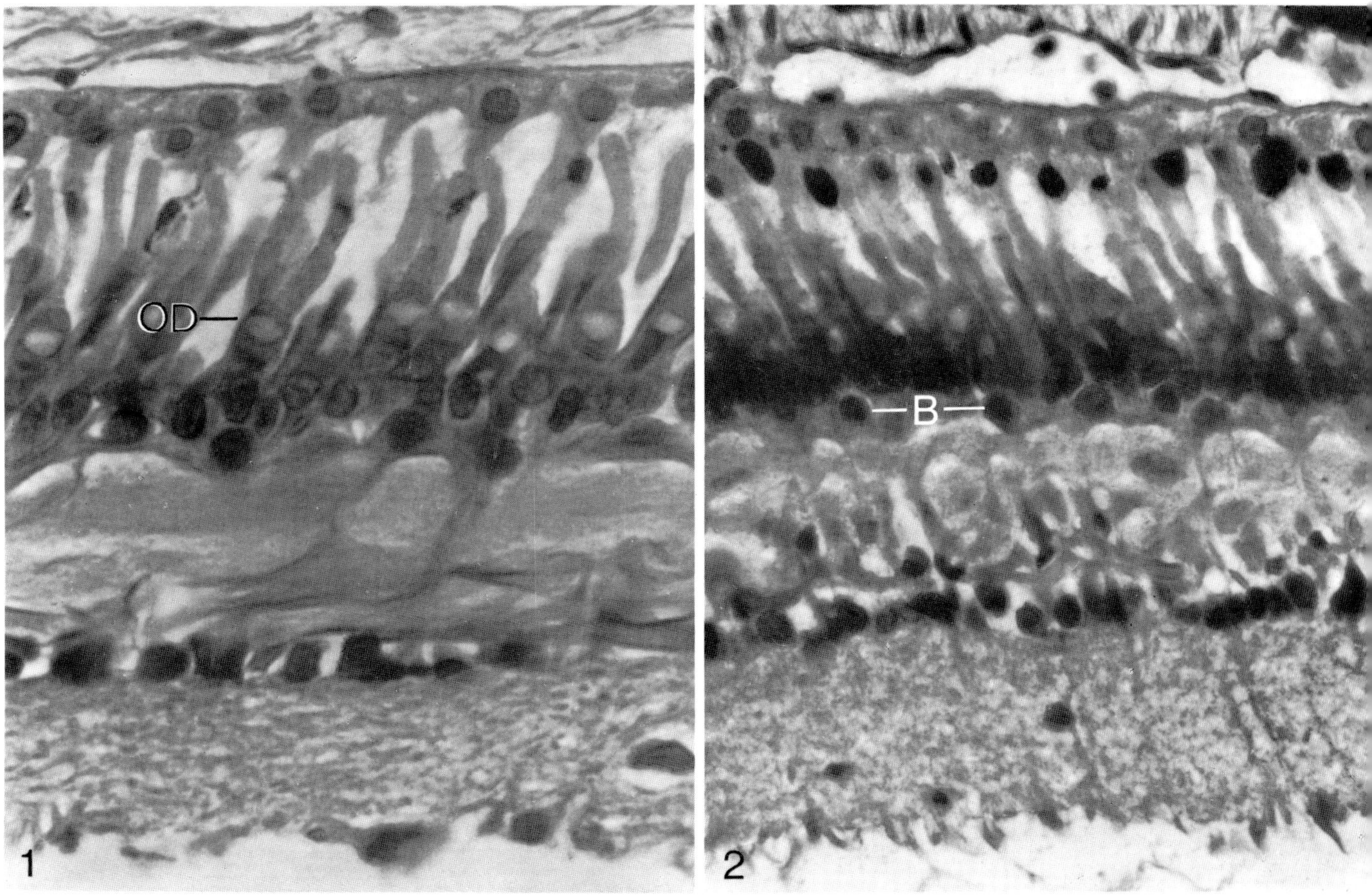

Plate 19.1, 2. Transverse sections of the ventral fundus-retina (1) and peripheral retina (2) of the Atlantic sturgeon, *A. oxyrhinchus*. ×990. *B* displaced bipolar cells, *OD* cone oil droplet

Polyodontidae (*Paddlefishes*)

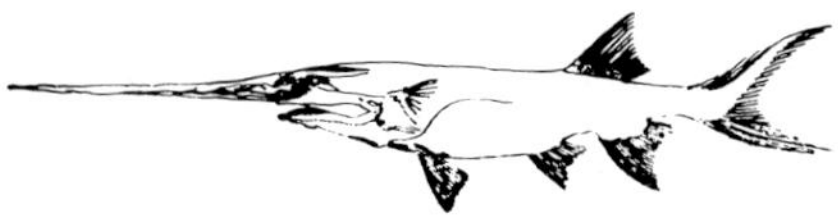

Teleostomi, Actinopterygii, Chondrostei, Acipenseriformes

Paddlefishes (*Psephurus gladius* and *Polyodon spathula*) have all the characteristics of the sturgeons, with the exception of their smooth, scaleless body and the monstrous paddle appendix attached to their nose. They have a large mouth and small eyes. The paddlefish catches its food by swimming with its mouth wide open; the long gill rakers strain the planktonic organisms thus engulfed. The paddlefish is strictly a freshwater fish.

The retina of *P. spathula* resembles that of the sturgeons in structure. Single cones with paraboloid and oil droplet, and long rods without these organelles constitute the visual cell layer.

Reference: MUNK (1969b)

Lepisosteidae (*Gars*)

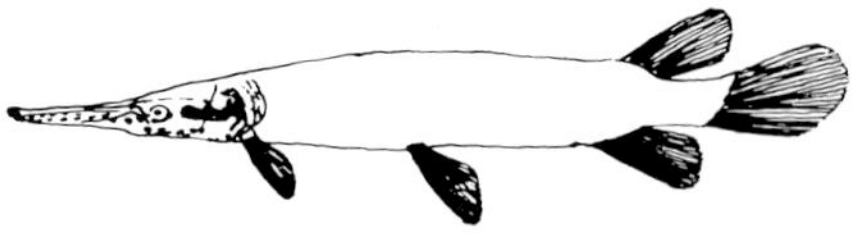

Teleostomi, Actinopterygii, Holostei, Lepisosteiformes

Gars are close relatives of the bowfin. They are characterised by an elongated body covered with ganoid scales arranged in rhombic flat plates which do not overlap. Gars have the same habitat and habits as those of *A. calva.*

The retina of *Lepisosteus productus* is essentially similar to that of the bowfin. The epithelial cell processes are more heavily pigmented in the ventral retina. In the dorsal retina, where the tips of the rod outer segments are in close contact with the epithelial cell bodies, the rods are very long.

References: MCEWAN (1938); MUNK (1968b)

Visual Pigments: A_2 max 523 (LYTHGOE, 1972)

Polypteridae (*Bichirs*)

Teleostomi, Brachiopterygii, Polypteriformes

Polypterids constitute a small African family of primitive freshwater fishes. They are characterised by an elongated body, ganoid scales, which are rhombic in shape, and a pair of gular plates. As in other primitive fishes, they possess a digestive tract with a spiral valve. As in the lungfishes they possess lungs that permit them to live out of water, but for a few hours only.

The retina of *Calamoichthys calabaricus* and *Polypterus* spp. is better developed than that of the lungfishes, and is made of smaller cells. The rods are large and have only the paraboloid, whereas the single cones have both the oil droplet and paraboloid. The principal element of the unequal double cone has both the oil droplet and a small paraboloid, while the accessory element has only a small paraboloid. Displaced bipolar cells form one layer immediately vitread to the visual cell nuclei. There is one layer of small, flattened horizontal cells and the bipolar and ganglion cells are numerous.

References: ROCHON-DUVIGNEAUD (1943); MUNK (1964c); PFEIFFER (1968)

Amiidae (*Bowfins*)

Teleostomi, Actinopterygii, Holostei, Amiiformes

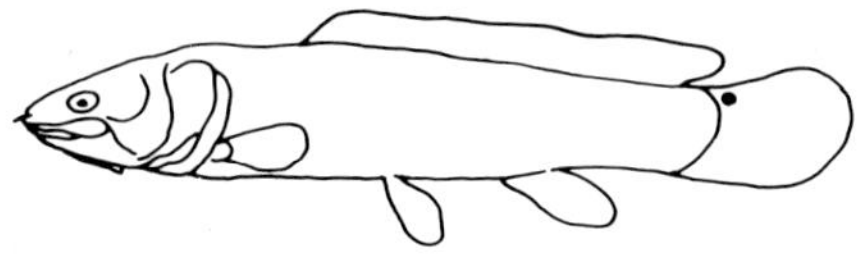

The bowfin is the only living representative of this family of primitive fishes, which flourished millions of years ago. It is restricted in distribution to some freshwater basins of North America. As in other primitive fishes, it possesses a spiral valve in the intestine and a gular plate on the lower jaw. The nostril barbels are prominent and the eyes are small. The bowfish inhabits shallow, weedy waters of bays, inlets, and lagoons. It is a relatively sluggish and carnivorous fish.

The retina of *A. calva* is better developed than the chondrostean retina and its cells smaller and more numerous (Plate 20.1). The epithelial cells are heavily pigmented and have relatively long processes. Rods and cones (single and unequal double) compose the visual cell layer; all of them possess a paraboloid of varying size (Plate 20.1), but no oil droplet. The single cone: double cone: rod ratio average 1:1:2. A row of displaced bipolar cells occupies the vitread margin of the external nuclear layer. There are two layers of small horizontal cells and the bipolar cells are numerous. The centre of cone abundance is in the ventro-temporal quarter (Plate 20.2).

References: McEwan (1938); Walls (1942); Munk (1968b); Gramoni and Ali (1970); Ali and Anctil (1974b); Ali and Wagner (1975a)

Visual Pigments: A_2 max 525 (Lythgoe, 1972)

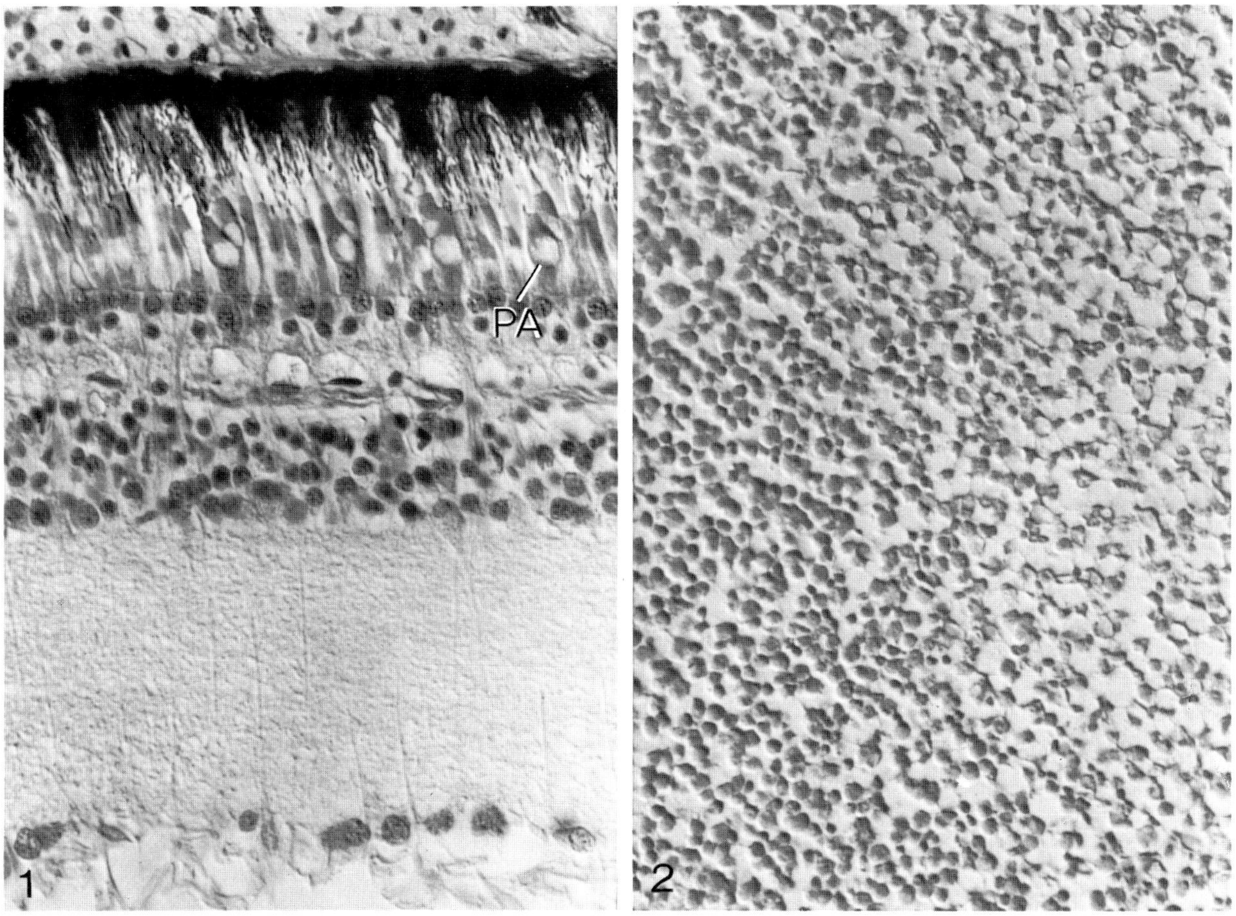

Plate 20.1, 2. Transverse (1) and tangential (2) section of the retina of the bowfin, *A. calva*. ×560. *PA* cone paraboloid. See also Plates 4 and 5 in Introduction

Elopidae and Megalopidae
(*Tenpounders, Tarpons*)

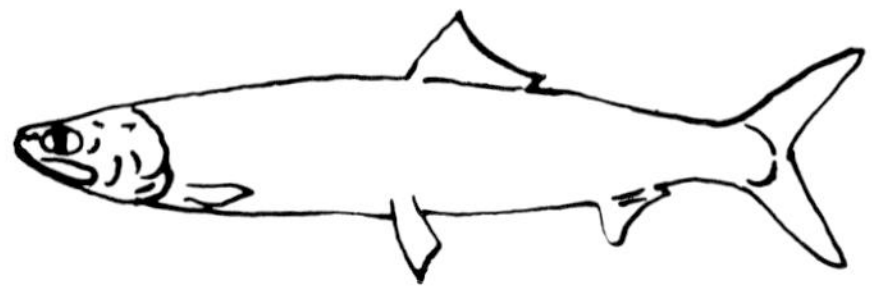

Elopomorpha, Elopiformes, Elopoidei

Elopomorphs constitute a primitive, divergent teleost group. Among them, the elopids and the megalopids are the most generalised forms as opposed to the more specialised Anguilliformes. They possess, however, a distinctive mark: a fairly large bony gular plate located under the mouth, between the two mandibles. These teleosts are primarily marine pelagic fishes.

In the retinas of the tenpounder (*Elops saurus*) and the Pacific tarpon (*Megalops cyprinoides*), only single cones have been reported. These cones are arranged in bundles ensheathed by a group of pigment cells. A tapetum lucidum system was reported in these pigment cells, possibly made of guanine.

Reference: McEwan (1938)

Albulidae (*Bonefishes*)

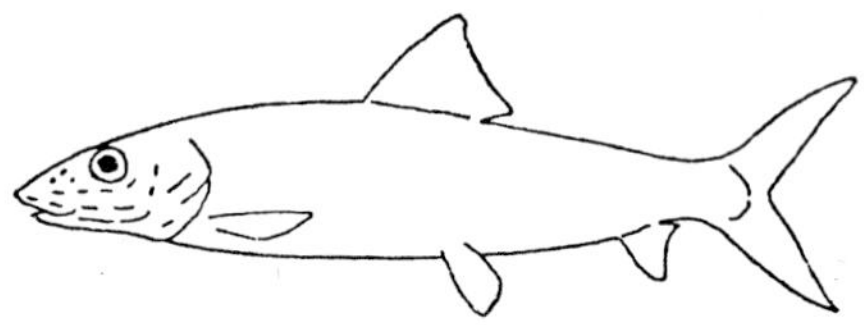

Elopomorpha, Elopiformes, Albuloidei

The silvery, pelagic bonefish is also a generalised elopomorph teleost, and like all the members of this superorder it has an eel-like, leptocephalus larval stage.

Single and double cones have been reported in the retina of *Albula conorhynchus*. The distal (sclerad) parts of the double cone ellipsoids are separate.

Reference: McEwan (1938)

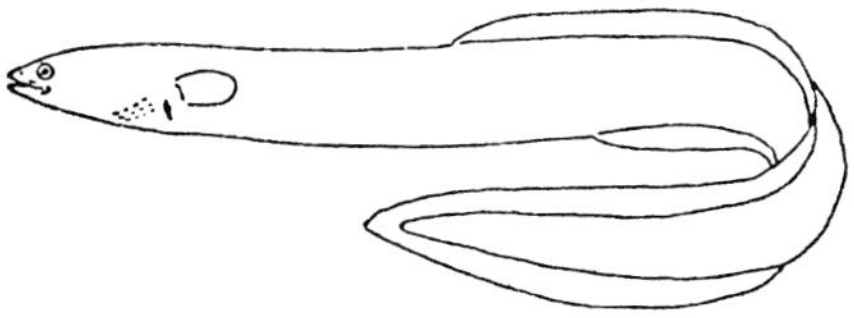

Anguillidae (*Freshwater Eels*)

Elopomorpha, Anguilliformes, Anguilloidei

These eels are specialised elopomorphs sharing a number of characteristics with some of their more generalised relatives such as the tarpons and tenpounders. They are elongate in shape and the pelvic girdle and fins are absent; they are also characterised by a leptocephalus larval stage as in all the elopomorph teleosts. The American eel is a catadromous fish, inasmuch as it lives in fresh water as an adult and undertakes seaward migrations to spawn. The American eels eat voraciously, chiefly at night; the main items are small fishes and invertebrates.

In their retina, the epithelial cell processes are well developed and filled with melanin pigment (Plate 21). The rods are numerous, whereas the cones are fewer and very small (Plate 22.1). Single cones, and probably double cones, are present (Plate 21; Plate 22.1). The cone nuclei are placed sclerally to the external limiting membrane and the cone ellipsoids are about the same size as the nuclei. Only two or three poorly defined layers of bipolar and amacrine cells are present, and the ganglion cells are few (Plate 22.2).

References: GARTEN (1907); WUNDER (1925); VERRIER (1928a); VILTER (1951b); MIRZALIEV and KOLOSS (1964); ALI et al. (1968); BEATTY (1975)

Visual Pigments: A_1 max 487, A_2 max 523. (ALI and WAGNER, 1975b)

Plate 21 ▷ Plate 22 ▷ ▷

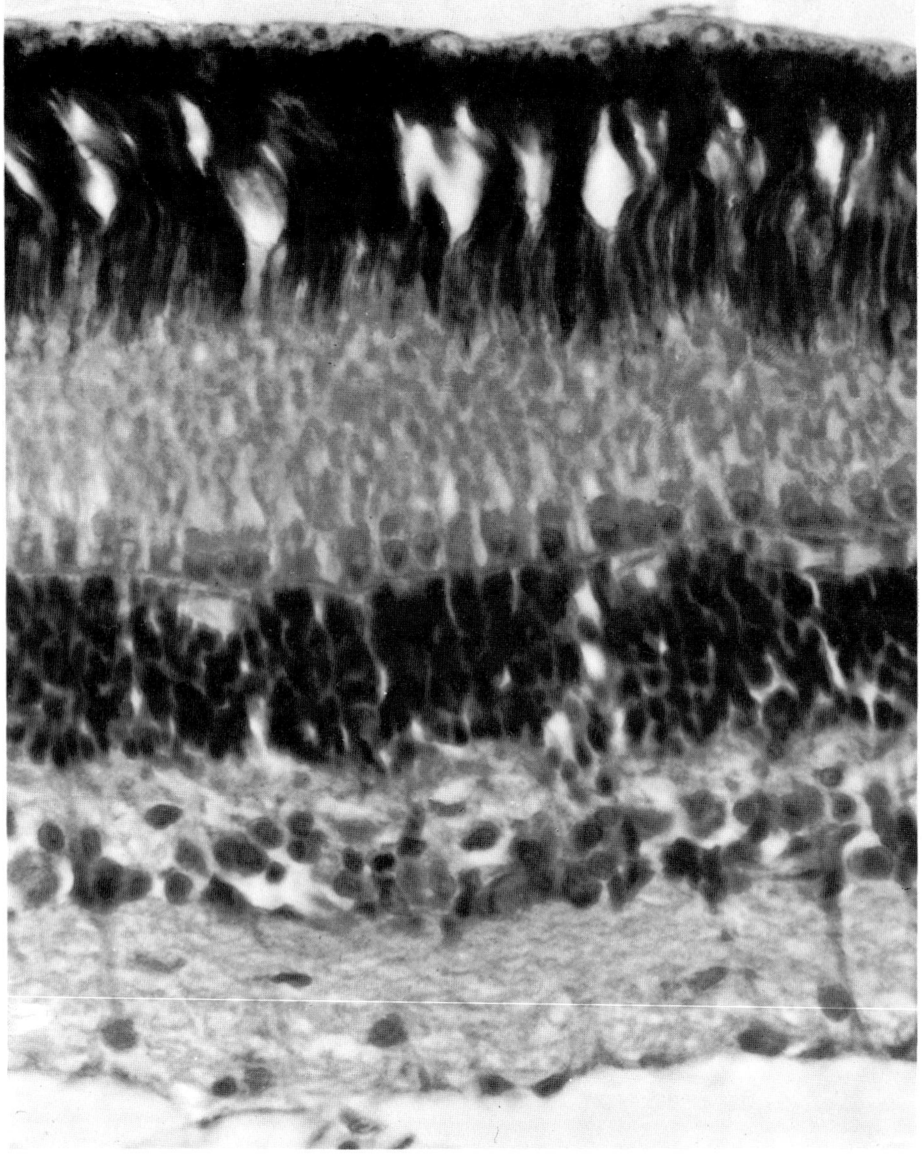

Plate 21. Transverse sections of the retina of *Anguilla rostrata*. ×650. Plate 22.1, 2. *A. anguilla*. Note the presumed double cone, *(arrow)* in (1). *B* bipolar cell layer, *RBV* retinal blood vessel. ×1,000

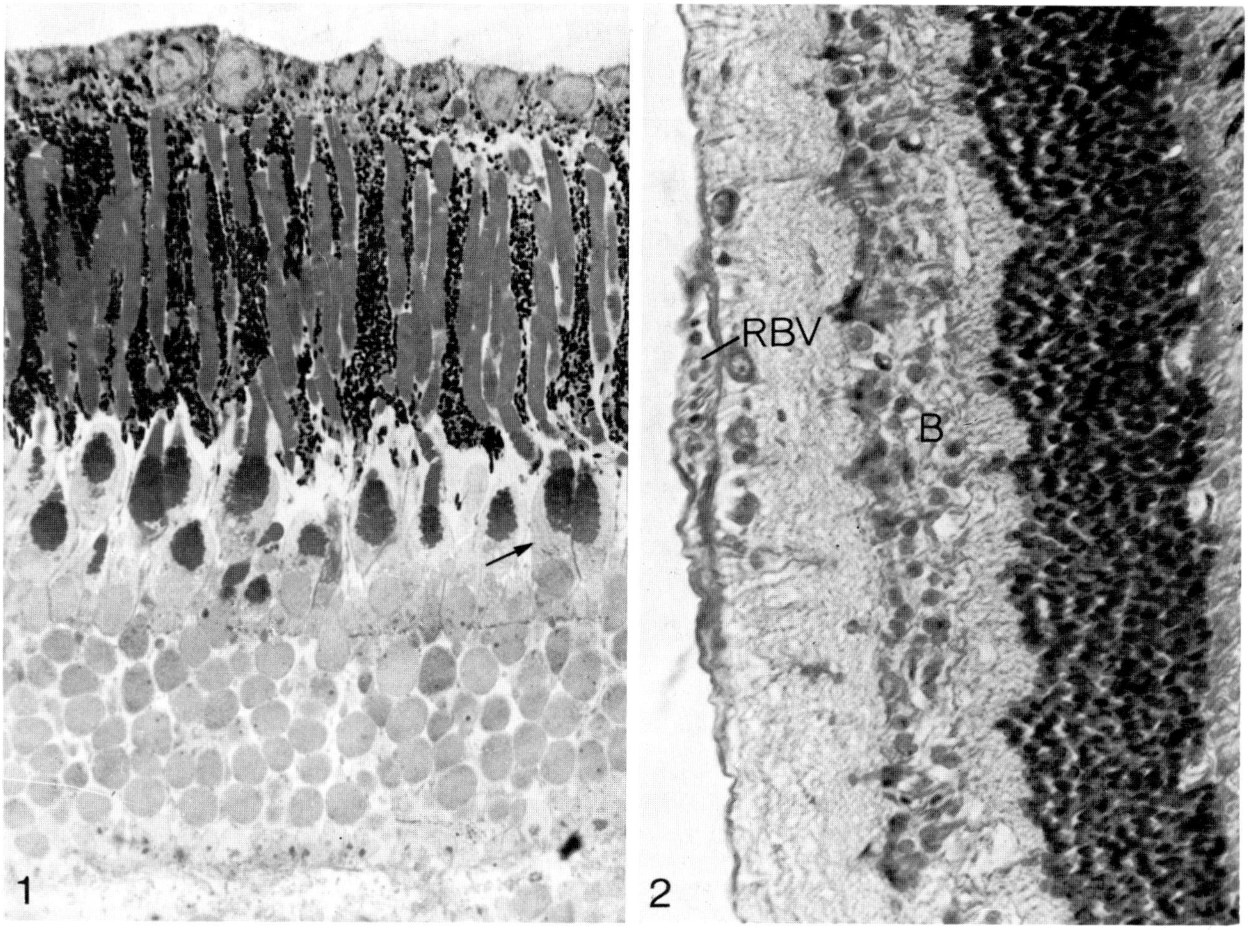

Plate 22.1, 2. (Legend see opposite page)

Muraenidae (*Moray Eels*)

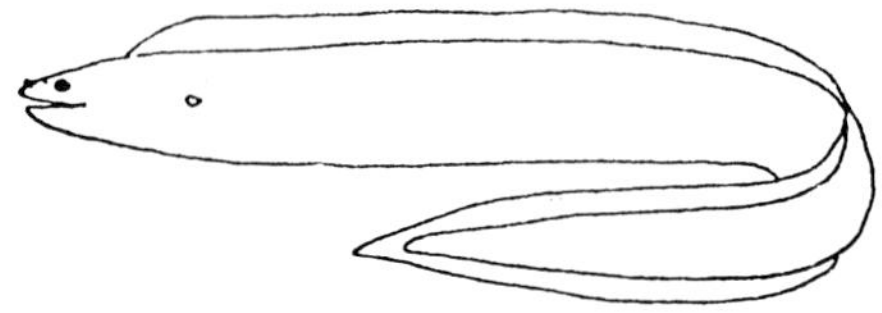

Elopomorpha, Anguilliformes, Anguilloidei

Moray eels are the most common of the rock-and-reef dwelling eels. Nearly all have strong jaws and fang-like teeth; they differ from most of the other eels inasmuch as they lack pectoral fins and lateral-line pores on the body and in the gill openings, which are small and round. Morays are very secretive, usually hiding during the day and coming out at night to forage. They are known to locate their food largely by smell.

The retina is characterised by abundant epithelial pigment, a large number of rods and very small cones with short outer segments (Plate 23.1). The mosaics are square with central single cones (Plate 23.2). There are apparently both single and double cones in the retina of the moray eel (Plate 23.2). The bipolar and ganglion cells are fairly numerous.

Visual Pigments: A_1 (ALI and HEUMANN, 1970)

Congridae (*Conger Eels*)

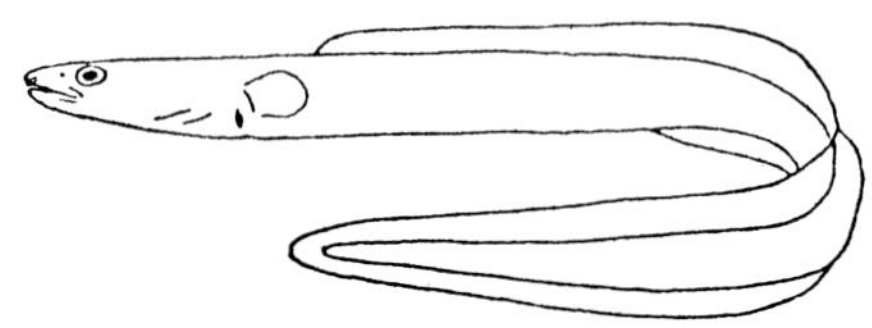

Elopomorpha, Anguilliformes, Anguilloidei

Congers are oceanic eels usually having light-coloured pectoral fins. Although congers are often found in shallow waters, they all apparently spawn in deep water.

The retinas of the following species have been studied: "*Leptocephalus*" *lacrymatus*, *L. mirabilis*, *Coloconger raniceps* and *Conger conger*. *L. mirabilis* possesses tubular eyes. All of these species have pure-rod retinas. The rods are slender, elongate, and very numerous. Some pigment granules are present in the epithelial cells and there is a single layer of well-defined horizontal cells.

References: BRAUER (1908); FRANZ (1910); VERRIER (1928a)

Visual Pigments: A_1 max 488–487 nm. (MUNZ and MCFARLAND, 1973; ALI and WAGNER, 1975b)

2. (Legend see opposite page)

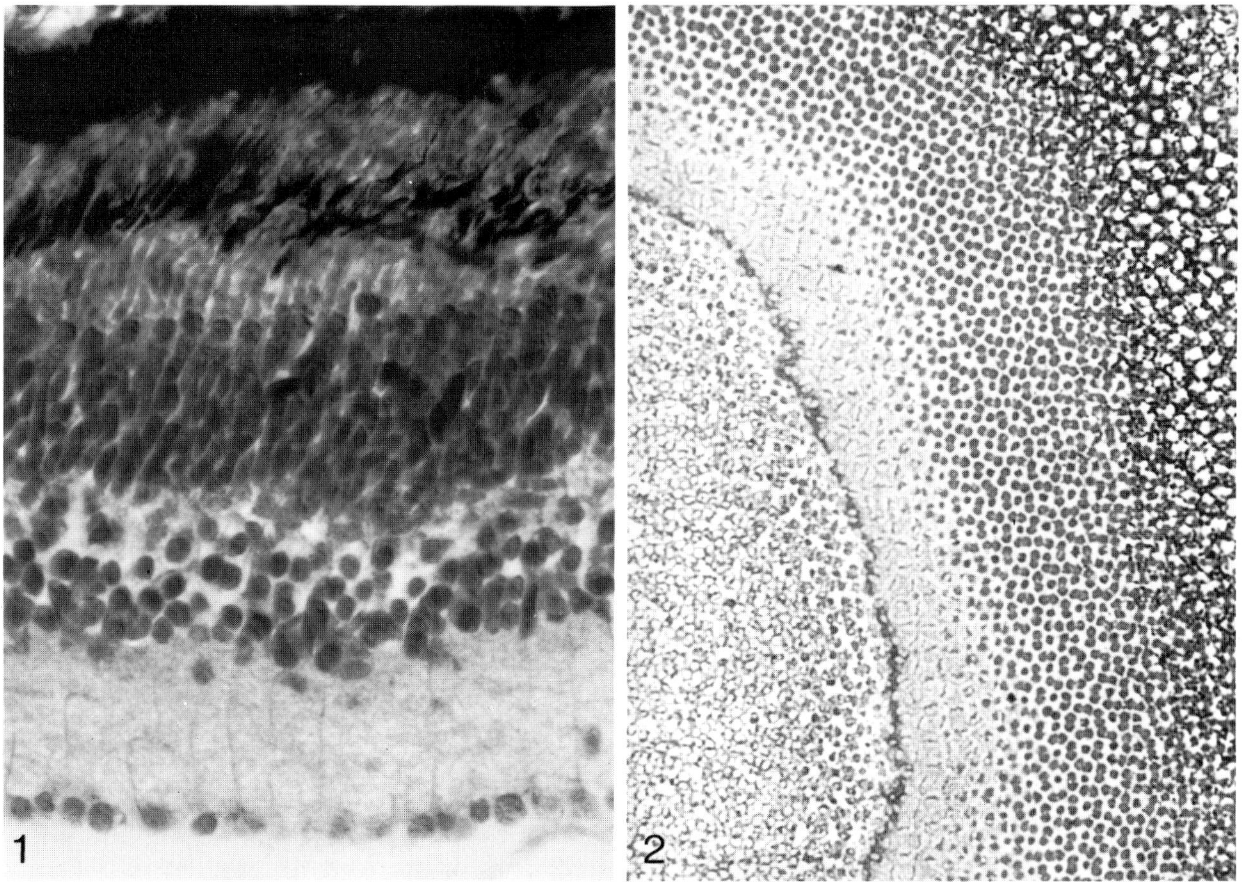

Plate 23.1, 2. (1) Transverse section of the retina of the moray eel, *Gymnothorax richardsoni*. ×700. (2) Tangential section of the dorsal retina of *G. ocellatus*. ×420

Ophichthidae (*Snake Eels*)

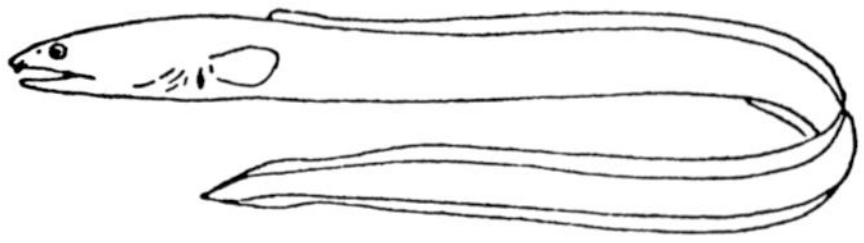

Elopomorpha, Anguilliformes, Anguilloidei

Snake eels are small and very slender. They are tail burrowers, having accordingly a very sharp, strong, spikelike tail. Snake eels are worldwide in distribution, however they are found principally inshore in the tropical seas.

The retina of the shrimp eel, *Ophichthus gomesi,* follows essentially the structural model encountered in freshwater eels (Anguillidae) and moray eels (Muraenidae). The rods are very slender and densely packed (Plate 24). The single and double cones are tiny and considerably less numerous than the rods. The external nuclear layer, mainly composed of rod nuclei, is extremely thick. The horizontal cells are few and scattered, while the internal nuclear layer is poorly developed.

Synaphobranchidae (*Deep-Sea Eels*)

Elopomorpha, Anguilliformes, Anguilloidei

Synaphobranchid eels appear to be intermediate in morphology between the moray eels and the conger eels. The pectoral fins are present and the gill openings are placed between them. They are mostly found in deep water, but the larval forms commonly appear near the surface.

The retina of *Synaphobranchus brevidorsalis,* caught in the bathypelagic zone, has been described. The retinal structure strikingly resemble that of *C. raniceps* (Congridae), except that the rods are less abundant and the horizontal cell layer appears to be missing.

Reference: Brauer (1908)

Nemichthyidae (*Snipe Eels*)

Elopomorpha, Anguilliformes, Anguilloidei

Snipe eels are elongate, deep-sea fishes with extremely long, needlelike upper and lower jaws; the upper jaw is bent upwards and the lower jaw downwards, therefore they cannot be closed. The tail usually tapers out to a long filament. The eyes of these eels are quite small.

Avocettina infans has a pure-rod retina; the rods are short, slender and very numerous. The epithelial cells are pigmented. There is one layer of horizontal cells, and the bipolar and amacrine cells each form one layer.

Reference: Brauer (1908)

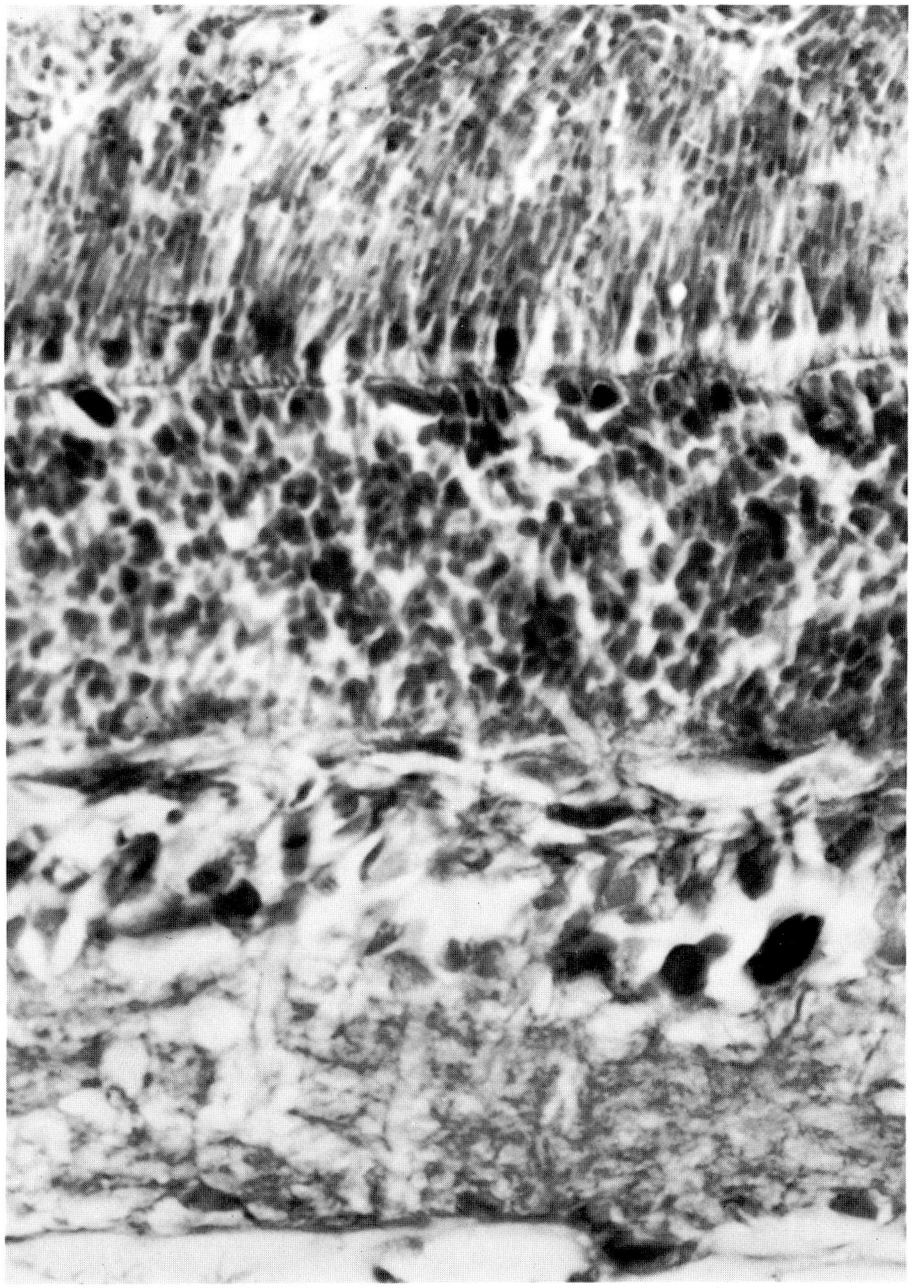

Plate 24. Transverse section of the retina of the snake eel, *O. gomesi.* ×825

Clupeidae (*Herrings*)

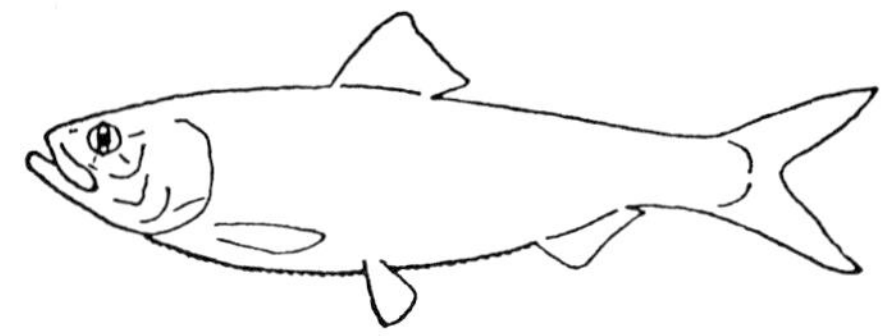

Clupeomorpha, Clupeiformes, Clupeoidei

Herrings comprise a large family of primarily marine fishes. Although the clupeomorpha constitute a clearly defined teleost group on morphological grounds, their relationship to other fishes remains obscure.

The herring is a coastal pelagic fish forming large schools and large aggregations. These fishes usually swim close to the surface and feed on planktonic forms.

The retina of the herring conforms to the well-developed teleostean retina. The epithelial pigment is moderately abundant, the rods are very numerous and slender, the cones are large, the horizontal cells are present in three layers, and the bipolar and ganglion cell layers are more or less well represented (Plate 25.1). The double cones constitute almost the unique cone type in most retinal regions (Plate 25.2), but in the ventro-temporal area retinae the single cones are about half as many as the double cones.

The alewife (*Alosa pseudoharengus*) is an anadromous fish, being found close to shore. It feeds chiefly on plankton. It resembles the herring in that it is a schooling fish.

The retina of the alewife is quite similar to that of the herring, except that the area retinae is ventral (Plate 25.3) and that some triple cones are seen in the fundus-retina (Plate 25.4, *arrow*). The retina of *Harengula jaguana,* a tide pool fish, is quite different. Note the prominent amacrines, Müllerian fibres and the difference between dorsal and ventral regions (Plate 26.1-4).

References: McEwan (1938); Vilter (1950); Baburina (1955); Baburina and Kovaleva (1959); Verheijen (1959); Engström (1963b); O'Connell (1963); Blaxter and Jones (1967); Anctil (1969); Wagner (1972); Zyznar and Nicol (1973); Ali and Wagner (1975a); Blaxter (1975); Wagner et al. (1976)

Visual Pigments: A_1 max 500–508; A_2 max 521–528 (Ali and Heumann, 1970, 1972; Ali and Wagner, 1975b)

Plate 25.1-4. Transverse (1) and tangential (2–4) sections of the retinas of the herring, *Clupea harengus,* and alewife, *A. pseudoharengus.* (1) ×320; (2) and (3) ×400; (4) ×530 ▷

Plate 26.1-4. Transverse sections (×550) of dorsal (1) and ventral (2) retinas of *H. jaguana*. Tangential section (×1,380) of dorsal and ventral retinas through the cones are shown in (3) and (4) respectively. *A* amacrine cell layer ▷ ▷

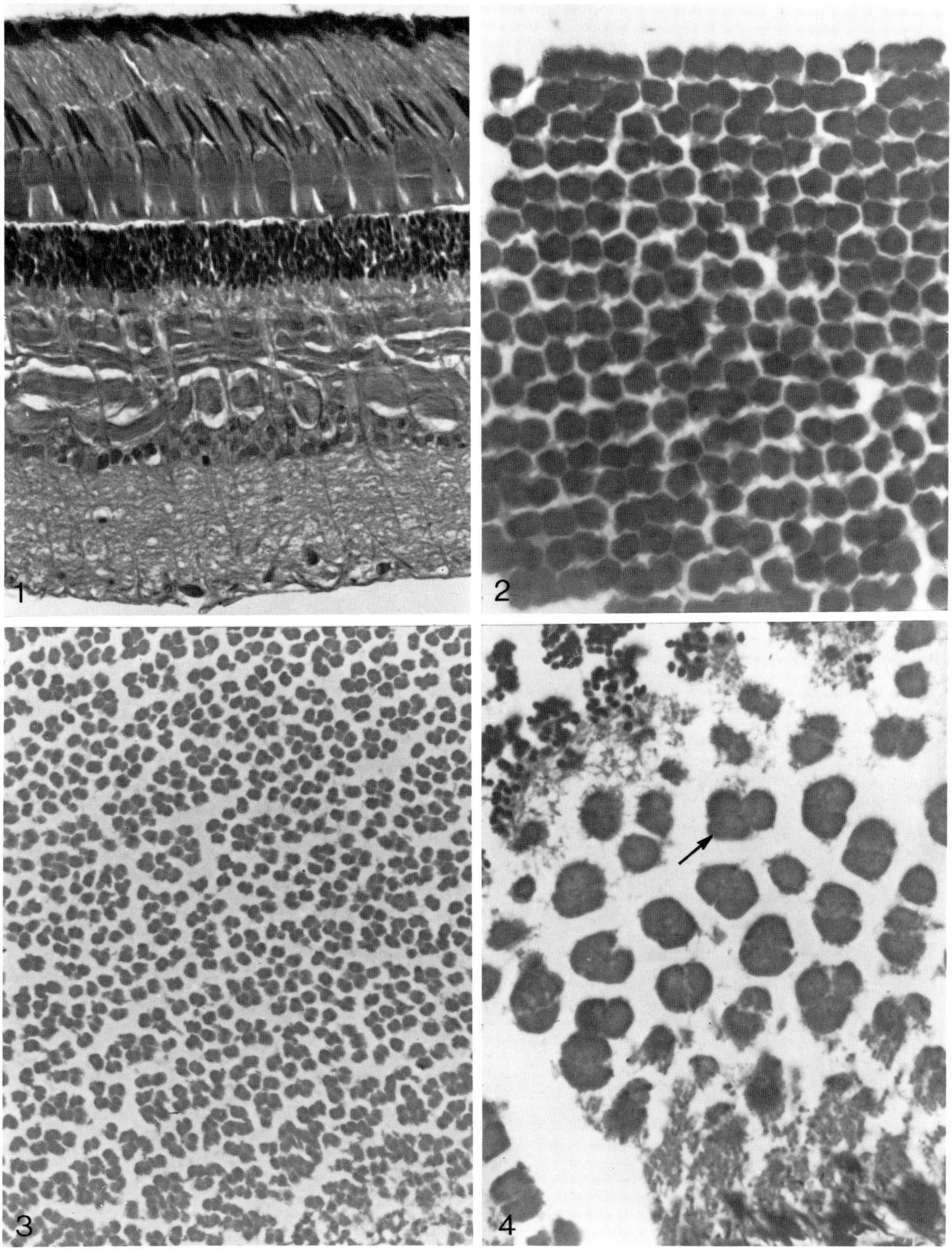
1
2
3
4

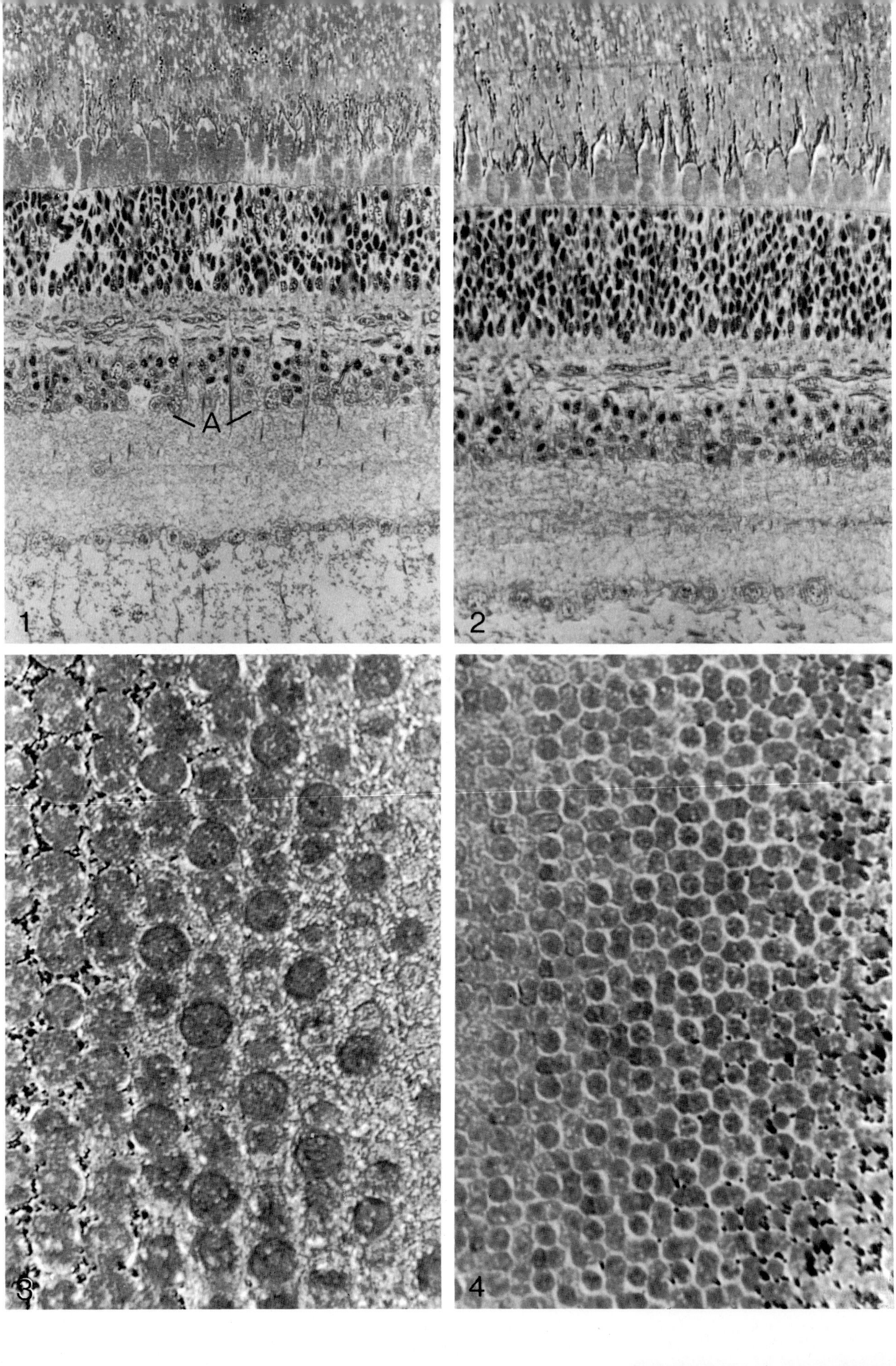
A
1
2
3
4

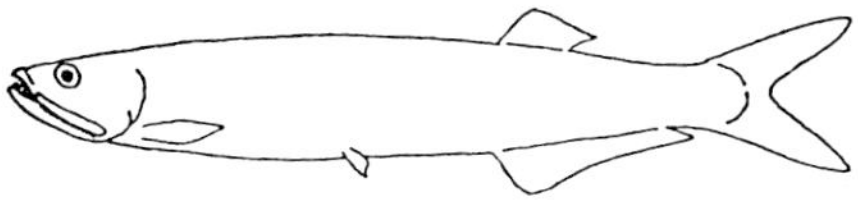

Chirocentridae (*Wolf Herrings or Dorabs*)

Clupeomorpha, Clupeiformes, Clupeoidei

This large clupeoid fish is characterised by the presence of a spiral valve in the intestine, a rare feature among the teleosts. This pelagic fish is a voracious carnivore, eating chiefly microplankton.

The only known mention of the retina of *Chirocentrus dorab* emphasizes the well-developed internal nuclear layer.

Reference: McEwan (1938)

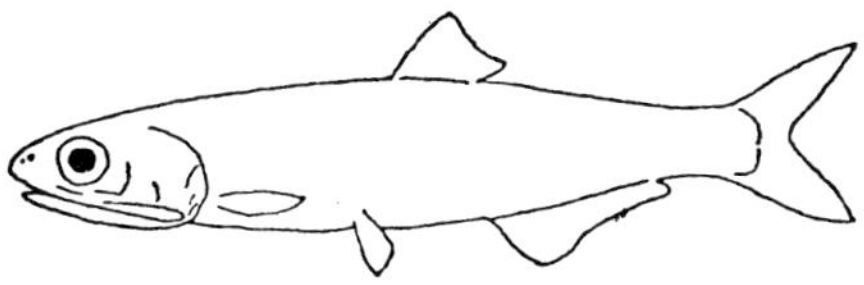

Engraulidae (*Anchovies*)

Clupeomorpha, Clupeiformes, Clupeoidei

Anchovies are marine pelagic fishes closely related to the herring family. They differ from herrings by the following characteristics: (1) the mouth of the anchovy is set far back on the underside of the head, (2) the lower jaw is small and inconspicuous, and (3) the adipose fin and lateral line are absent. Some species are distributed in coastal turbid waters.

Very slender cones (unidentified) and numerous rods have been reported in the retina of *Engraulis encrasicholus*. The retinas of *E. mordax* and *Anchoa compressa* are characterised by the presence of an occlusible tapetum lucidum system in the epithelial cell processes. The rods are slender and numerically well represented, however only equal double cones with no single cones have been reported in these species. The area retinae is located in the ventro-temporal region.

References: McEwan (1938); Sverdlick (1940, 1955); O'Connell (1963); Zyznar and Nicol (1973); Ali and Wagner (1975a)

Visual Pigments: A_1 (Ali and Heumann, 1972)

◁ Plate 26.1-4. (Legend see page 58)

Hiodontidae (*Mooneyes, Goldeyes*)

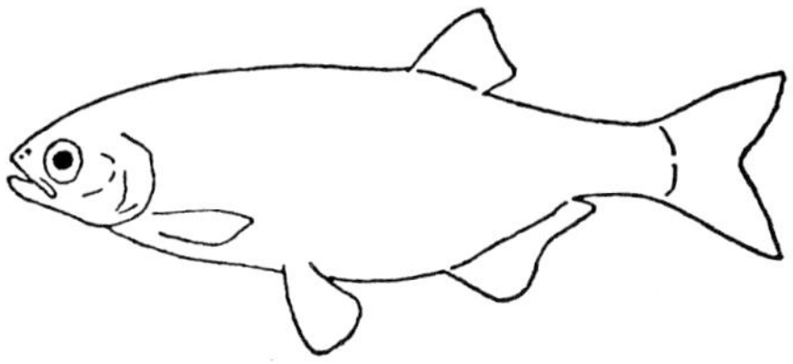

Osteoglossomorpha, Osteoglossiformes, Notopteroidei

The living osteoglossomorph teleosts are all freshwater fishes and appear to be a separate group, not closely allied to any other teleostean group. The mooneye family comprises three species of herringlike fishes, which are limited in distribution to central and eastern North America. The hiodontids are related to the Asiatic and African featherbacks (Notopteridae). *Hiodon tergisus* (mooneye) has a very large eye, a silvery colour, and inhabits mostly clear water. *H. alosoides* (goldeyes) is similar to *H. tergisus* in appearance, but differs by the golden colour of the eye and can tolerate muddy and turbid waters.

In both species mentioned above, the retina is characterised by a visual cell layer composed of rods and cones (Plate 27.1,4). These are arranged in bundles ensheathed in epithelial cells, like the cone bundles of the Elopiformes and Mormyriformes. The epithelial cells are apparently filled with reflecting tapetal material (Plate 27.2, 4).

References: Moore (1944); Moore and McDougal (1949)

Visual Pigments: A_2 max 535 (Beatty, 1973)

Plate 27.1-4. Transverse (1) and (2) and tangential (3) and (4) sections of the mooneye, *H. tergisus* (1) and (3) and the goldeye; *H. alosoides* (2) and (4) retinas. (1) ×292; (2) ×228; (3) ×182; (4) ×571. *ET* epithelial tapetum lucidum, *P* epithelial pigment, *R* rod bundle ▷

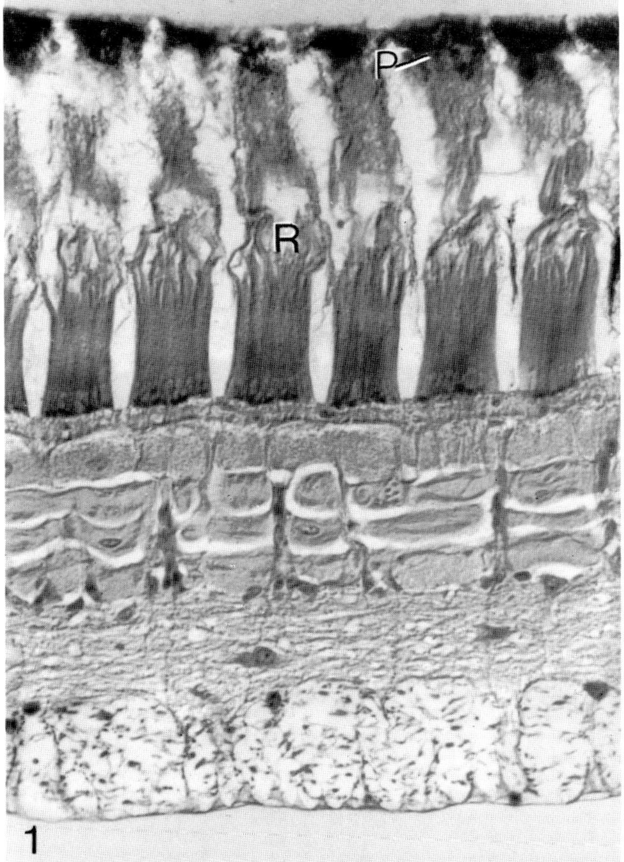

1

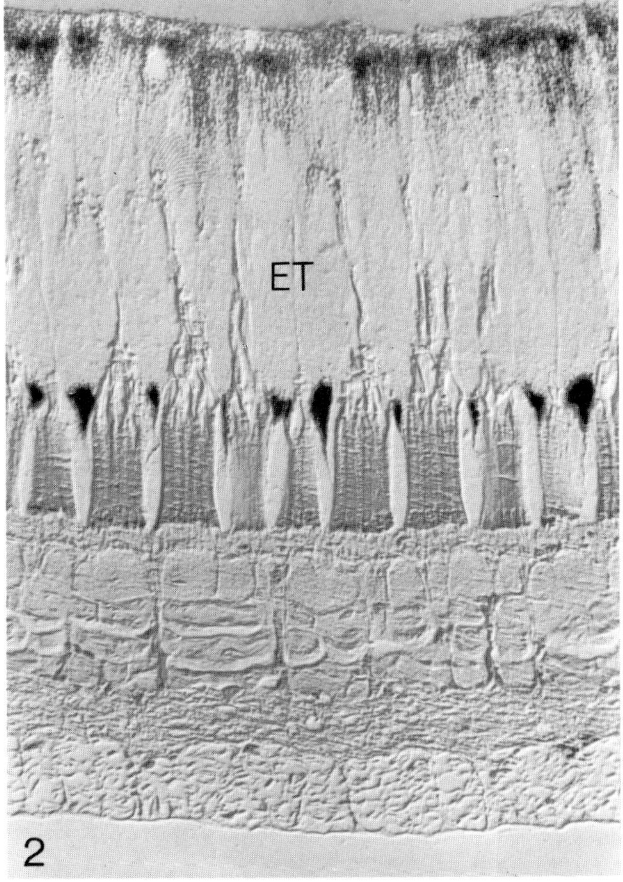

2

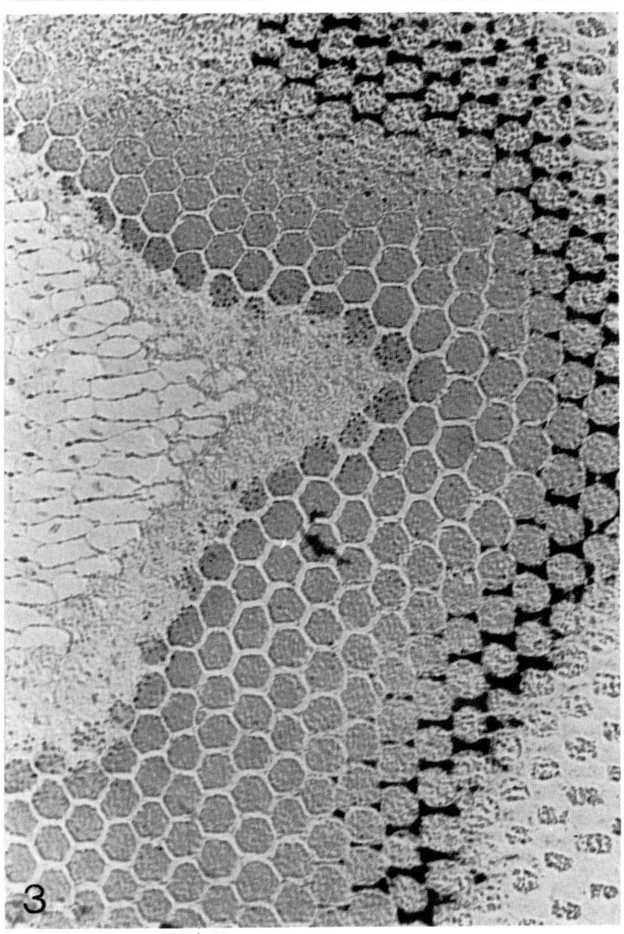

3

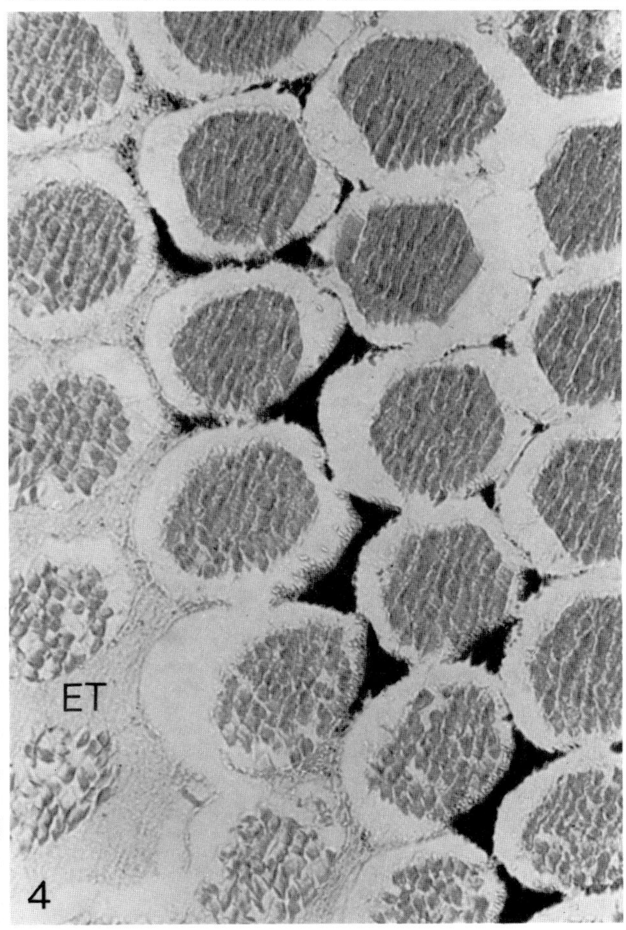

4

Notopteridae (*Featherbacks*)

Osteoglossomorpha, Osteoglossiformes, Notopteroidei

This is a small family of freshwater fishes from the Indo-Australian and West African regions. Featherbacks are easily identified by the very long anal fin extending from behind the head up to the tip of the tail. They possess small bones and a small, slender dorsal fin on the middle of the back, or, as in the false featherfin, *Xenomystus,* the dorsal fin is entirely missing.

The retinas of the featherback, *Notopterus notopterus,* and the featherfin, *X. nigri,* are characterised chiefly by the presence of a reflecting tapetal material in the large processes of the well-developed epithelial cells (Plate 28.1,3), and by the grouping of the visual cells into bundles (Plate 28.2,4). In *X. nigri,* the photoreceptor bundles comprise about 15–20 cones and 20–25 rods; rods and cones are segregated, rods forming a sclerad "bouquet" and the cones forming a distinct, vitread bunch with strikingly small outer segments (Plate 28.2,4). The horizontal cells form two or three layers in *N. notopterus* (Plate 28.1), but they are not so clearly defined in *X. nigri.* The bipolar and amacrine cells are relatively few, and the ganglion cells are moderately abundant.

Visual Pigments: A_1 max 503; A_2 max 527 (ALI and WAGNER, 1975b)

Plate 28.1-4. Transverse sections of the retinas of the false featherfin, *X. nigri* (1) and (2) and the featherback, *N. notopterus* (3) and (4). (1) and (4) ×350; (2) ×560; (3) ×225. *C* cone bundle, *ET* epithelial tapetum lucidum, *R* rod bundle

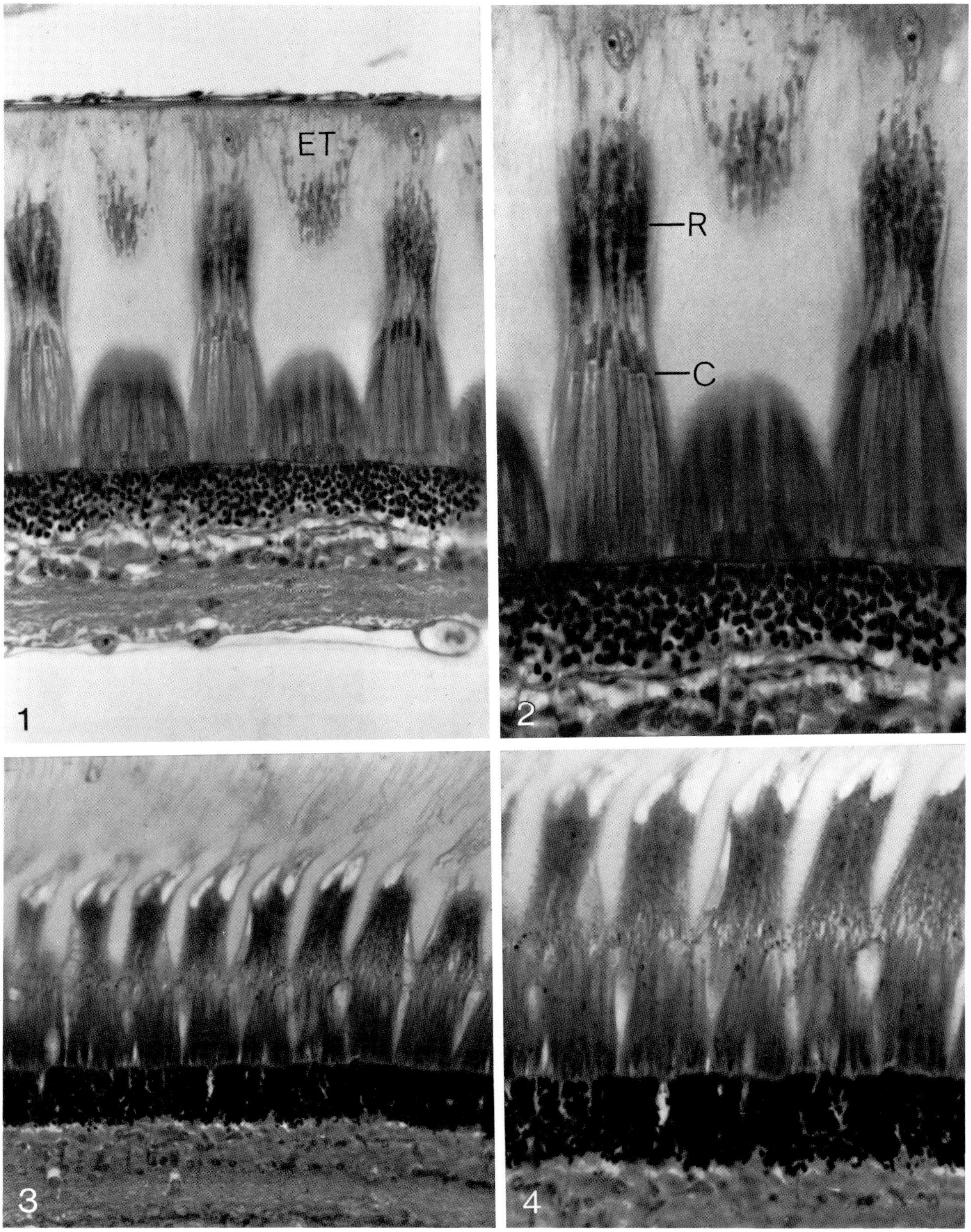

Plate 28.1-4. (Legend see opposite page)

Mormyridae (*Elephantfishes*)

Osteoglossomorpha, Mormyriformes

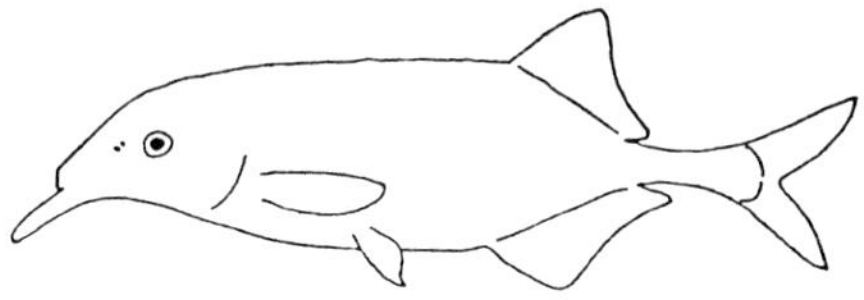

These freshwater fishes are confined to Africa. Some mormyrids have very atypical forms while others have peculiar appearances. Some species have a tuberous process on the jaw serving as a feeler, others have a greatly elongated proboscislike snout, both adaptations to a bottom life. The eyes are small and many species possess weak electric organs. Most of these species are solitary and crepuscular in habits, and live in muddy, slow-flowing waters.

The retinas of *Marcusenius longianalis, M. rudebeckii, Petrocephalus stuhlmanni, P. brevipedunculatus* and *Gnathonemus macrolepidotus* have been described. Rods and cones are grouped in bundles, each bundle being ensheathed by a tube formed by the processes of the pigmented epithelial cells. Only the single cones are present. Their ellipsoid and myoid segments cannot be differentiated. Several rods surround each cone in the bundle. The epithelial cells are packed with reflecting tapetal granules and little melanin pigment is present.

References: Franz (1920); McEwan (1938); Engström (1963b)

Gymnarchidae (*Gymnarchids*)

Osteoglossomorpha, Mormyriformes

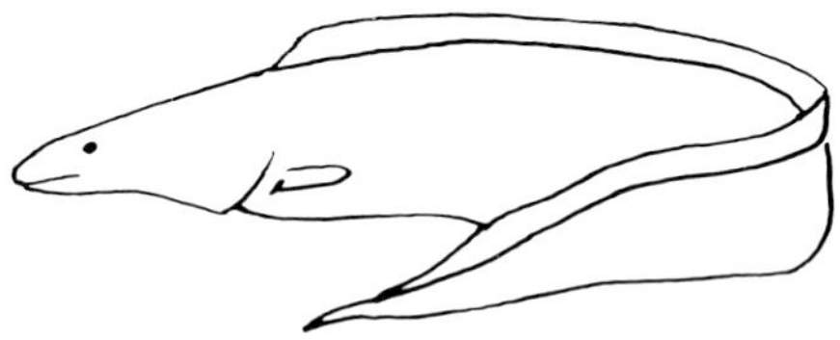

These fishes are very closely related to the elephantfishes, but they lack the pelvic, anal, and caudal fins and have no teeth on the tongue or palate. They also possess weak electric organs, and inhabit the same muddy waters. They are carnivorous and eat small fishes when adult.

The retina of *Gymnarchus niloticus* has been described and is identical, in most morphological characters, with that of the elephantfishes.

Reference: Engström (1963b)

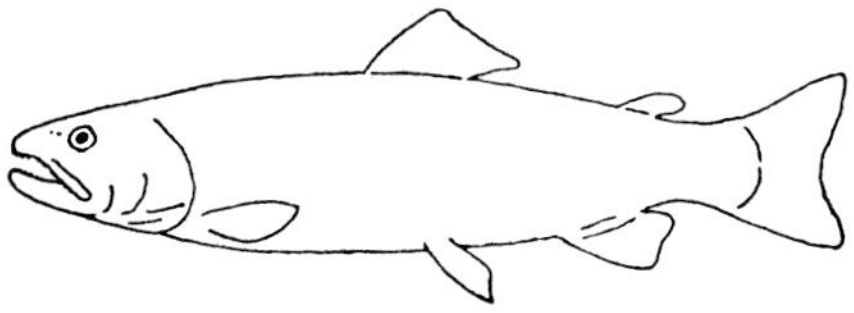

Salmonidae (*Salmons, Trouts, Chars*)

Protacanthopterygii, Salmoniformes, Salmonoidei

The salmonids constitute a family of generalised teleosts morphologically speaking. These are soft-rayed fishes with an adipose fin and posteriorly placed pelvic fins. The salmonids, in general, are anadromous (*Salmo, Oncorhynchus*), passing their adult life in open sea and spawning in fresh water. They are swift swimmers, travelling close to the surface, and feed on fishes and crustaceans. The Pacific salmon is a schooling fish, which depends on vision for the location and capture of its food.

The retina is of the classical teleostean type in its morphological characteristics. The epithelial pigment fills the long processes and migrates in response to light- or dark-adaptation (Plate 29.1, 2). The visual cells comprise numerous rods, single, and equal double cones which are arranged in regular patterns when viewed tangentially (Plate 29.3). These cells undergo retinomotor responses (Plate 29.1, 2). Two or three layers of horizontal cells are present (Plate 29.1, 2) and the bipolar and ganglion cells are relatively abundant. The retinas of the Sockeye (*O. nerka*), Coho (*O. kisutch*), Chum (*O. keta*) and Pink (*O. gorbuscha*) are essentially similar.

References: Brett and Ali (1958); Ali (1959, 1971, 1975); Ali and Wagner (1975a, 1975b)

Atlantic Salmon

The retina of the Atlantic salmon (*S. salar*) resembles that of the Pacific salmon in morphological characters and retinomotor responses (Plate 29.4, 5). However, the retina of the former is more sensitive to light than that of the latter.

References: Ryder (1895); Fürst (1904); Lyall (1957a); Ali et al. (1961); Engström (1963b); Ali (1963, 1964d); Wagner (1972)

Brook Trout

The brook trout (*S. fontinales*) lives in cool, clean streams and lakes; it eats a large variety of food (insect larvae, adult insects, fish).

The retina of this trout resembles closely that of the salmons. Plate 30.1 shows a semi-adapted retina in which the epithelial pigment, rods, and cones are found in transitional positions. Plate 30.2 shows a regular cone mosaic of double cones arranged in squares with central single cones.

References: Verrier (1928a); Rochon-Duvigneaud (1943); Polyak (1957); Ali and Kobayashi (1968a, b); Kobayashi and Ali (1971); Ali and Wagner (1975a)

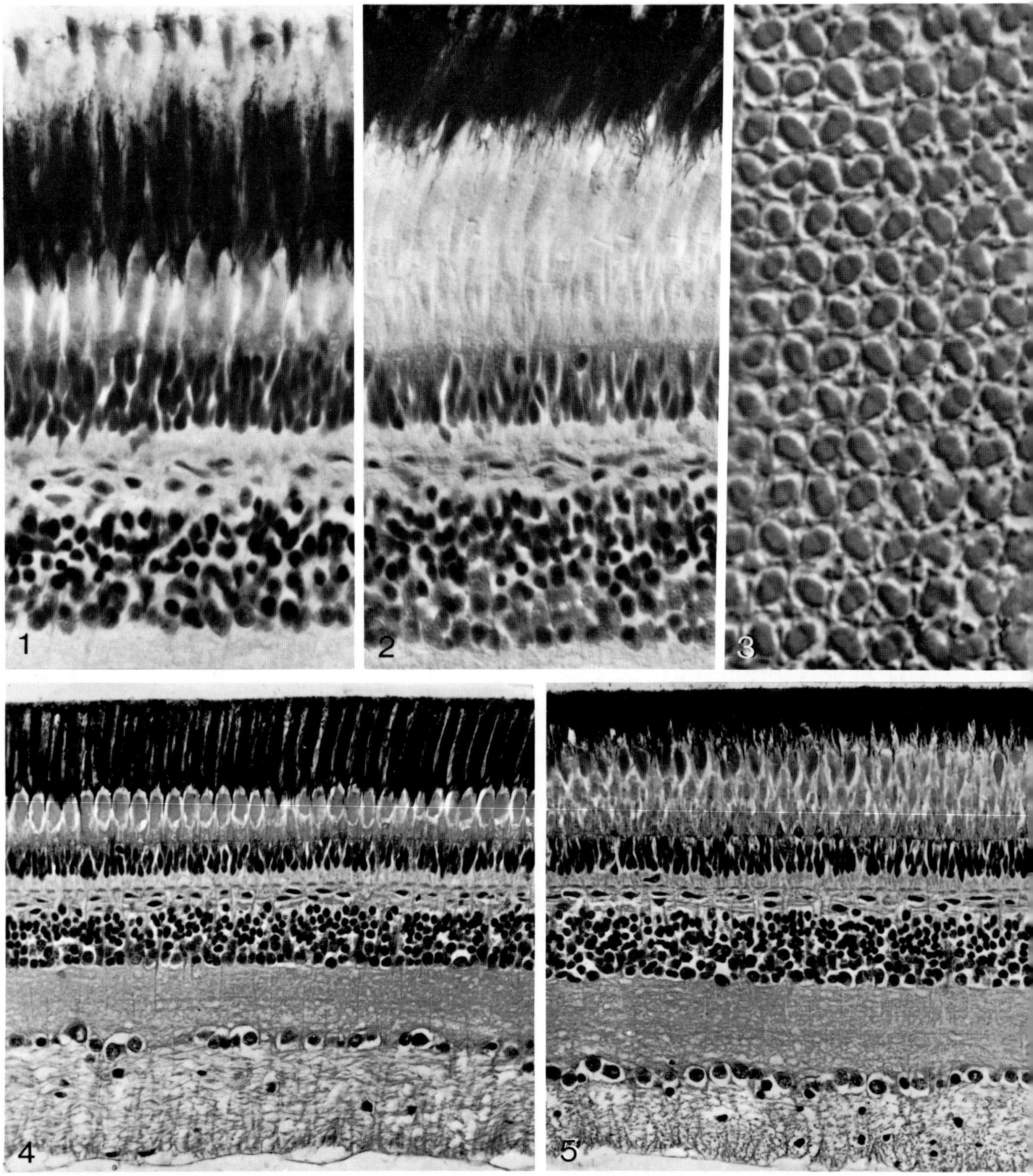

Plate 29.1-5. Transverse (1), (2), (4) and (5) and tangential (3) sections of the retinas of the sockeye salmon, *O. nerka,* and the Atlantic salmon, *S. salar*. Retinas in (1) and (4) are light-adapted, those in (2) and (5) are dark-adapted. (1)–(3) × 560; (4) and (5) × 300

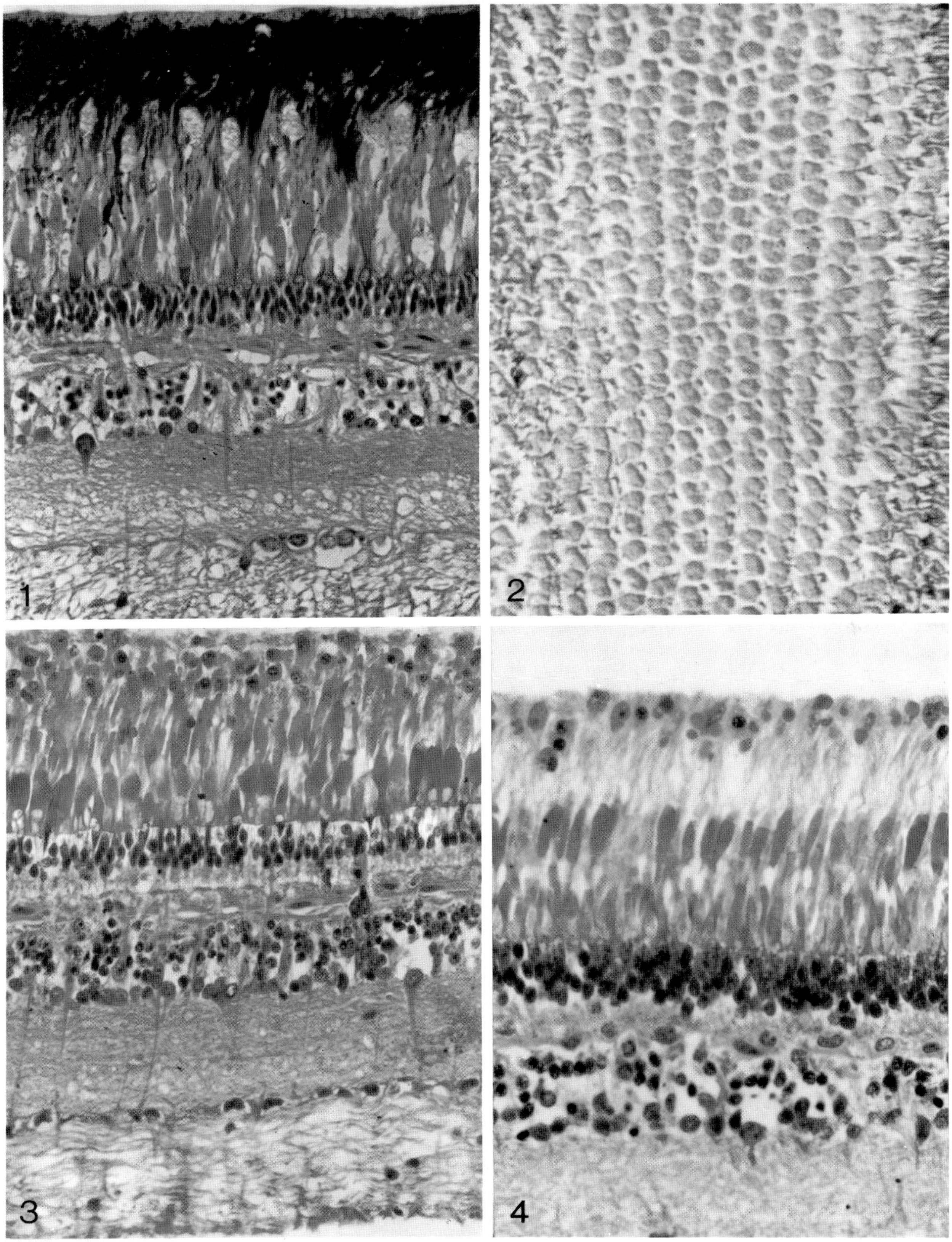

Plate 30. Transverse (1), (3) and (4) and tangential (2) sections of the retinas of the brook trout, *S. fontinalis,* and the albino splake F_1 hybrid between *S. fontinalis* and *S. namaycush*. (1), (3) and (4) ×420; (2) ×440

Albino Splake

The retina of the albino splake (cross between *S. fontinalis* and *S. namaycush*) is devoid of epithelial pigment, thus permitting better observation of the retinomotor changes of rods and cones. Plate 30.3 shows a light-adapted splake retina in which the cones are closely applied to the external limiting membrane, and a dark-adapted one (Plate 30.4) in which the cones have moved sclerally and the rods are in turn found close to the limiting membrane.

References: ALI (1964c); ALI and WAGNER (1975a)

Arctic Char

Plate 31, shows a dark-adapted retina of the Arctic char (*S. alpinus*) in which the double cones appear unequal in shape.

Reference: ALI (1965)

Ciscos

Ciscos are freshwater salmonids with adaptations for plankton-feeding, such as a small mouth and a protruding lower jaw. Ciscos apparently look at their prey from below (silhouetting), and are found at various depths in lakes, according to species.

The retinas of *Coregonus acronius, C. albula, C. pennantii* and others are well developed. The rods are of moderate size and more or less numerous. The cones are numerous especially in the ventro-temporal area retinae. Very small single cones and unequal double cones with long, light-adapted myoids are present; they are arranged in square mosaics in the temporal half of the retina, and in row mosaics in the nasal half. There are two or three layers of horizontal cells, and the internal nuclear layer is thick and densely populated in the ventro-temporal part of the retina.

References: FRIIS (1879); EIGENMANN and SHAFER (1900); WUNDER (1926, 1933); LYALL (1957a); AHLBERT (1969)

Visual Pigments: A_1 max 502–520; A_2 max 526–545 (BEATTY, 1973; ALI and WAGNER, (1975b)

(See also Plates 1, 3 and 5 in Introduction)

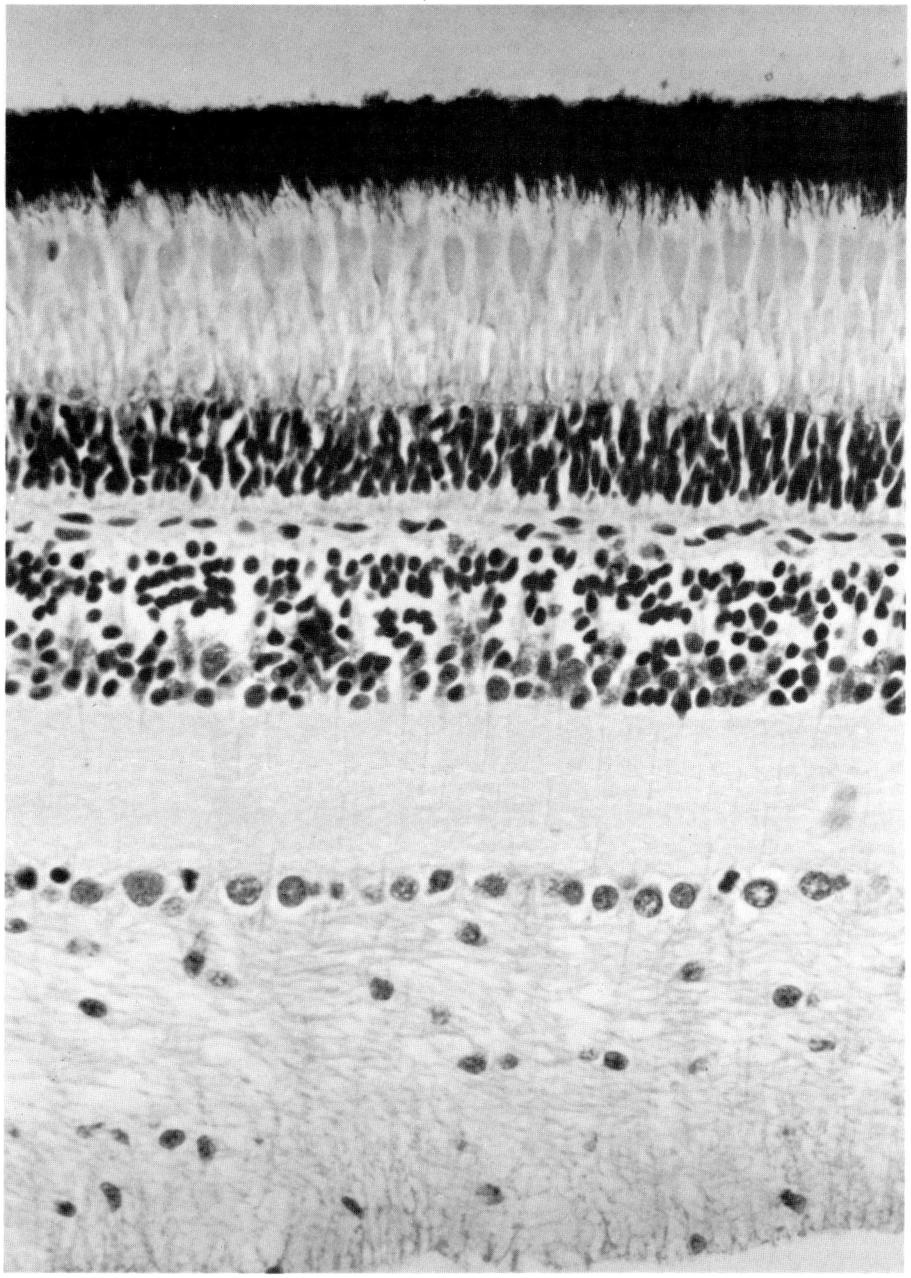

Plate 31. Transverse section of the dark-adapted retina of the Arctic char, *S. alpinus*. $\times 800$

Osmeridae (*Smelts*)

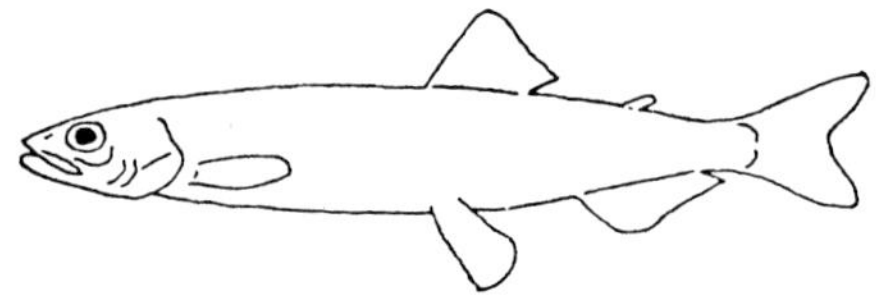

Protacanthopterygii, Salmoniformes, Salmonoidei

Smelts are small salmoniform fish from the continental shelf differing from the salmonids by the reduction or absence of pyloric caeca. Smelts live in shallow coastal waters and ascend streams to spawn. They feed mainly on small crustaceans.

The retina of the smelt, *Osmerus mordax,* is characterised by a large number of small cones, which are of the single and double types; and the rod nuclei form only one or two layers transversely (Plate 32,1). The internal nuclear layers are well developed and there are two or three layers of horizontal cells (Plate 32.1). The cones are smaller and more abundant in the ventro-temporal region (Plate 32.2) than in other retinal regions (Plate 32.3). They form alternating rows of single and double cones.

Reference: ANCTIL (1969)

Capelin

The capelin (*Mallotus villosus*) is very similar in appearance to the smelt, however differs in its habits and other characteristics. It is usually found in deeper shelf waters, except during the reproduction period when it comes close to the tideline to spawn on the shore.

The retina of the capelin differs strikingly from that of the smelt. The epithelial pigment is less abundant and retracted sclerally (Plate 32.4); the rods are very long, extremely abundant, and only equal double cones are found.

References: ANCTIL (1969); ALI and WAGNER (1975a)

Plate 32.1-4. Transverse (1) and (4) and tangential (2) and (3) sections of the retina of the smelt, *O. mordax*; and capelin, *M. villosus*. (1) and (2) *O. mordax*; (2) and (4) *M. villosus*. (1) ×310; (2) and (3) ×765; (4) ×600 ▷

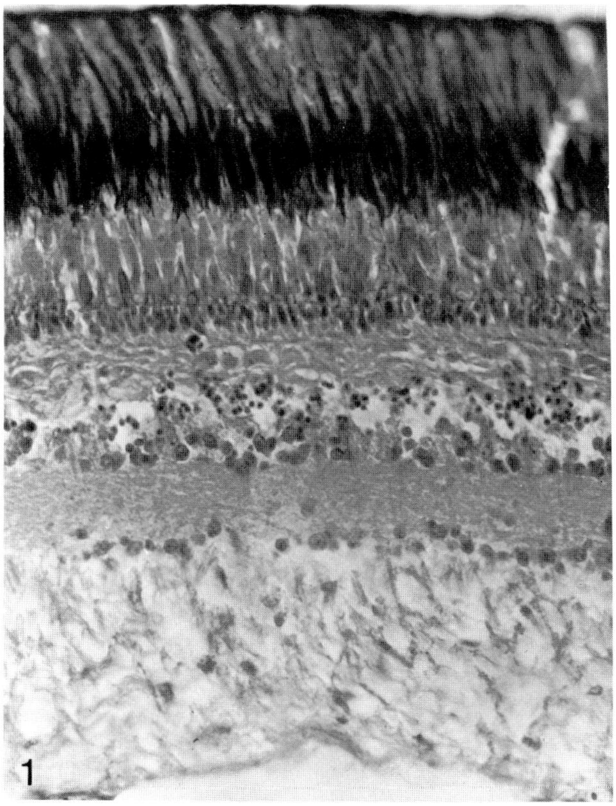
1

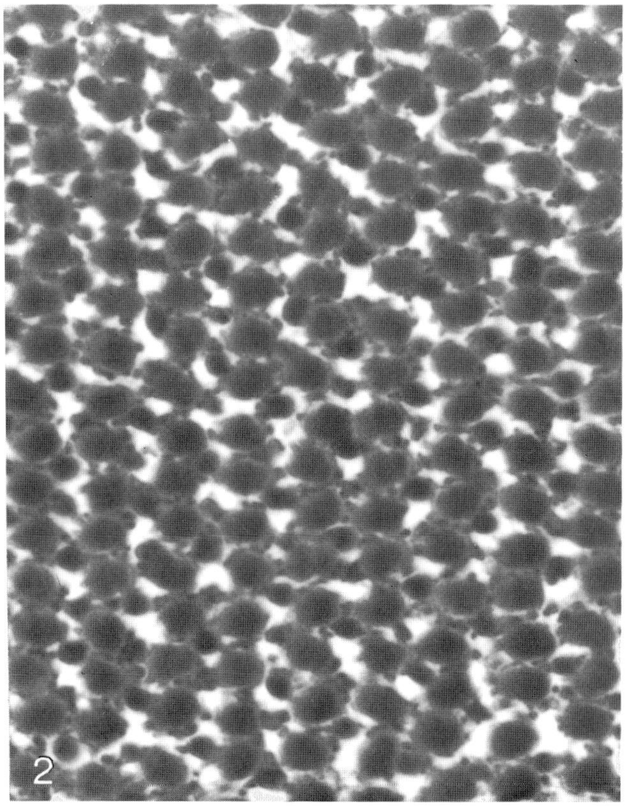
2

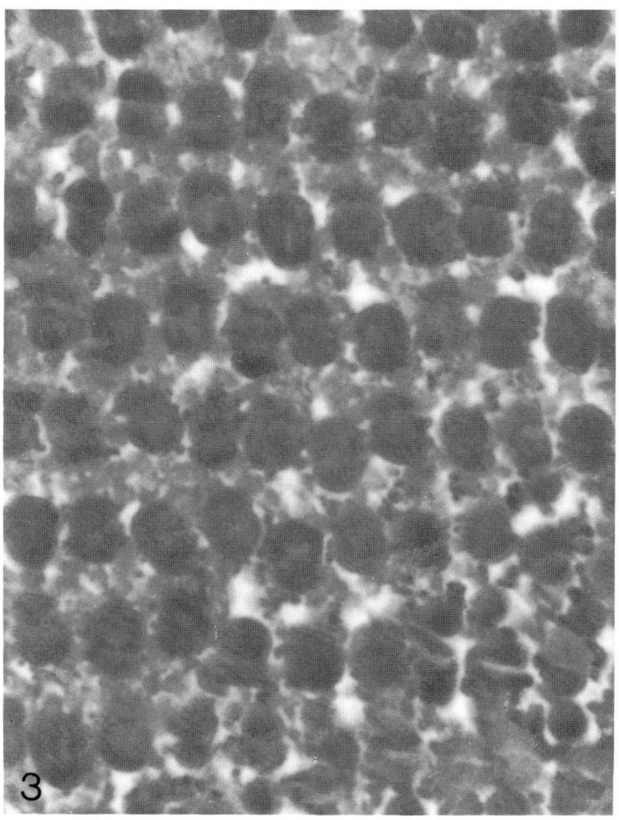
3

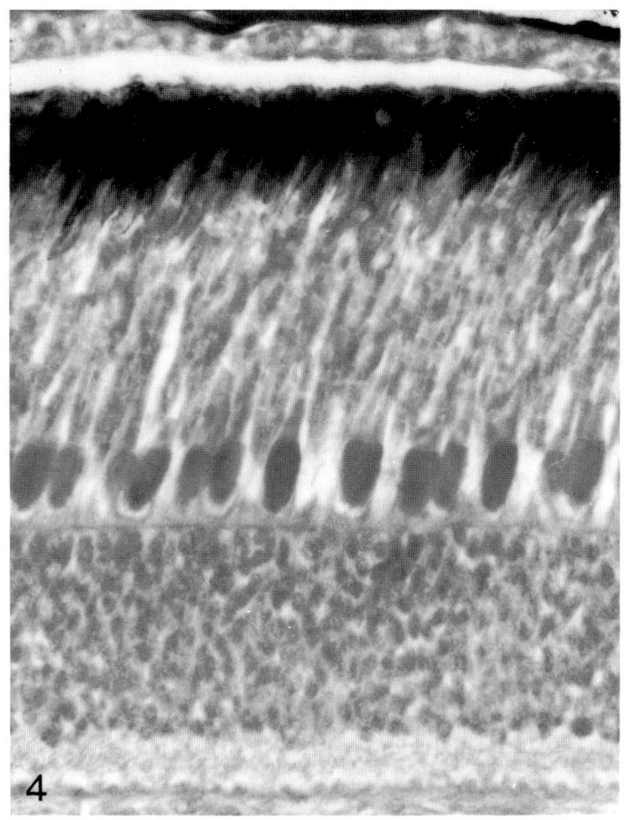
4

Argentinidae (*Argentines*)

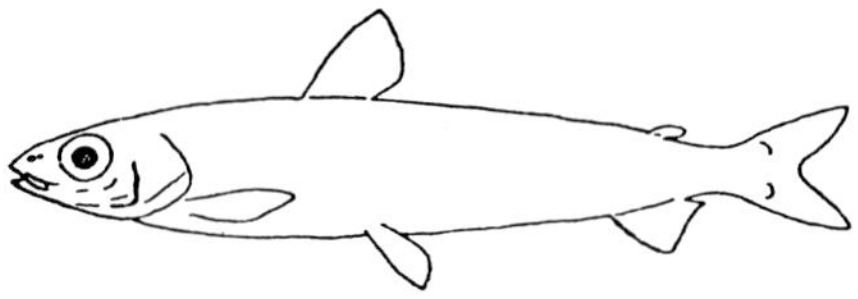

Protacanthopterygii, Salmoniformes, Argentinoidei

The argentines resemble the smelts (osmerids) in most of their external characters, but their mouths are much smaller and the entire base of their rayed dorsal fin is in front of their pelvic fins. They have eyes much larger than either the smelt or the capelin, a characteristic probably associated with their deepwater habitat. They live in marine shelf water as deep as 80 to 300 fathoms.

The Atlantic argentine has a pure-rod retina. The epithelial pigment is packed sclerally (Plate 33.1, 2); the rods are very numerous and arranged in two or three superimposed layers (Plate 33.1). Large horizontal cells are also present in two or three layers (Plate 33.2). Note the blunt epithelial processes devoid of pigment in Plate 33.2; this feature is not present in the ventral retina and possibly constitutes a retinal tapetum lucidum.

References: TAMURA (1957); ALI and HANYU (1963); MUNK (1966a)

Visual Pigments: A_1 max 497 (ALI and WAGNER, 1975b)

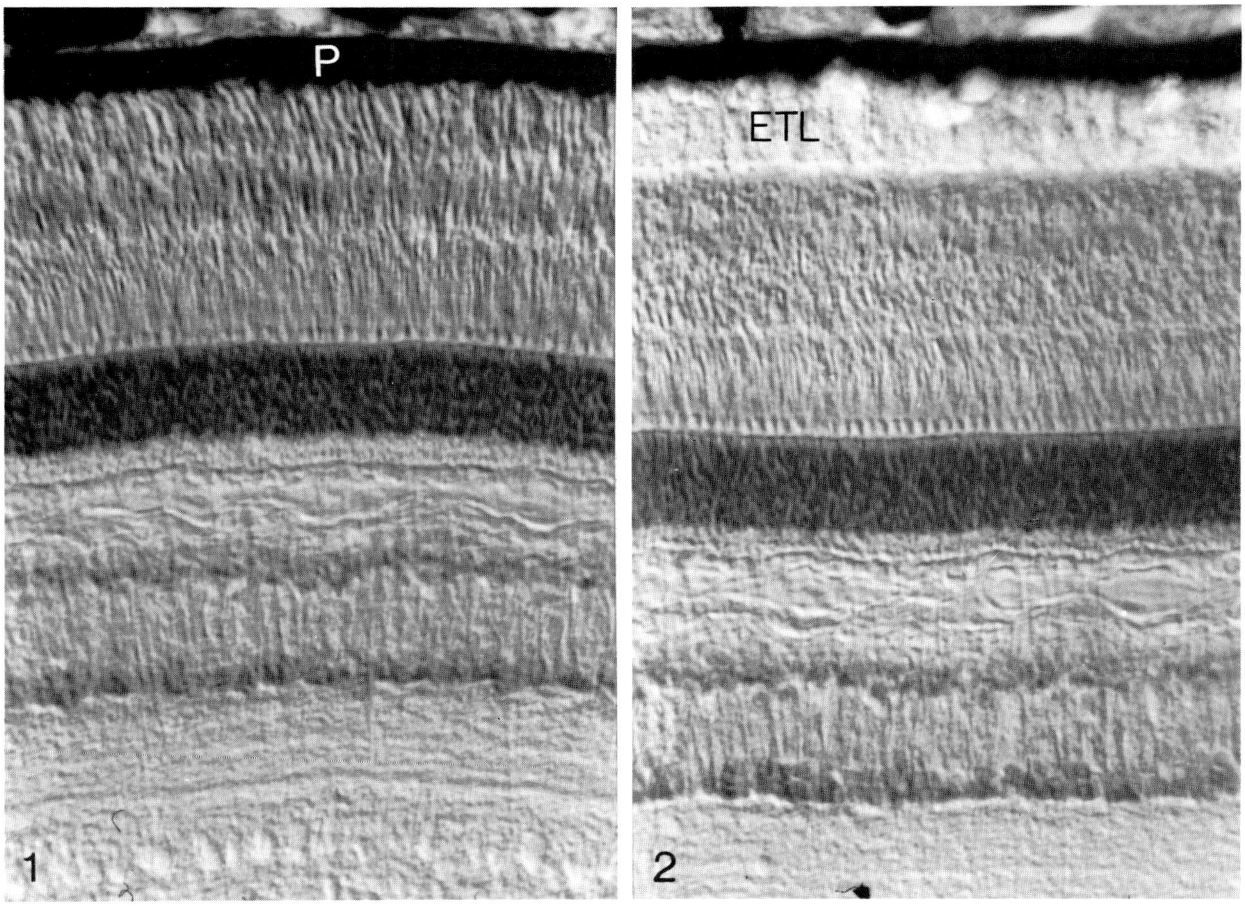

Plate 33.1, 2. Transverse section of the ventral (1) and dorsal (2) retinas of the Atlantic argentine, *Argentina silus*. ×440. *ETL* epithelial tapetum lucidum, *P* pigment epithelium

Bathylagidae (*Deep-Sea Smelts*)

Protacanthopterygii, Salmoniformes, Argentinoidei

Deep-sea smelts are mesopelagic or bathypelagic fishes characterised by a slender body, rarely stout, covered with large cycloid scales. They have a small mouth and large eyes, which are sometimes tubular and directed upwards.

Deep-sea smelts *Bathylagus* spp. and *Microstoma microstoma* have a typical pure-rod retina like their relatives, the argentinids. However, in many species of the genus *Bathylagus* the rods are set in several (3–6) superimposed layers probably in order to increase sensitivity by increasing visual pigment concentration (Plate 34). The area retinae, in this case a retinal region of higher rod density with a foveal depression, is temporal (Plate 34).

References: Brauer (1908); Vilter (1953, 1954a, b, c); Munk (1966a)

Visual Pigments: A, max 467–501 (Ali and Wagner, 1975b)

Opisthoproctidae (*Barreleyes, Spookfishes*)

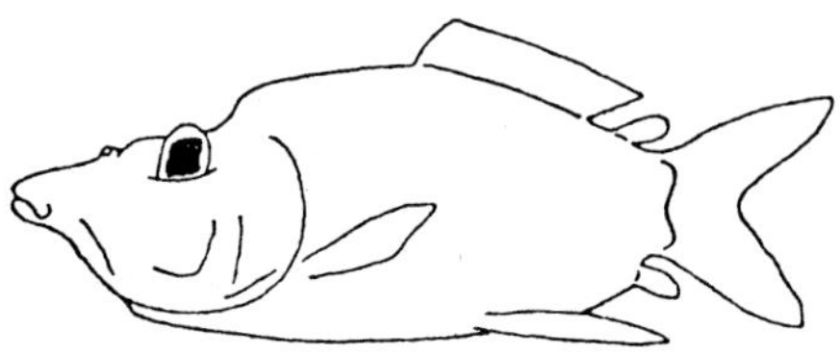

Protacanthopterygii, Salmoniformes, Argentinoidei

This small family of mesopelagic and bathypelagic fishes is characterised by a ventral surface broadened as in the soles, supported anteriorly by an expansion of the cleithra; other features include large, tubular eyes and the elongate, doglike snout. Some species possess luminescent organs whose luminescence is of bacterial origin.

Since most of the opisthoproctids (*Dolichopteryx, Opisthoproctus, Rhynchohyalus, Winteria*) have tubular eyes which are directed upwards, a main (bottom) and accessory (lateral) retinas are differentiated, the main one being better developed. In *Rhynchohyalus,* a third specialisation, the diverticulum retina projecting rostro-laterally, is present. All these retinal specialisations are pure-rod; rods are notably fewer and shorter in the accessory and diverticulum retinas compared to the main one. In the nontubular eye of *Bathylychnops,* the diverticulum retina is developed spectacularly and possesses its own lens; several superimposed rod layers are found in the main and diverticulum retinas of this "four-eyed" fish.

References: Brauer (1908); Bertelsen et al. (1965); Munk (1966a)

Plate 34. Photomicrograph of the transversely cut temporal retina of the deep sea smelt, *B. pacificus.* ×290. *R1–R6* superimposed layers of rod acromeres ▷

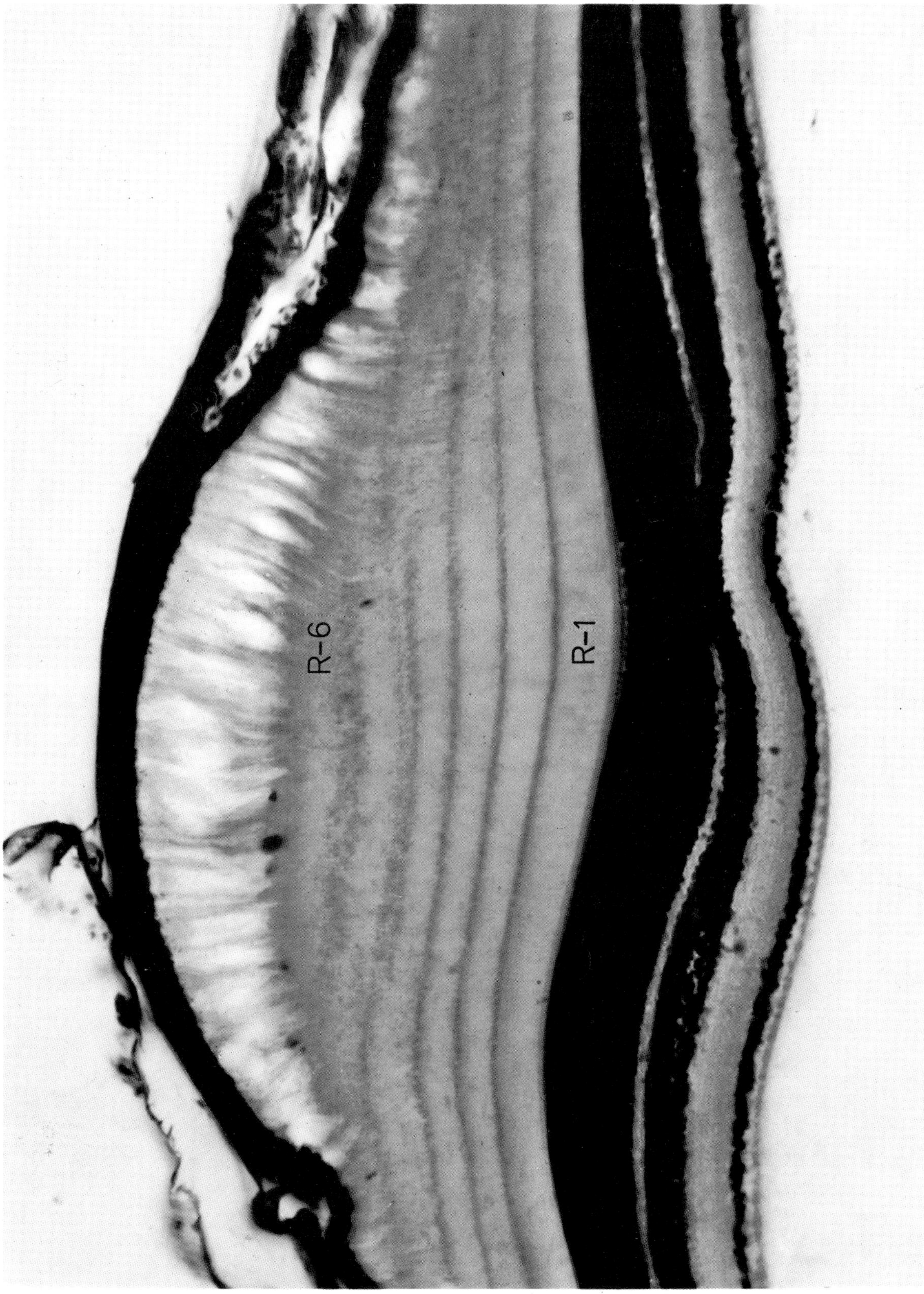

Plate 34. (Legend see opposite page)

Galaxidae (*Puyes*)

Protacanthopterygii, Salmoniformes, Galaxoidei

Galaxiids or puyes form a small family of fishes with primitive characters, and are distributed in Chile, New Zealand, Australia, and South Africa. They have an elongate and almost cylindrical body, a naked skin and a single dorsal fin inserted close to the caudal peduncle. They are found in lakes, rivers, and estuaries. The young are surface forms and eat planktonic animals (copepods). As they grow, the adults live in increasingly deeper water, and finally move close to the bottom, where they feed on benthic invertebrates.

In the retina of the juvenile *Galaxias platei* (Plate 35.1), the epithelial cells possess long and poorly pigmented processes. The visual cell layer is thick and composed of bulky rods and large equal double cones; single cones may be present, but are not clearly defined. The rod outer segments are in close contact with the epithelial cell bodies (Plate 35.1, 2). The cones have long ellipsoids but their outer segments are as cylindrical as those of the rods (Plate 35.1, 2). The bipolar and ganglion cells are present in large numbers.

Reference: SCHOEBITZ et al. (1973)

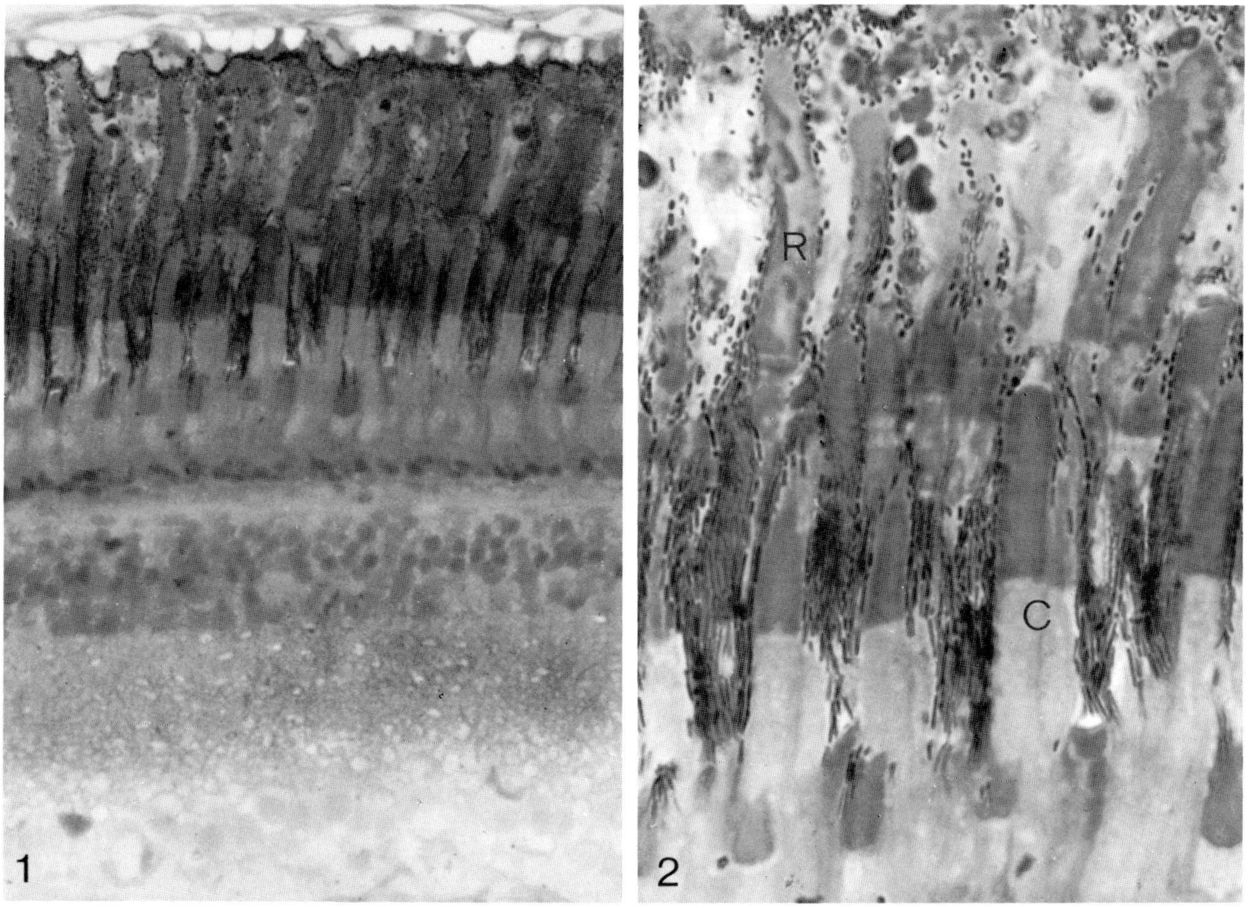

Plate 35.1, 2. Transverse section of the retina of the puye, *G. platei*. (1) × 560; (2) × 1,400. *C* double cone, *R* rod

Esocidae (*Northern Pikes*)

Protacanthopterygii, Salmoniformes, Esocoidei

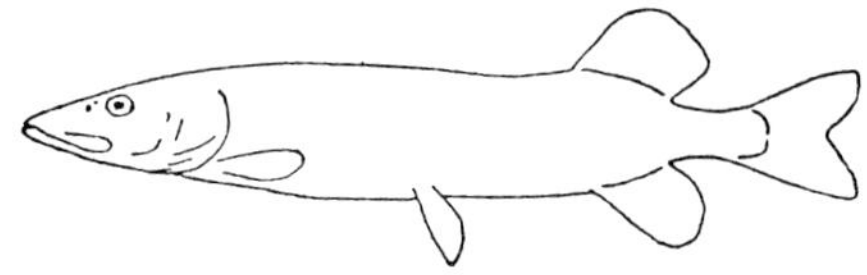

These predaceous freshwater fishes are characterised by the soft-rayed fins of the primitive teleosts, the posterior placement of the dorsal fin, the presence of cycloid scales, the elongated and well-toothed jaws, and the large gape. They are sluggish in general, though capable of rapid movements. They feed chiefly on other fishes.

The main characteristics of this retina are the small number of rods present, the large single and equal double cones with bulky foot-pieces (Plate 36.1), the large horizontal cells arranged in three layers (Plate 36.1, 2), and the presence of displaced bipolars (Plate 36.3), a unique feature among the teleosts. The area retinae is ventral, and the peculiar triangular cone mosaic is better resolved in this region (Plate 36.4) than elsewhere in the retina (Plate 36.5). This type of cone mosaic has not been found in other teleosts.

References: EIGENMANN and SHAFER (1900); WUNDER (1925); ROCHON-DUVIGNEAUD (1943); LYALL (1957b); POLYAK (1957); ENGSTRÖM (1963b); WAGNER (1972); BYZOV et al. (1972); ALI and WAGNER (1975a)

Visual Pigments: A_2 max 533 (ALI and WAGNER, 1975b)

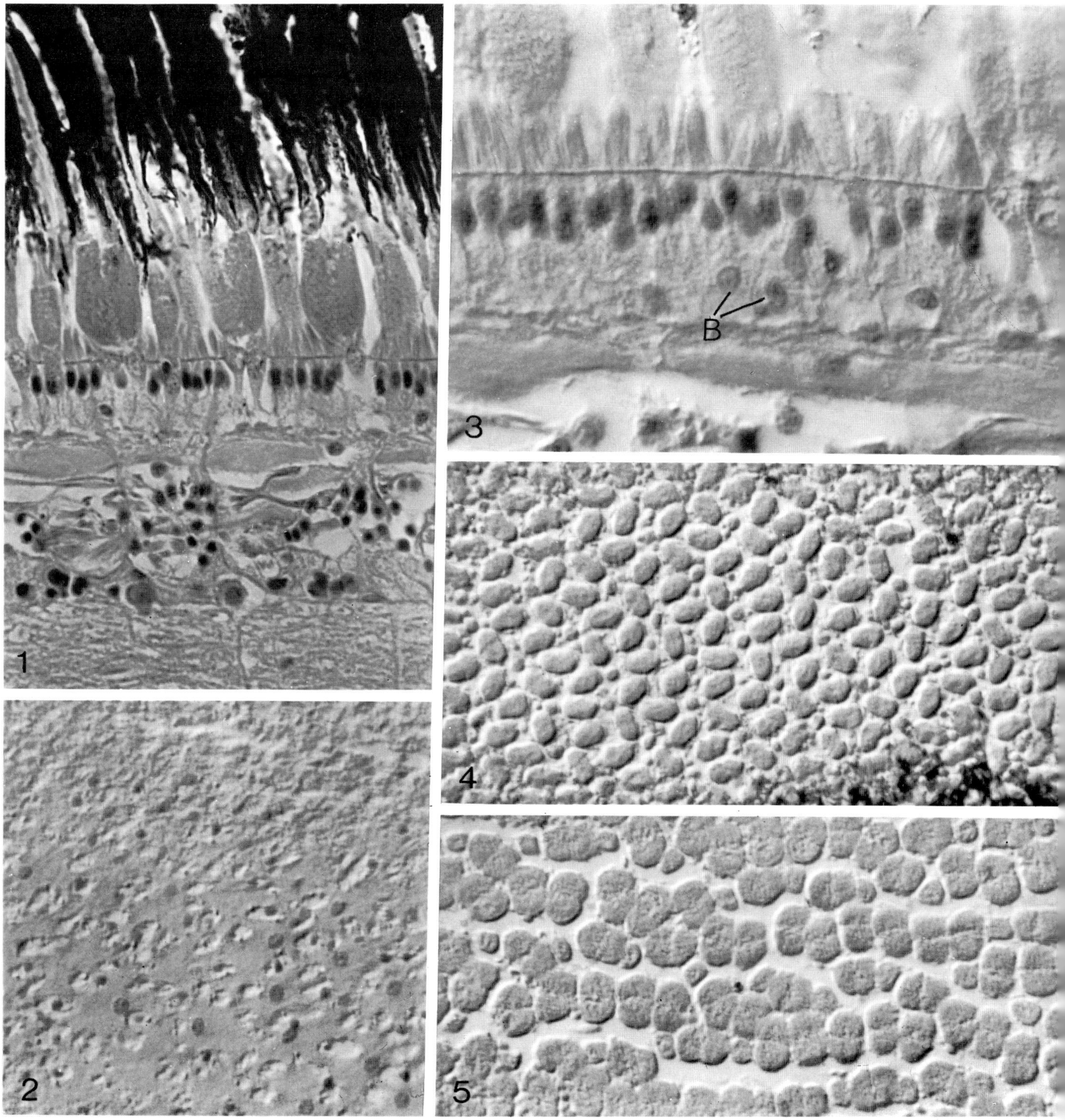

Plate 36.1-5. Transverse (1) and (3) and tangential (2), (4) and (5) sections of the retina of the pike, *Esox lucius*. (1) ×560; (2), (4) and (5) ×440; (3) ×1,100. *B* displaced bipolar cells

Gonostomatidae (*Lightfishes*)

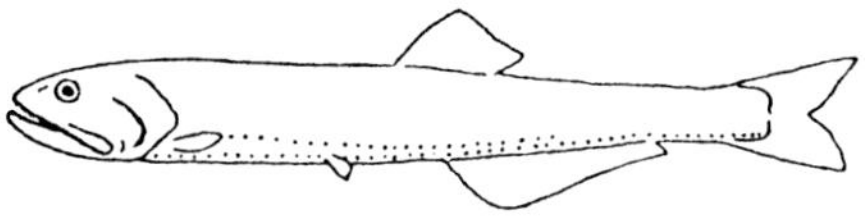

Protacanthopterygii, Salmoniformes, Stomiatoidei

Gonostomatids are small mesopelagic fishes, which are distinguished from other stomiatoids by the possession of true gill rakers and small teeth, and by the absence of postorbital photophores. Müller's pearlside is an oblong fish with large eyes, a relatively small mouth, and rows of small ventro-lateral photophores. They feed on plankton, migrate vertically in the ocean and form schools at the surface during the night.

The retinal epithelium is moderately pigmented and the pigment is packed sclerally (Plate 37.1). This is a pure-rod retina. The rod inner segment is relatively long and the rod diameter is large; there is only one row of rod nuclei. The ventral retina (Plate 37.2) is thicker and better developed than the dorsal retina (Plate 37.1); the rods are longer and the bipolar and ganglion cell populations are denser in the ventral retina. There are two layers of horizontal cells in the ventral retina and only one in the dorsal retina (Plate 37.1).

References: BRAUER (1908); VERRIER (1930b); MUNK (1966a)

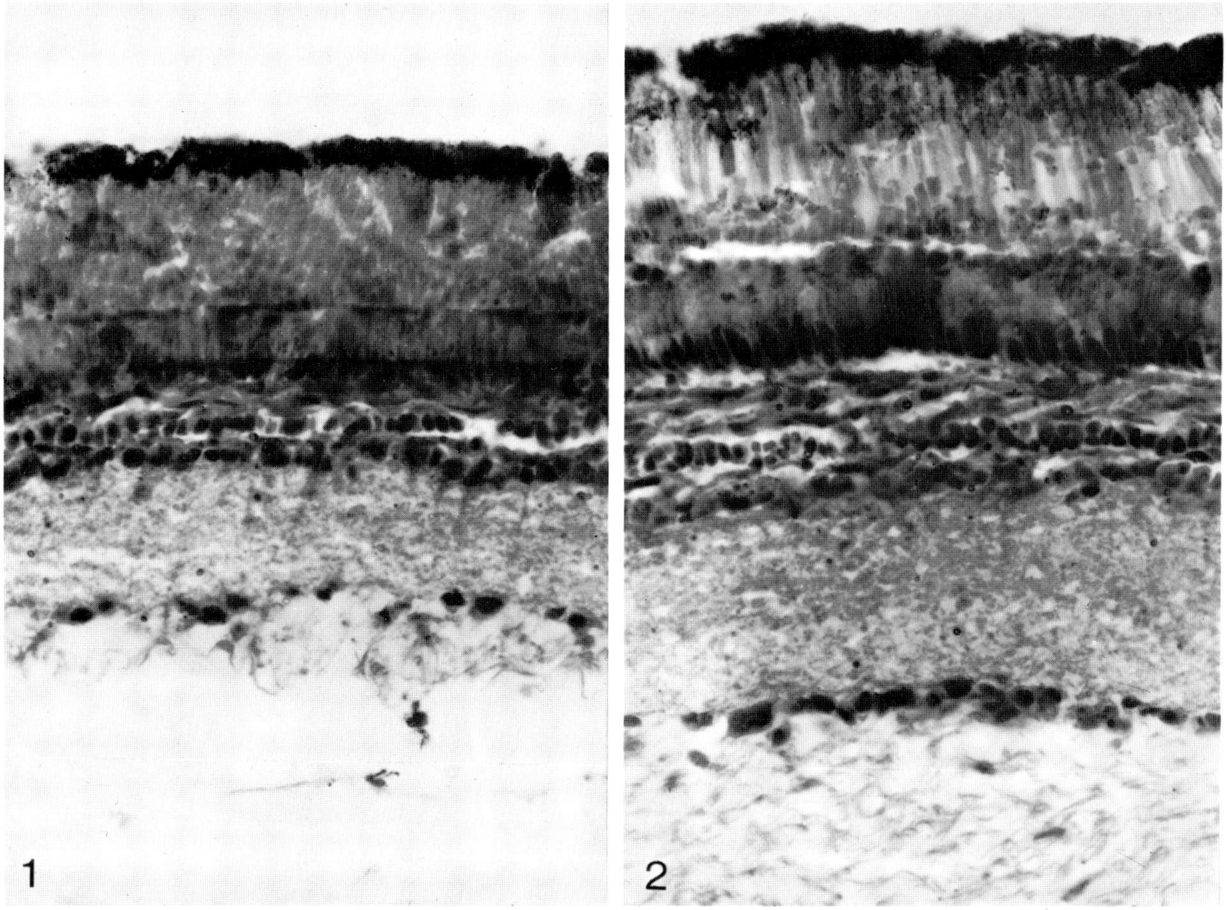

Plate 37.1, 2. Transverse sections of the dorsal (1) and ventral (2) retinas of the pearlside, *Maurolicus muelleri*. × 560

Sternoptychidae (*Hatchetfishes*)

Protacanthopterygii, Salmoniformes, Stomiatoidei

These small, silvery fishes are closely related to the gonostomatids (deep-bodied). All have luminescent organs along the ventro-lateral margins of the body, and some have tubular eyes directed upwards. The hatchetfishes, like the gonostomatids, are mesopelagic fishes and feed mainly on planktonic crustaceans.

Sternoptyx diaphana and *Polyipnus spinosus* have a simple, pure-rod retina, with few rods (a single layer of rod nuclei) and one layer each of horizontal, bipolar, amacrine, and ganglion cells (Plate 38). In the tubular eye of the fishes of the genus *Argyropelecus* the dorsal retinal region of the larva is modified into a rudimentary "accessory retina", with short rods and devoid of horizontal cells, while the larval ventral retina developed into the main retina with long-rod outer segments and a layer of horizontal cells.

References: Brauer (1908); Verrier (1930b); Contino (1939); Munk (1966a); Locket (1970a)

Visual Pigments: A_1 max 478, 485 (Ali and Wagner, 1975b)

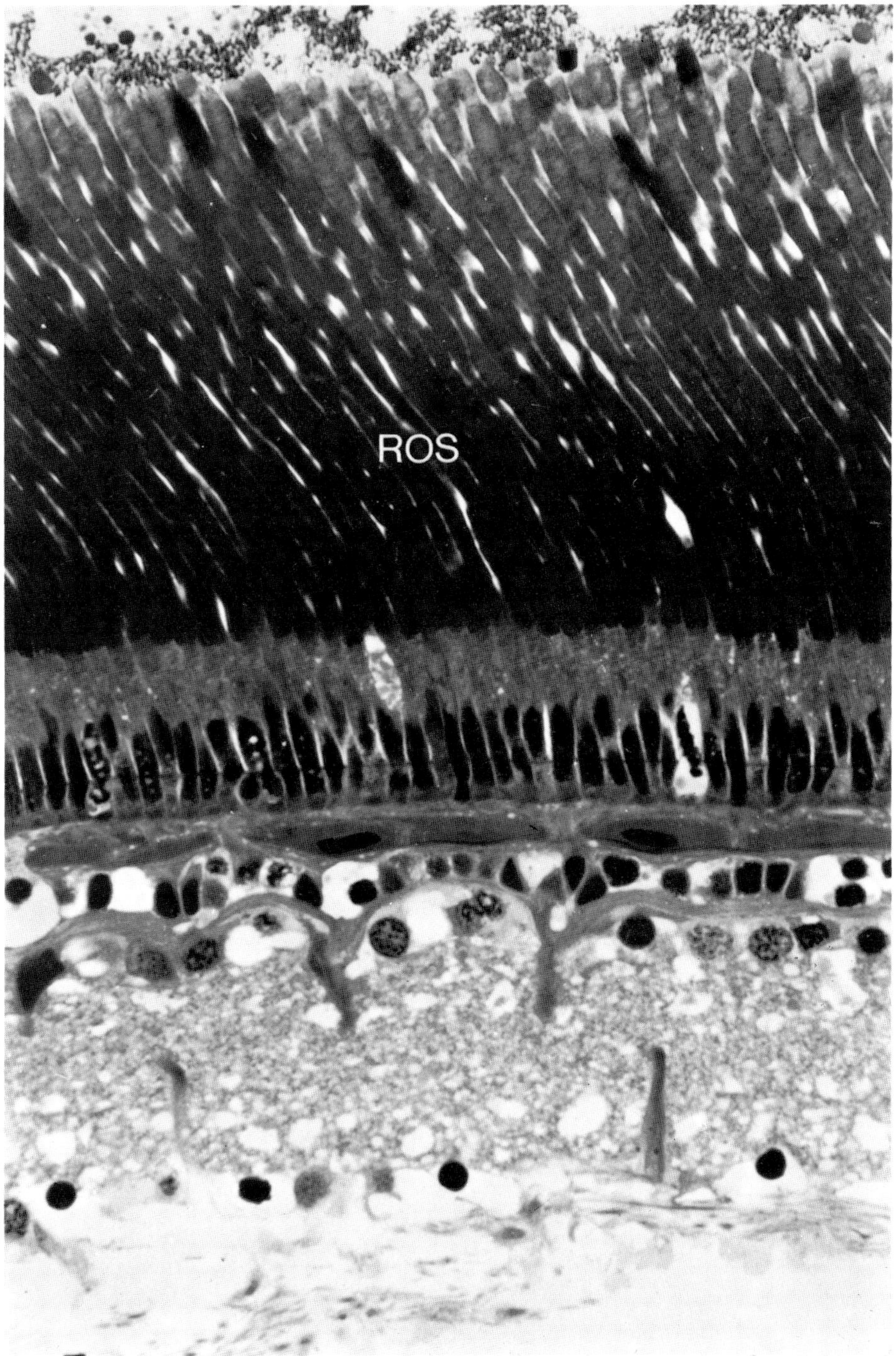

Plate 38. Photomicrograph of a transverse section of the retina of the hatchetfish, *S. diaphana* (about ×800). *ROS* rod outer segments

Chauliodontidae (*Viperfishes*)

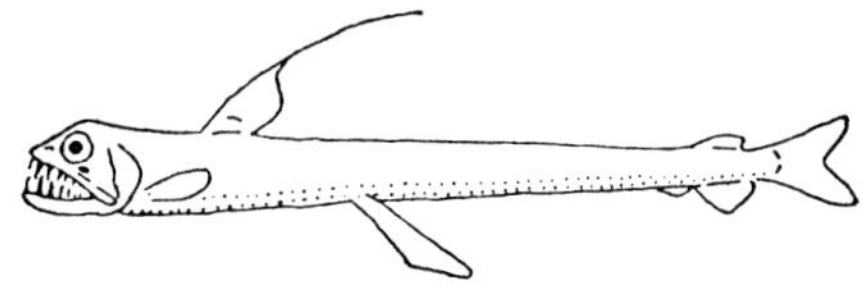

Protacanthopterygii, Salmoniformes, Stomiatoidei

This family of deep-sea (mesopelagic) fishes is characterised by an elongate body, a large gape with many large fanglike teeth, a first dorsal ray which is modified as a long sensory filament, and a ventro-lateral series of photophores. The body is covered with large and thin cycloid scales which do not overlap, a specialisation associated with a distensible body wall, and an ability to ingest large prey.

The retina of the viperfish possesses a dense amount of pigment distributed in epithelial cells without processes. This is a pure-rod retina, the rods being numerous and arranged in a single layer (three such layers in *Chauliodus sloani*). The internal layers are poorly represented.

References: Brauer (1908); Haffner (1952); Munk (1966a)

Idiacanthidae (*Chin-Whisker Fishes or Dragonfishes*)

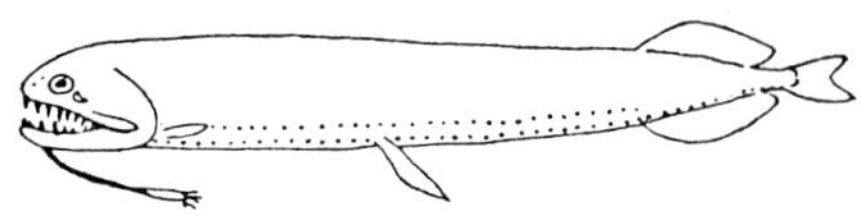

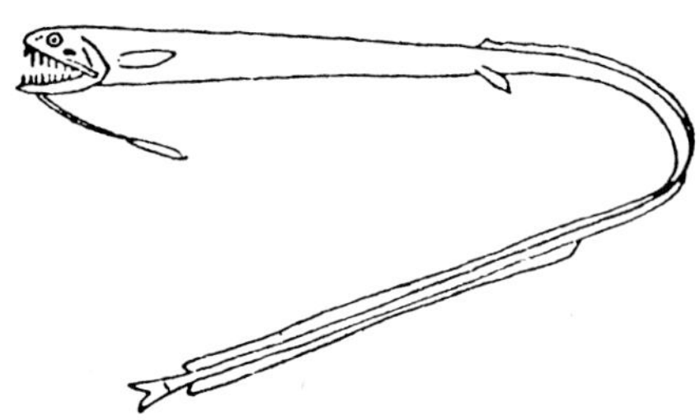

(including Astronesthidae, Melanostomiatidae, Malacosteidae, Stomiatidae)

Protacanthopterygii, Salmoniformes, Stomiatoidei

These include several world-wide families of mesopelagic and bathypelagic teleosts possessing a single whiskerlike barbel attached to the underside of the chin. They also possess serial luminescent organs on the body, and the barbel itself is luminous. All these fishes have daggerlike teeth, some with small hooks at the tip. The dragonfishes (Melanostomiatidae, Stomiatidae, Idiacanthidae) have very elongate snakelike bodies, whereas the astronesthids and malacosteids have bodies of moderate length. They are active predators, eating nectonic and planktonic organisms. All have normally shaped eyes with a simple pure-rod retina. The epithelium is pigmented and sends short processes vitreally (Plate 39.1, 2). The rods are numerous and arranged in two superimposed layers [(Plate 39.1, 2) Melanostomiatidae], except in *Idiacanthus* where there is only one such layer. A retinal tapetum lucidum has been reported in *Malacosteus indicus*. The internal nuclear layers are poorly developed (Plate 39.1, 2).

References: Brauer (1908); Munk (1966a)

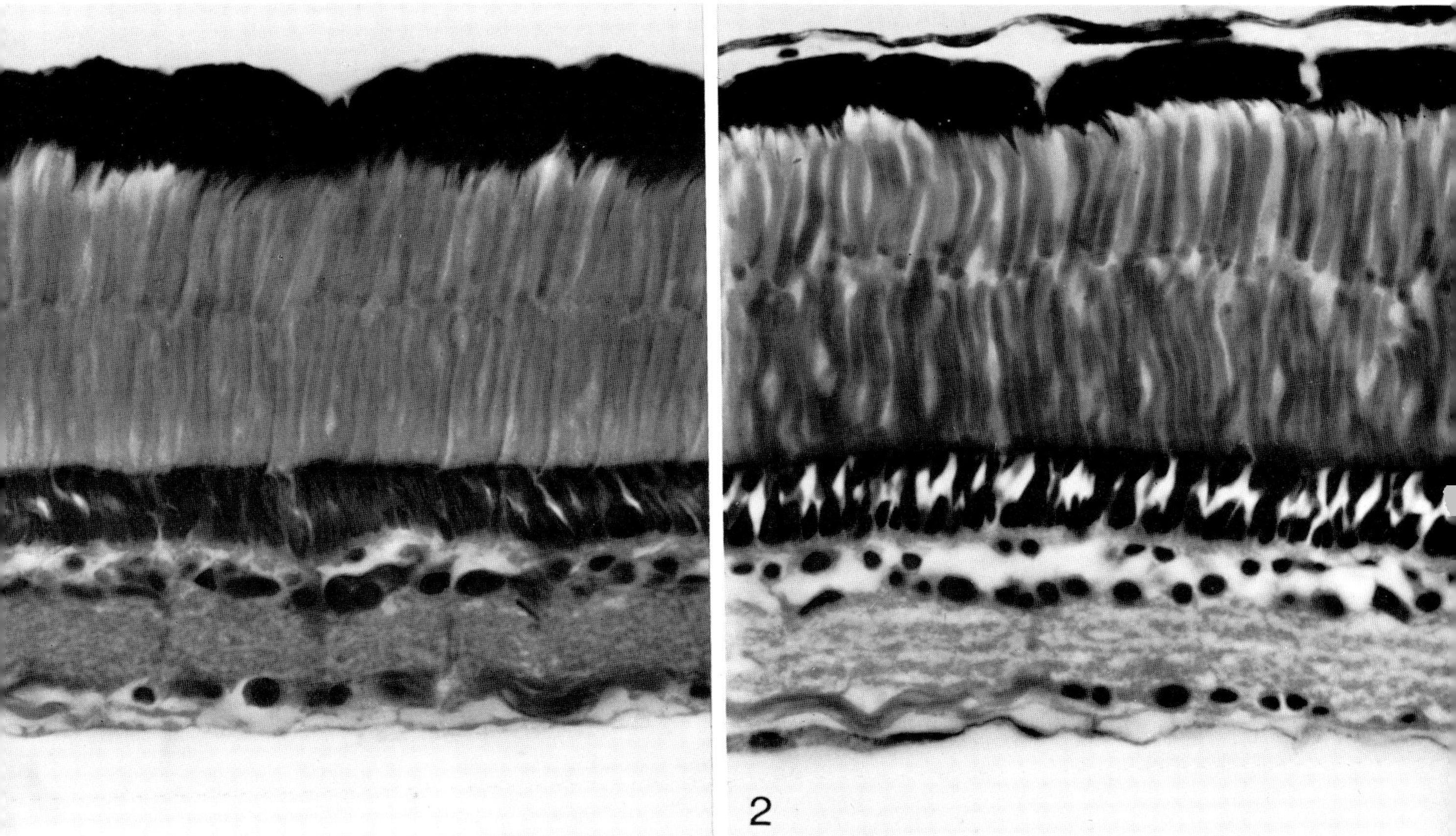

Plate 39.1, 2. Photomicrographs of the retinas of two stomiatoid fishes (Melanostomiatidae), cut in cross-sections. (1) Transverse section of the retina (fundus) of *Melanostomias spilorhynchus*. ×945. (2) Transverse section of the retina (fundus) of *Flagellostomias boureei*. ×945

Alepocephalidae (*Slickheadfishes*)

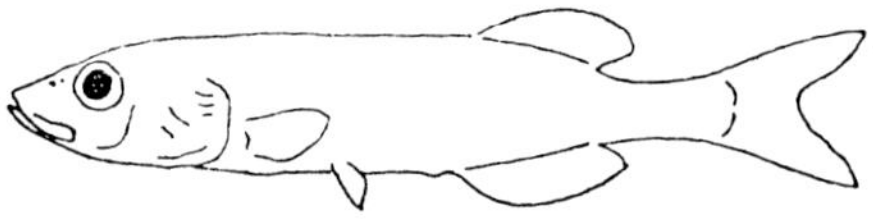

Protacanthopterygii, Salmoniformes, Alepocephaloidei

This is a large group of small, slender-bodied mesopelagic fishes. Most of them possess a lateral line and lack both the swim bladder and adipose fin. Some species possess luminescent organs.

These fishes have pure-rod retinas. In some species (*Searsia koefoedi, Platytroctes procerus, P. apus, Platytroctegen mirus*), the retina is specialised into a main retina with long rods (Plate 40.1) and a foveal depression (Plate 40.2), and an accessory retina with short rods. The epithelial cells have short processes and are pigmented. The horizontal cells are present only in the main retina. *P. apus* is a grey, bathypelagic fish seldom found shallower than 900 metres. It has large eyes with a wide pupil only partly occupied by the spherical lens.

References: Brauer (1908); Munk (1966a); Locket (1971a)

Visual Pigments: A_1 max 471–479 (Ali and Wagner, 1975b)

Bathylaconidae (*Bathylaconids*)

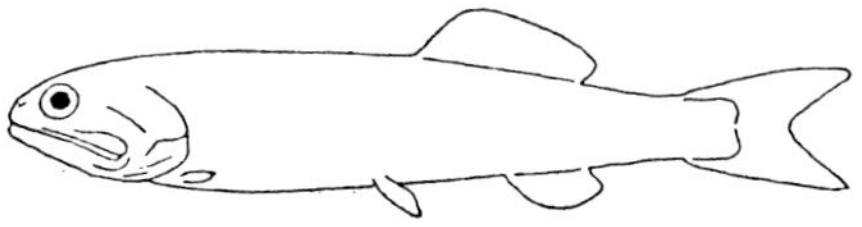

Protacanthopterygii, Salmoniformes, Bathylaconoidei

These are bathypelagic fishes, which resemble the alepocephalids externally. They have a short snout and large lateral eyes. These fishes are considered to be transitional forms between the stomiatoids and esocoids.

Bathylaco nigricans has a pure-rod retina with short pigmented epithelial processes. There are three superimposed layers of rods in the visual cell layer, except in the temporal fovea where only one rod layer is present. The horizontal cells can be observed. The retinal anatomy of this fish is similar to that of the alepocephalids.

Reference: Munk (1968a)

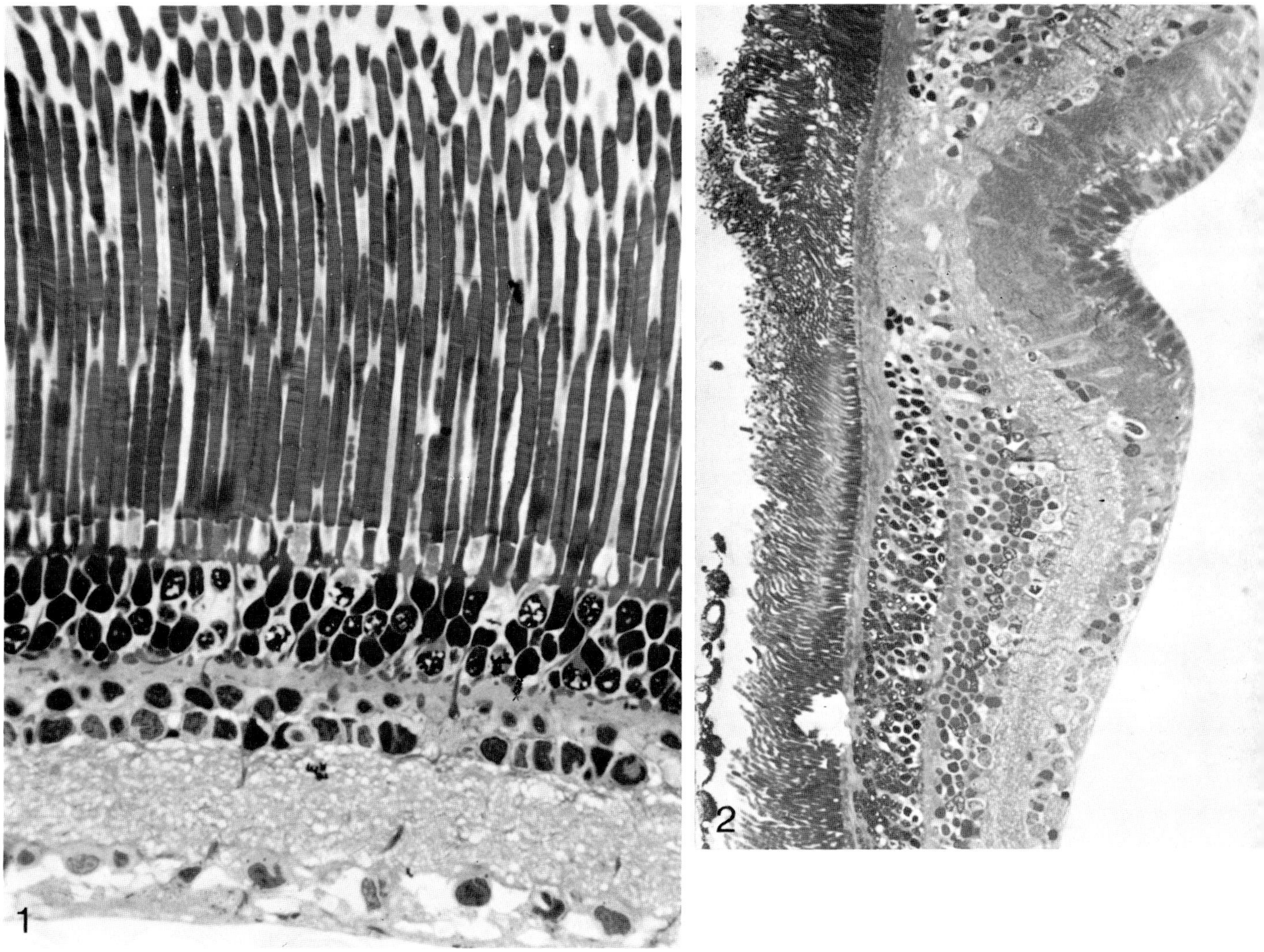

Plate 40.1, 2. (1) Photomicrographs of a transverse section (about × 550) and the foveal depression (2) of the retina (about × 250) of the searsid fish, *P. apus*

Chlorophthalmidae (*Greeneyes*)

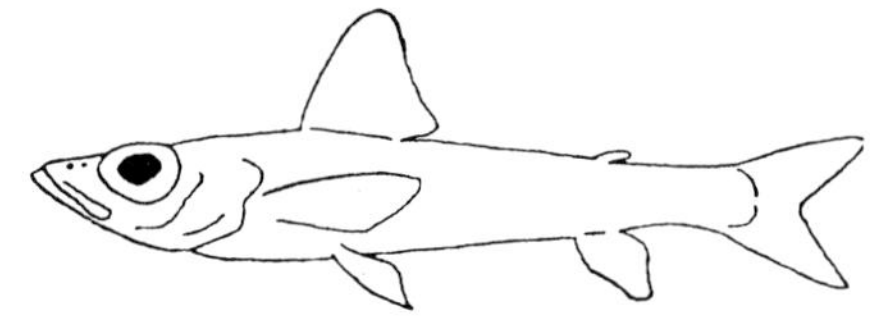

Scopelomorpha, Myctophiformes

This is a small family of teleosts inhabiting deep shelf waters. In appearance greeneyes are much like the thread-sail fishes (Aulopidae) but their fins are more normal in size and they are usually found in much deeper water. *Chlorophthalmus albatrossis* lives at depths of 200–400 m on the continental shelf. The retina of this fish presents interesting characteristics. Only rods and equal double cones are present in the visual cell layer. The cones are mainly confined to the ventral part of the retina. The remaining parts of the retina are composed of rods present in the form of bundles of 50–100 slender cells, except in a horizontal band slightly dorsal to the optic disc, where rods are grouped in columns with a few intercalated double cones.

References: TAMURA (1957); SOMIYA and TAMURA (1971)

Visual Pigments: A_1 max 485–490 (ALI and WAGNER, 1975b)

Bathypteroidae (*Spiderfishes*)

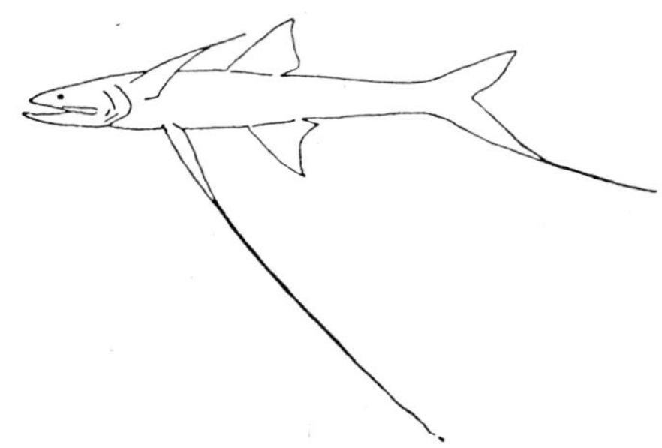

Scopelomorpha, Myctophiformes

This is a specialised family of deep-sea teleosts with elongate snouts. The first pelvic fin rays are greatly elongated and stiffened; these form a landing or resting tripod with the tips of the caudal fin rays. This is an adaptation for a benthic life. Spiderfishes are found on abyssal bottoms (2,000 m and deeper).

Varying degrees of retinal degeneration are reported in the eyes of *Bathypterois longipes* (Plate 41.2,3), *Benthosaurus grallator* (Plate 41.1), and *Bathymicrops regis*. These are all pure-rod retinas. Only the sclerad part of the epithelial cells are pigmented (Plate 41.1). The rod outer segments are fragmented and their tips include a considerable amount of macrophage cells with pigment granules (Plate 41.2). The retina of *B. longipes* is best preserved at the periphery (Plate 41.3). The internal nuclear layer is poorly developed in all cases (Plate 41.1-3).

Reference: MUNK (1964a, 1965a, 1966a)

Plate 41.1-3. Photomicrographs of the transverse sections of the retinas of two spiderfishes *B. grallator* (1) and *B. longipes* (2) fundus; (3) peripheral retina. (1) ×878; (2) ×945; (3) ×715. *EP* epithelial process, *MA* macrophage cell, *R* rod layer ▷

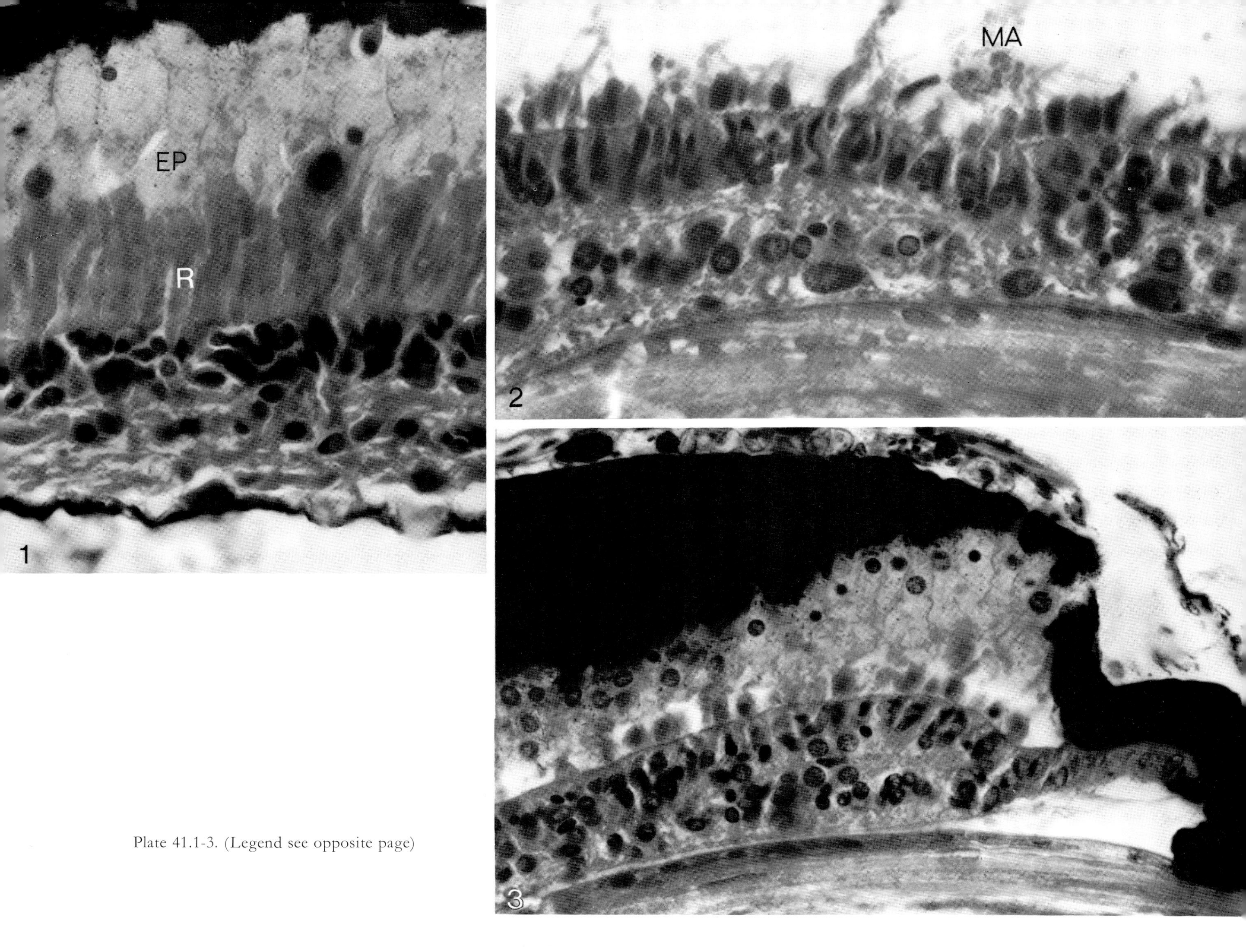

Plate 41.1-3. (Legend see opposite page)

Ipnopidae (*Grideye Fishes*)

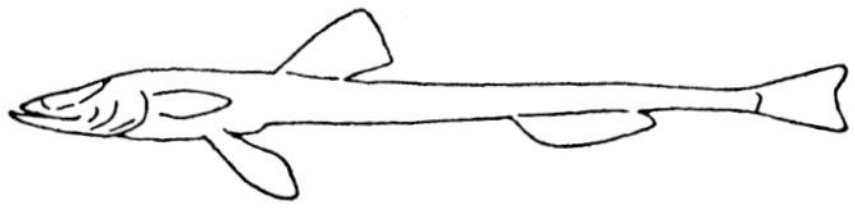

Scopelomorpha, Myctophiformes

Bathybenthic ipnopids are remarkable in that in some species the eye structure is greatly modified. The cornea is flattened and the retina enlarged so that its sensitivity is potentially enhanced; this would be efficient in perceiving ambient luminescence, but luminous organisms seldom occur below depths of 1,000 m and the ipnopid eye presents some degenerative characters, as shown below.

In *Ipnops meadi* the flattened retinas (Plate 42.1) cover a large part of the upper surface of the broad head, and are covered by a bony plate. Only the scleral parts of the epithelial cells are pigmented (Plate 42.2). The coarse rod outer segments are irregularly swollen and surrounded by macrophages (Plate 42.2). There is a single layer of rod nuclei. The nuclei are rare and scattered in the thin internal nuclear layer.

Reference: MUNK (1959, 1965a, 1966a)

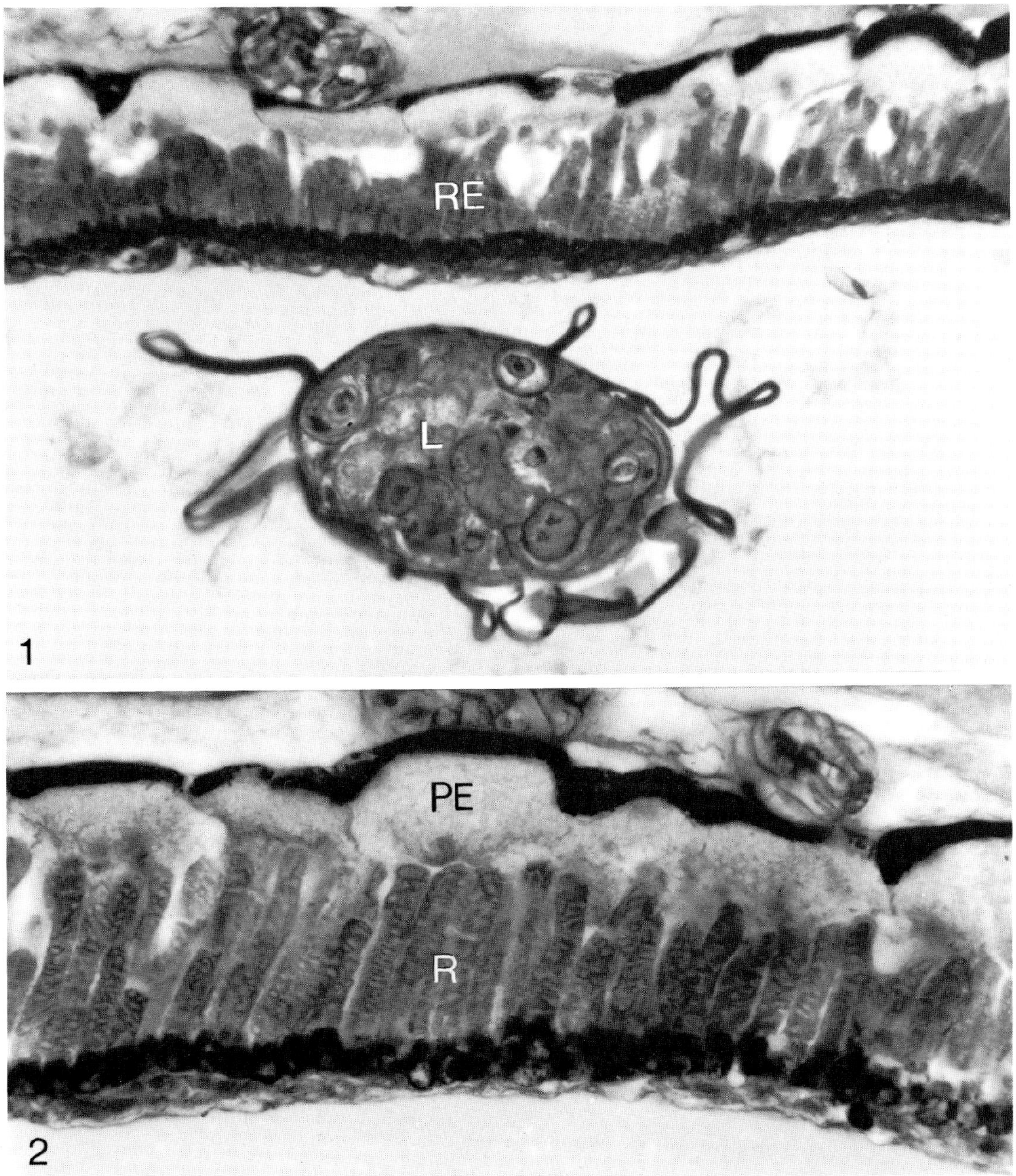

Plate 42.1, 2. Photomicrographs of the transversely sectioned retina of the grideye fish, *I. meadi*. (1) ×575; (2) ×945. *L* rudimentary lens, *PE* epithelial cell, *R* rod layer, *RE* retina

Omosudidae (*Hammerjaws*)

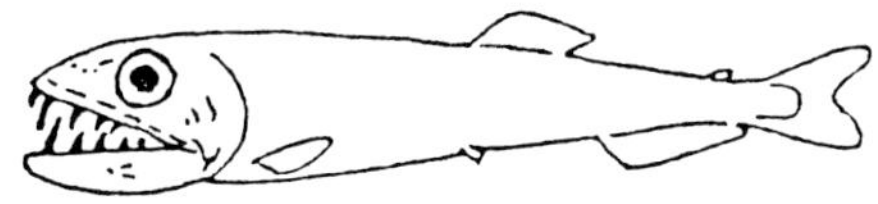

Scopelomorpha, Myctophiformes

This family is represented by a single deep-sea species, *Omosudis lowei*. The hammerjaw, which derives its name from the hammerlike lower jaw, resembles the alepisaurid (lancet-fish) to which it is closely related. The hammerjaw is a bathypelagic fish whose young undertake daily vertical migrations. Its diet includes squids and small fishes.

The retina of *O. lowei* possesses two remarkable features for a deep-sea fish: the long pigmented epithelial cell processes (Plate 43.1-3), and its almost pure-cone visual cell layer (Plate 43.1, 2, 4). The coarsest cones are found in the dorsal fundus-retina (Plate 43.2, 3), with slightly tapering outer segments and typically long, teleosteanlike ellipsoids. Transitional cones are present in the central fundus-retina. In the thickest ventral fundus-retina (Plate 43.1) the cone outer segments are actually rod-shaped. All are single cones and the area retinae is ventral (Plate 43.4). The few rods are found maximally in the dorsal fundus-retina (Plate 43.2), where the internal (vitread) layers are less developed.

Reference: MUNK (1965b, 1966a)

Evermannellidae (*Sabre-Tooth Fishes*)

Scopelomorpha, Myctophiformes

This is a family of small meso- and bathypelagic fishes, whose first pair of vomero-palatine teeth are modified into sabre-type structures. These fishes have either lateral or tubular eyes. They can swallow prey much larger than themselves. *Evermannella atrata* has lateral eyes and a pure-rod retina. The epithelial cell processes ensheath rod bundles. A reflecting tapetal system is apparently present in this epithelium, but only in the dorsal part of the retina. The eyes of *E. balbo* and *E. indica* are tubular, and the ventral (bottom) retina is very thick and specialised as an area retinae in which the epithelial cells are well pigmented, but devoid of tapetal material. The rods are slender and densely packed in the area retinae; they have long ellipsoids and relatively short, rod-shaped outer segments. The dorsal (accessory) retina of the tubular eye has coarser rods with long outer segments.

References: BRAUER (1908); MUNK (1966a)

Plate 43.1-4. Photomicrographs of the transversely cut ventral (1) and dorsal (2, 3) retinas, and the tangential (oblique) ventral retina (4) of the hammerjaw, *O. lowei*. *C* cone, *R* rod. (1) and (3) × 945; (2) × 1,900; (4) × 2,685 ▷

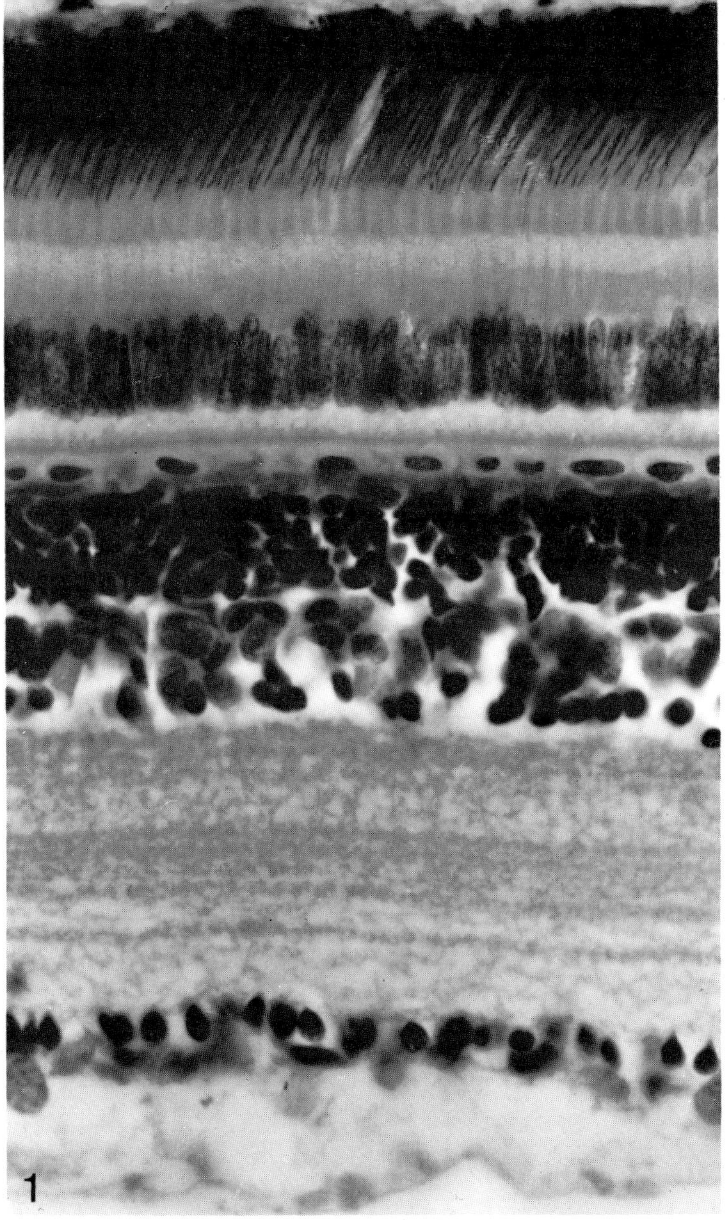
1

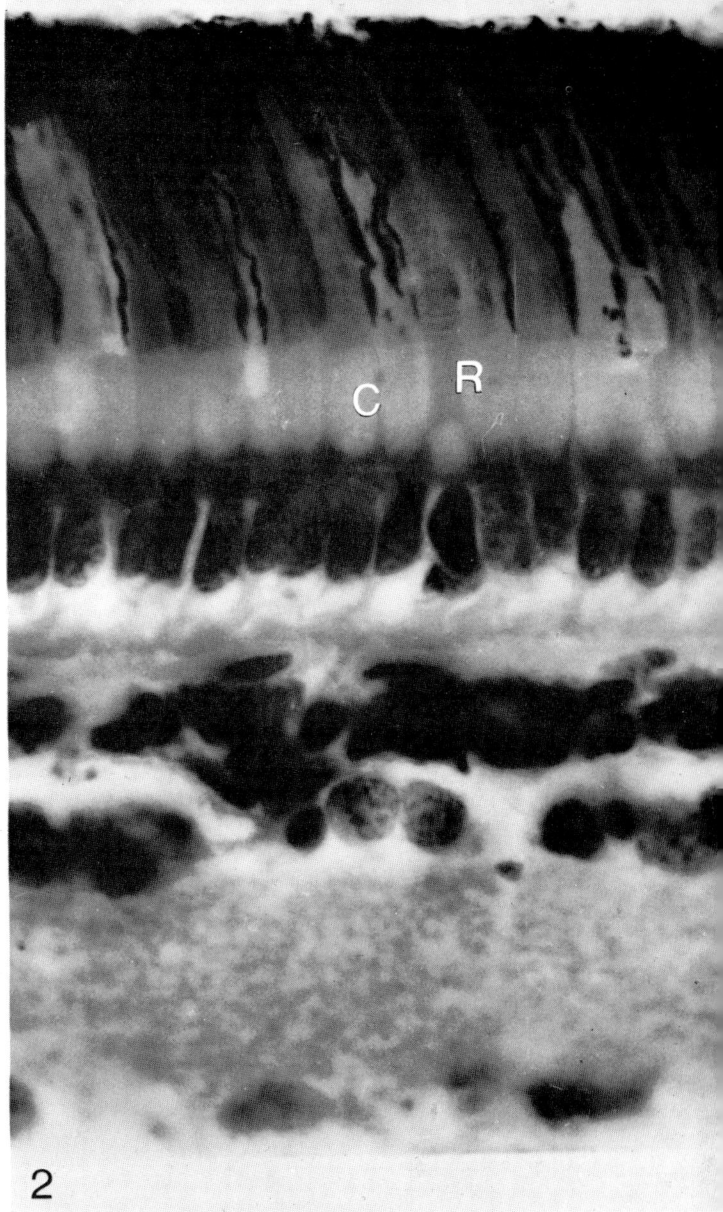
C
R
2

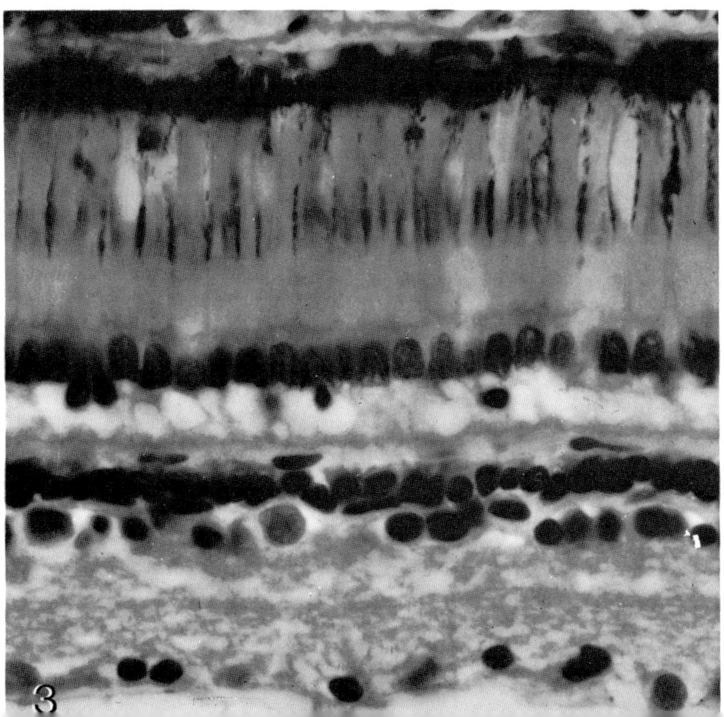
3

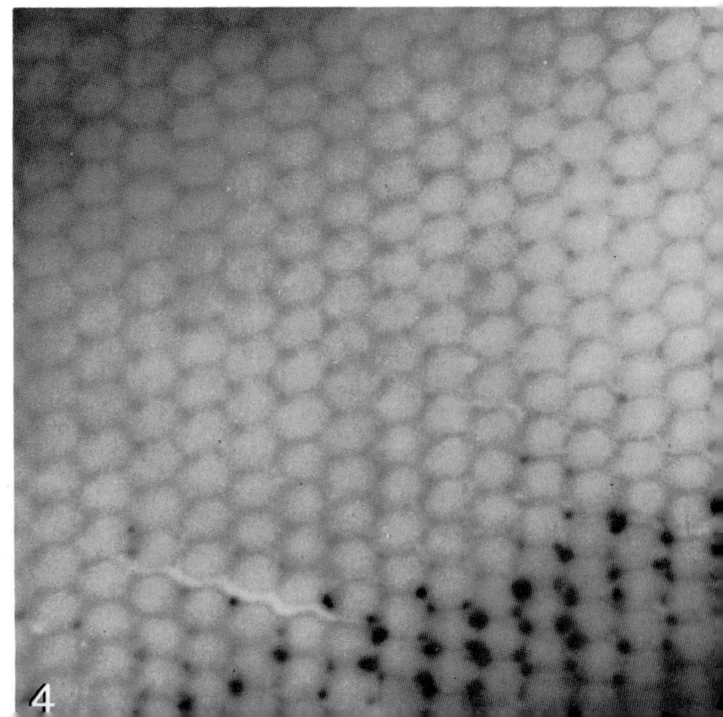
4

Scopelarchidae (*Pearleyes*)

Protacanthopterygii, Myctophiformes, Myctophoidei

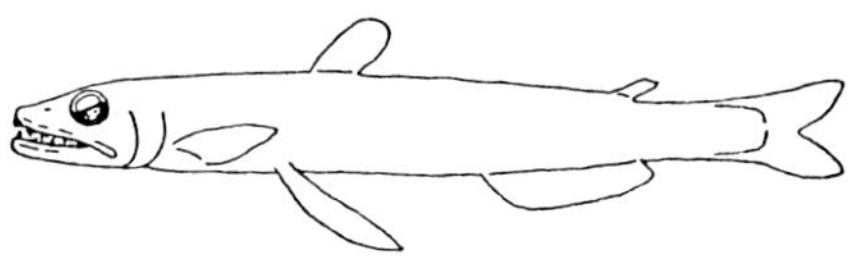

The pearleyes are small meso- and bathypelagic fishes, which are characterised by the presence of large, compressed, hooked teeth on the tongue, and tubular eyes. The tubular eyes are directed forwards in some species, upwards in some others, and slightly backwards in still others.

A main (bottom) and an accessory (dorsal) retina are differentiated in the tubular eyes of *Scopelarchus guentheri, S. sagax* and *Neoscopelarhoides* sp. All are pure-rod retinas, with well-developed internal nuclear layers. The main retina of *S. guentheri* includes an anterior part with a uniform population of rods associated with an unspecialised pigment epithelium (Plate 45.1), and a posterior part with bundles of about 23 rods inserted into pits in a reflecting tapetum lucidum formed by the epithelial cell processes (Plate 44; Plate 46). A third, anteriormost part is present in the main retina of *S. sagax,* with extremely long rod outer segments which are not grouped (Plate 46.2). The accessory retina possesses rod bundles of variable size (Plate 45.2).

References: BRAUER (1908); MUNK (1966a); LOCKET (1971b)

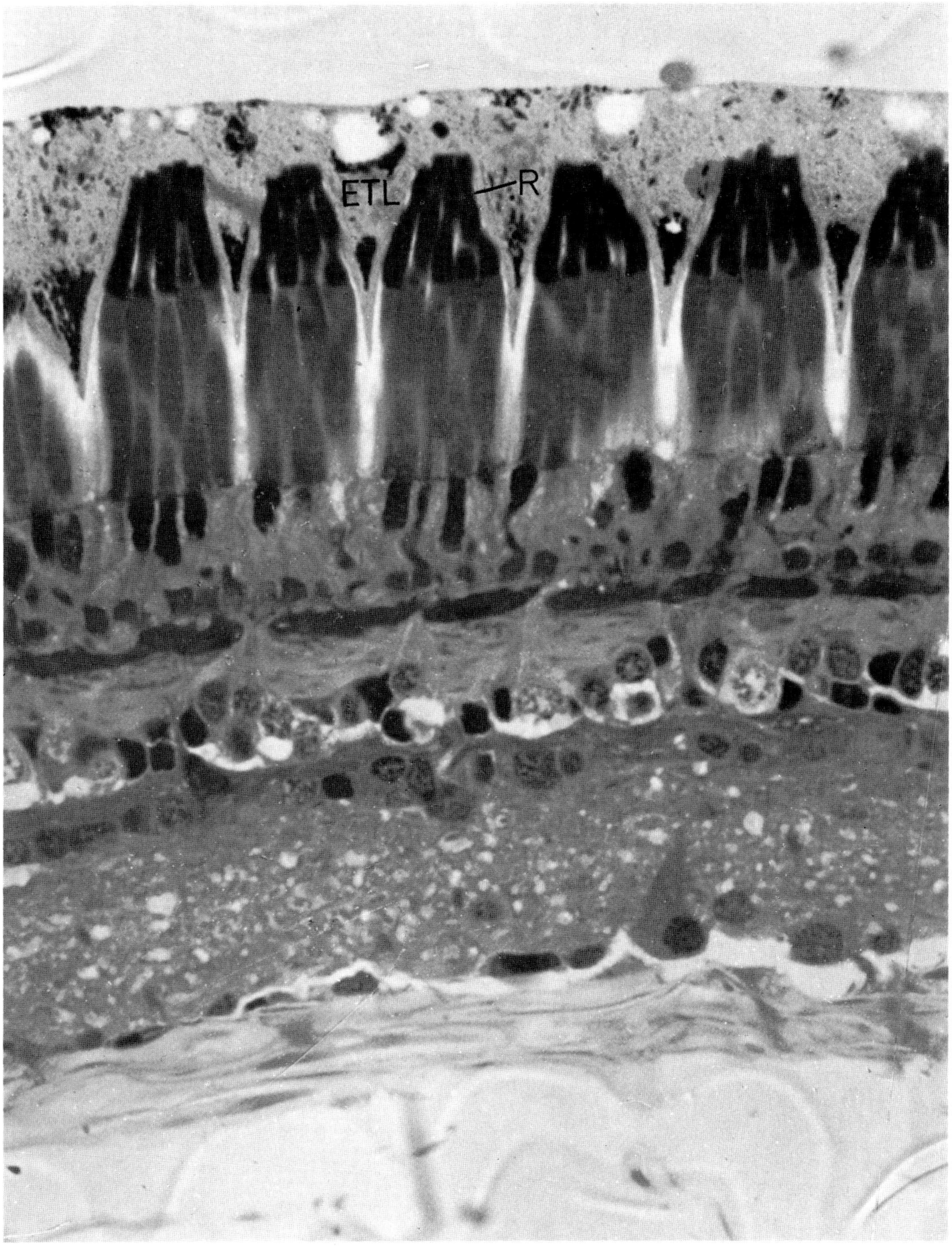

Plate 44. Photomicrograph of the grouped main retina of *S. guentheri,* seen in transverse section (about × 550). *ETL* epithelial tapetum lucidum, *R* grouped rods. Plate 45.1, 2. Photomicrographs of transverse sections of the nongrouped main retina (1) and the accessory retina (2) in *S. guentheri* (about × 550 and × 1,100). Plate 46.1, 2. Photomicrographs of a transverse section (about × 200) of the retinal step in *S. guentheri* (1) and a tangential section (about × 1,100) of the grouped main retina in *S. sagax* (2)

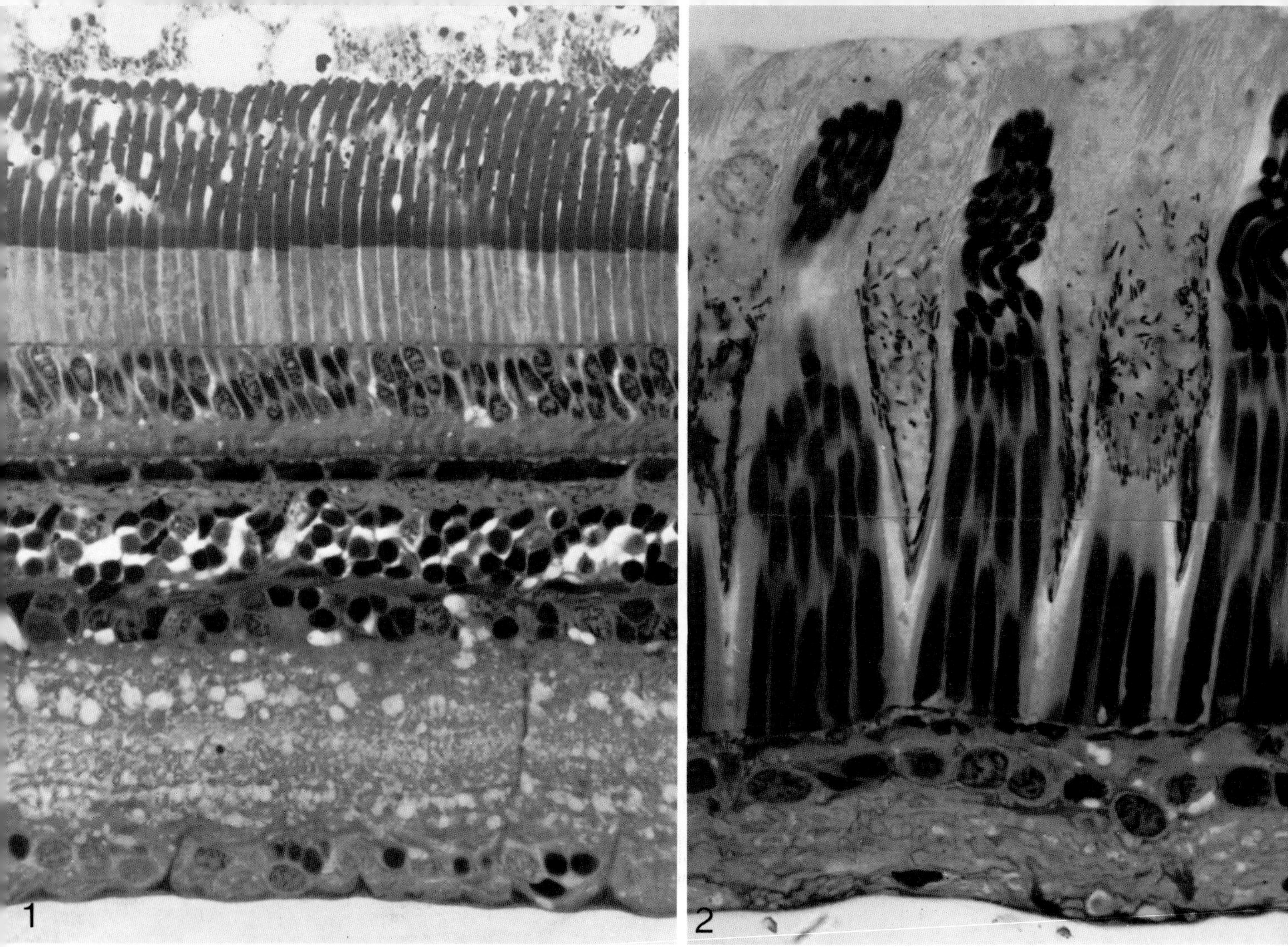

△ Plate 45 ▽ Plate 46 (Legends see page 97)

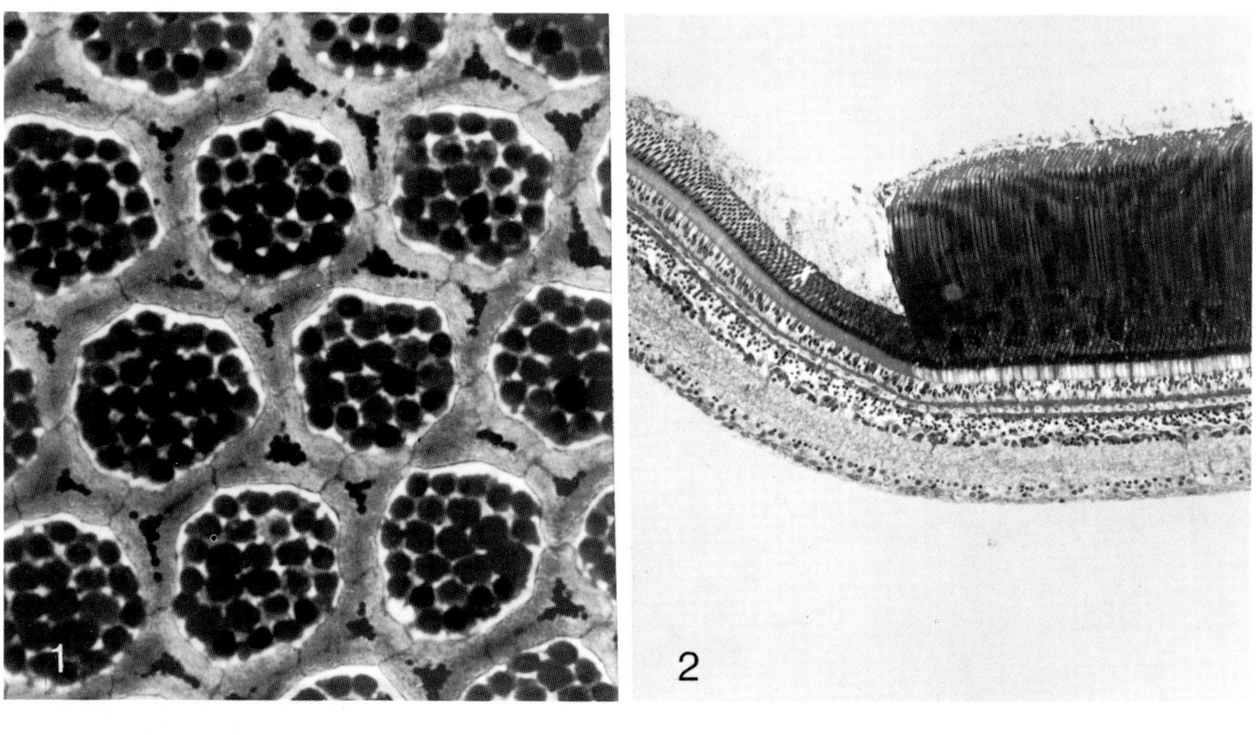

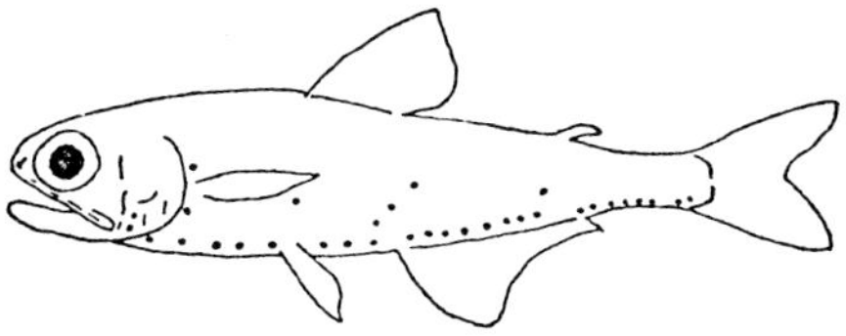

Myctophidae (*Lanternfishes*)

Scopelomorpha, Myctophiformes

Myctophoids constitute a large group of teleost families presenting some primitive, salmoniform characters as well as advanced morphological features; therefore it is thought that they may have contributed to the ancestry of certain major percomorph types. The lanternfishes are a major constituent of the mesopelagic fish fauna throughout the world. Their most distinctive external characters are the large eyes, the wide mouth gaping back beyond the eye, one soft-rayed dorsal fin, a deeply forked tail, and the presence of a series of luminescent organs on the head and trunk. These small and fragile fishes are in many cases active vertical migrants, living at considerable depths in daytime and ascending hundreds of metres by night, as they chase crustacean planktonic forms.

In general, the retina of the lanternfish is characterised by the degenerate pigment epithelium in which only remnants of melanin pigment are occasionally seen at the retinal periphery. This is a pure-rod retina (Plate 47.1, *Notoscopelus krøyeri*); the rods are extremely slender and densely packed, averaging around 500,000 cells per mm^2. Three types of retinal organisation may be observed among the myctophids: (1) the rod inner segment is very short in comparison with the outer segment, and the bipolar cells are relatively few (Plate 47.1, 2, *Stenobrachius leucopsarus*), (2) the rod inner segment is long in comparison with the outer segment, and the bipolar cells are numerically abundant (Plate 47.3 *Ceratoscopelus maderensis*), (3) an intermediary type in which the dorsal retina belongs to the A-type and the ventral region to the B-type retina (Plate 47.4, Plate 48.1 respectively, *Benthosema glaciale*). Note in Plate 48.2 (*Protomyctophum thompsoni*) what may be interpreted as macrophages, between the choroid pigment and the flat, degenerate pigment epithelium.

The retina of the larval lanternfish differs from the descriptions given above by the presence of a normal pigment epithelium with long processes and dense pigment inclusions (Plate 48.3, possibly *S. leucopsarus*). This pigment is less abundant in the dorsal retina than in the ventral part. The visual cells are fewer than in the adult and have filamentous nuclei, long inner segments, and tapering outer segments (Plate 48.3). In the retina of the postlarval lanternfish, degeneration of the pigment epithelium is under way (Plate 48.4) and the visual cells are more numerous.

References: Brauer (1908); Vilter (1951a); Jollie (1954); Ali and Hanyu (1963)

Visual Pigments: A_1 max 485–490 (Ali and Wagner, 1975b)

Plate 47 ▷ and 48 ▷▷

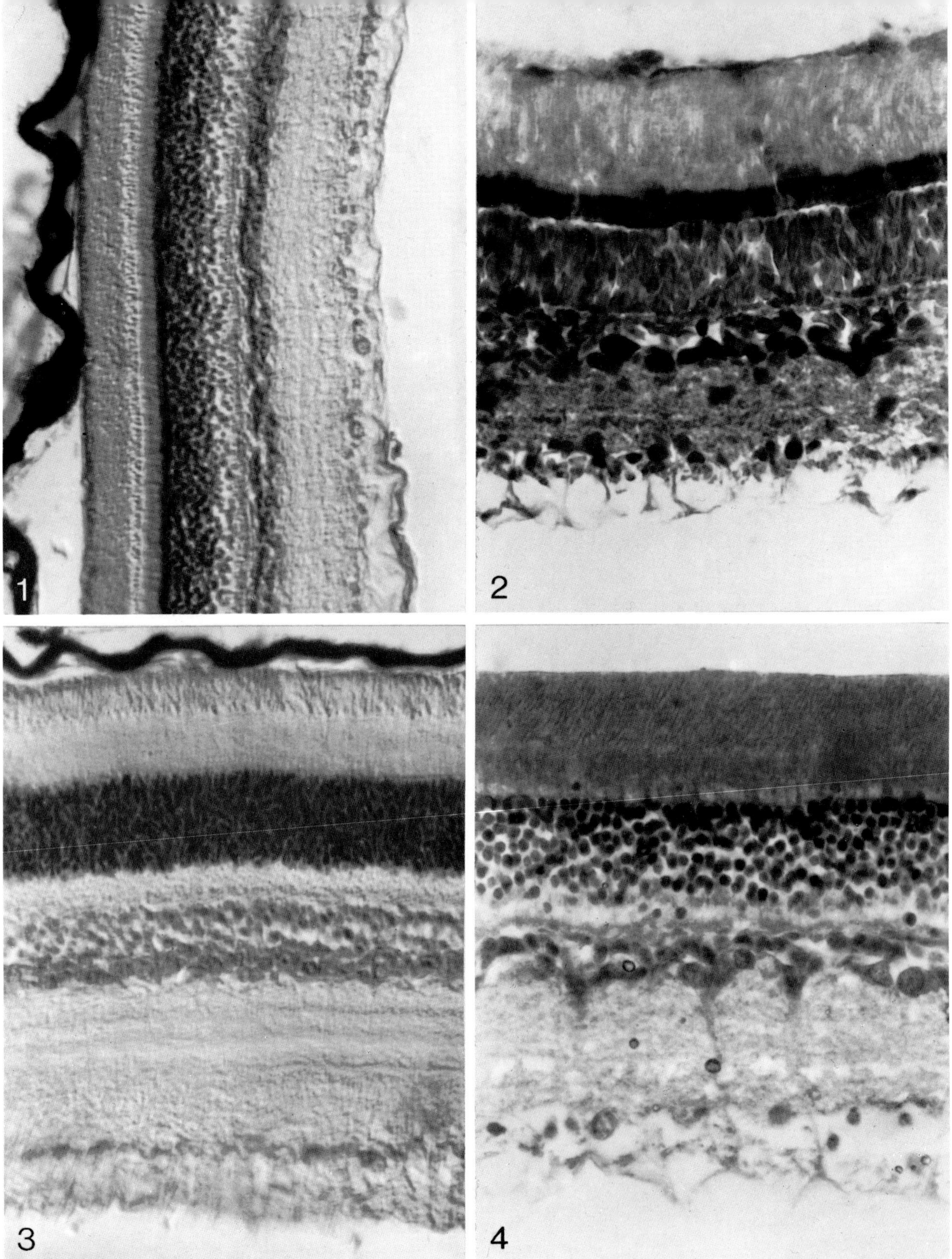

Plates 47.1-4, Plate 48.1-4. Transverse sections of the retinas of various lanternfish species (see text). Plate 47 (1) and (3) ×440; Plate 47 (2, 4), Plate 48 (1-4) ×720. *MA* macrophages

△ Plate 47

Plate 48 ▷

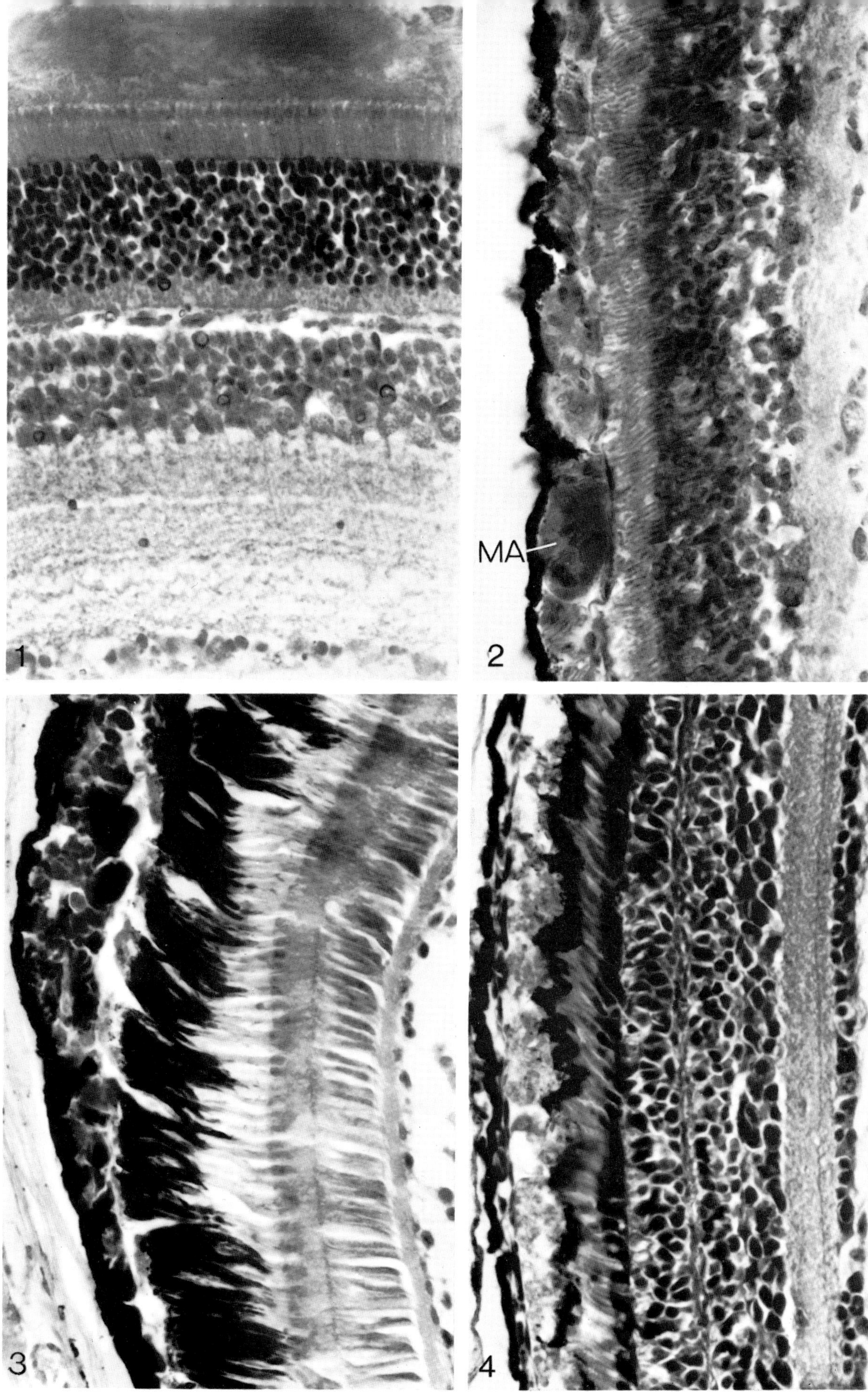
1
2
MA
3
4

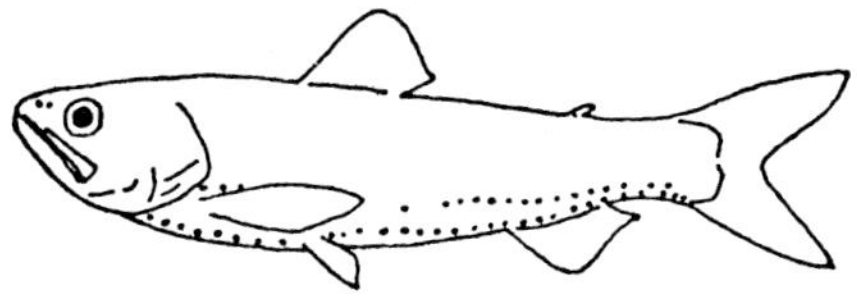

Neoscopelidae (*Neoscopelids*)

Scopelomorpha, Myctophiformes

Neoscopelids are deep-sea teleosts, which are closely related to the myctophids. They occur in deep shelf waters. The neoscopelids are noted for the advanced condition of the jaws, a hyoid and branchiostegal apparatus which is acanthopterygian in character. Some forms are luminescent.

The retina of *Neoscopelus macrolepidotus,* which is pure-rod, resembles very much that of the myctophids. The epithelial pigment is missing except near the iris region. The rods are numerous and long; the bipolar, amacrine and ganglion cells form one layer each. A tapetum lucidum has been described between the choriocapillaris and the retinal epithelium.

Reference: BRAUER (1908)

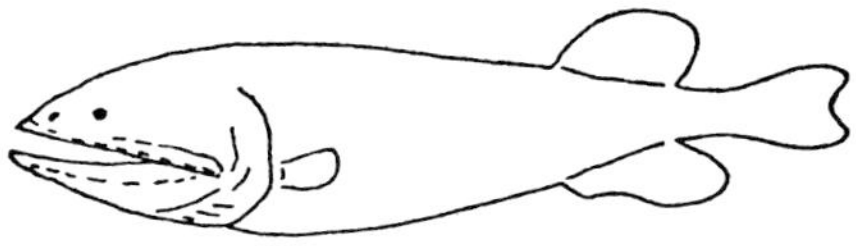

Cetomimidae (*Whalefishes*)

Acanthopterygii, Percomorpha, Beryciformes, Cetomimoidei

The rare, deep-sea whalefishes are undoubtedly the most interesting group among the Beryciformes, and the most specialised. They are very delicate and small, being remarkable due to the large mouth and highly distensible stomach which permit them to swallow other deep-sea fishes as large as themselves. The frontal and parietal bones are fused, the pelvic fins are absent, and the eyes are degenerate.

The retinas of *Cetomimus gilli, Ditropichthys storeri* and *Gyronomimus* sp. present various degrees of degeneration. The epithelial pigment has proliferated; the rods, which are the only visual cells present, are very small; the rod outer segment is swollen in many instances; in *Gyronomimus* sp. the rods are altogether absent. Macrophages accompanied by pigment granules are found between the outer segments. The inner retinal layers show marked cystoid degeneration.

References: Brauer (1908); Munk (1965a, 1966a)

Giganturidae (*Giganturids*)

Acanthopterygii, Percomorpha, Beryciformes, Giganturoidei

Like the whalefishes, giganturids are highly specialised fishes. They lack the pelvic fins, maxillary and palatine bones. They possess sharp and depressible teeth; the lower lobe of the tail is greatly extended. These small bathypelagic fishes are equipped with tubular eyes directed forwards.

The retina of *Gigantura indica* is divided into the main and accessory retinas. The main retina is located on the bottom of the tubular eye; it is a pure-rod retina in which horizontal cells are present and the bipolars, amacrines, and ganglion cells form one layer each. The accessory retina is located on the lateral wall of the tubular eye; the rods are considerably fewer and shorter in this area, and the horizontal cells are missing.

Reference: Brauer (1908)

Characidae (*Characins*)

Ostariophysi, Cypriniformes, Characoidei

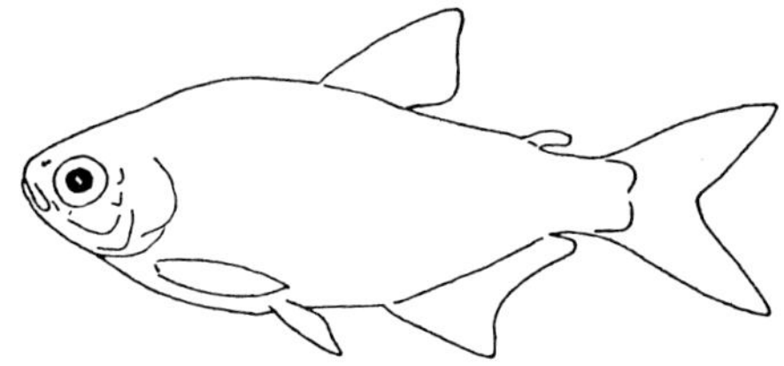

This is a large group of teleosts with great diversity of body forms, as is well indicated by the conflicting family classifications proposed so far. These freshwater fishes are distinguished from the cyprinoids by the presence of teeth in the jaws and a small fleshy adiopose fin on the back. Some are herbivorous or omnivorous, while others such as the piranhas are dangerous carnivores, especially when in schools.

The retina of the piranha is characterised by the long processes of the epithelial cells and the rods, which constitute an important part of the retinal thickness. The epithelial processes are filled with granular reflecting material in the dorsal retina (Plate 49.1), while they contain primarily melanin pigment in the ventral retina (Plate 49.2). Single cones and unequal double cones are present (Plate 49.1, 2), and they are arranged in square mosaics containing only accessory single cones (Plate 49.3, 4). The area retinae is ventral (Plate 49.4). The internal layers (bipolars, ganglion cells) are poorly developed.

References: Lüling (1955b); Cahn (1958); Engström (1963b); John and Haut (1964); John (1969); Ali and Raymond (1972); Ali and Wagner (1975a)

Visual Pigments: A_1 max 503; A_2 max 527–530 (Ali and Wagner, 1975b)

Plate 49.1-4. Transverse (1) and (2) and tangential (3) and (4) section of the dorsal (1) and (3) and ventral (2) and (4) retina of the piranha, *Serrasalmus marginatus*. × 560 ▷

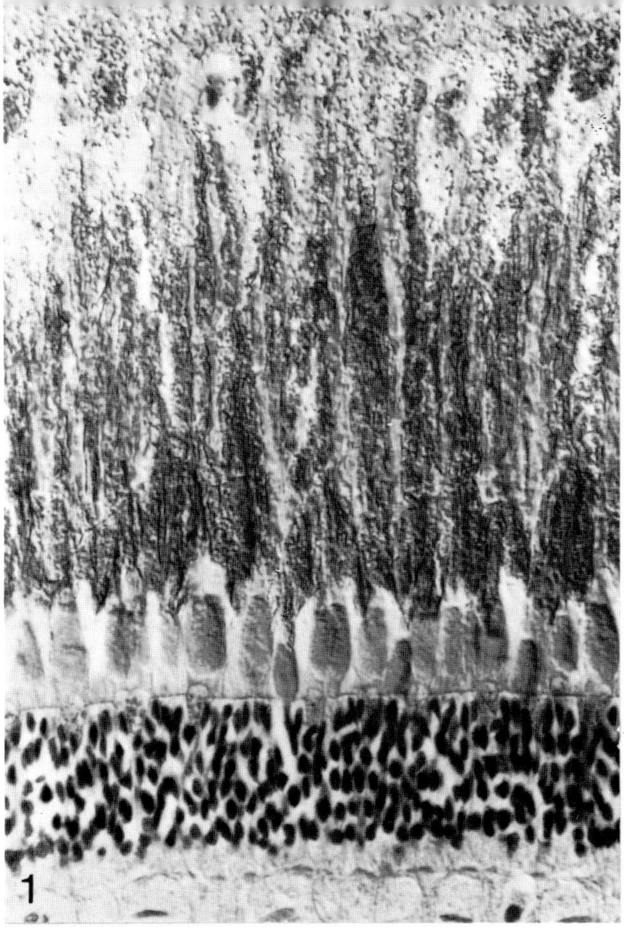
1
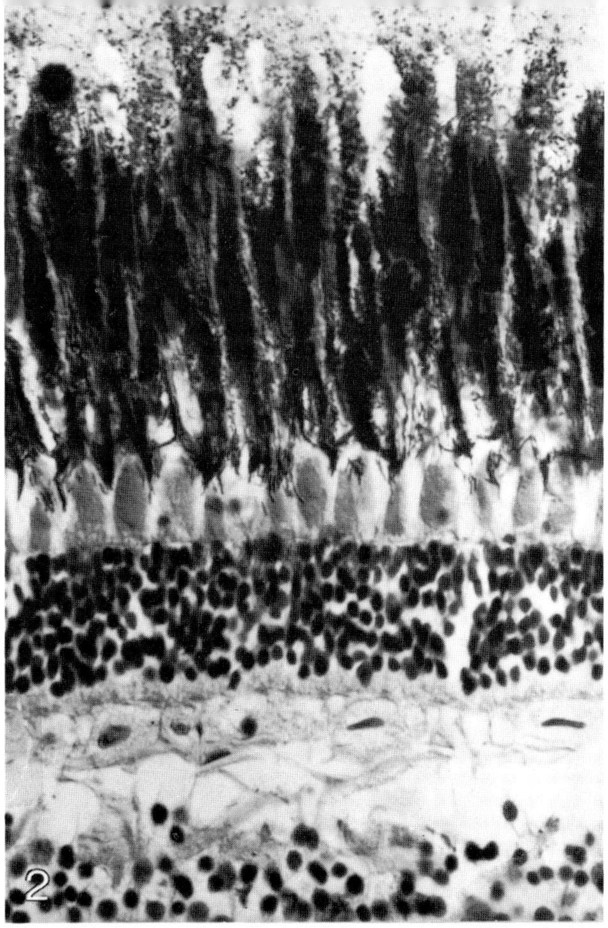
2
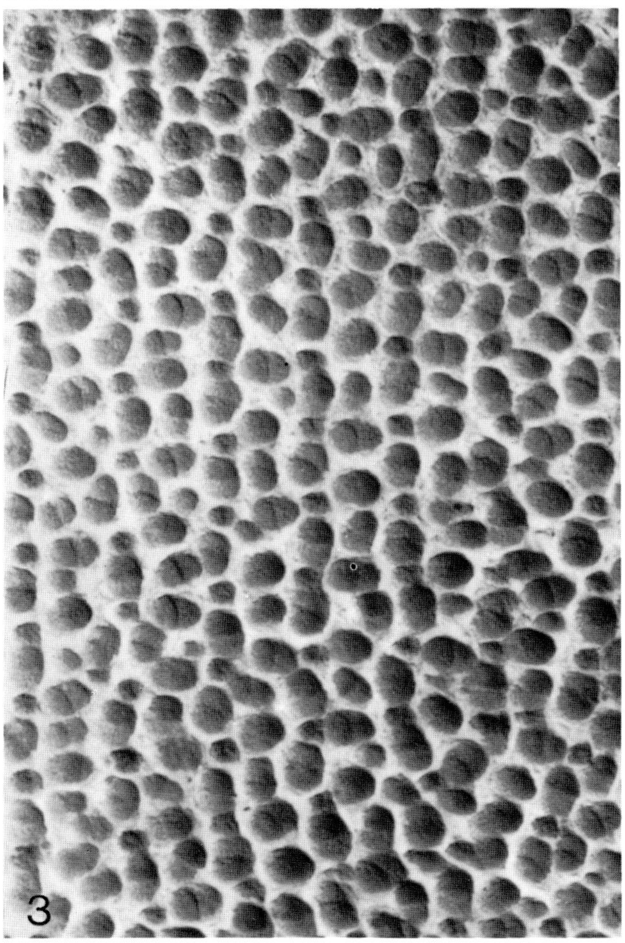
3
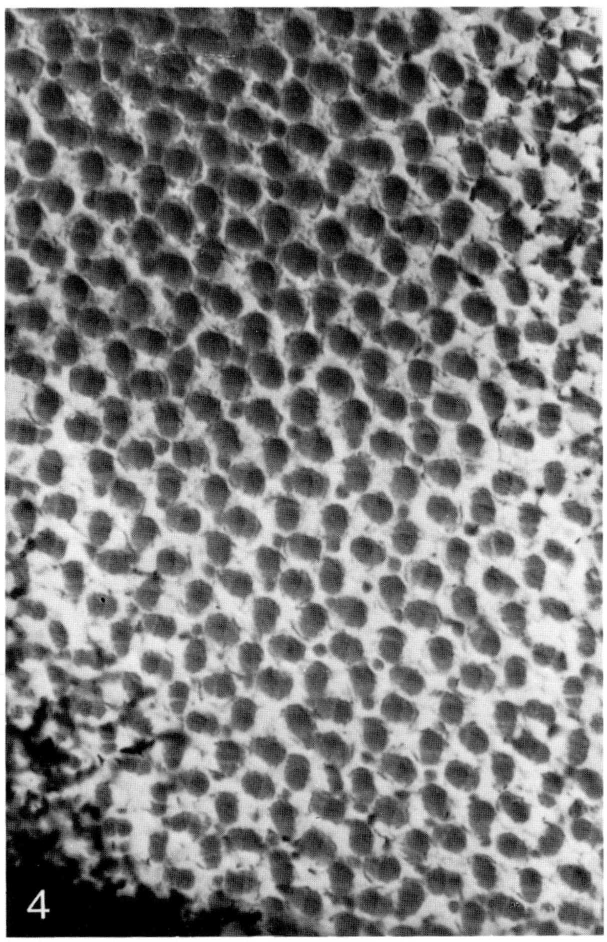
4

Lebiasinidae (including Nannostomidae) (*Pencilfishes*)

Ostariophysi, Cypriniformes, Characoidei

Pencilfishes are characinlike Ostariophysi of Central and South America. They are distinguished from typical characids by their dentition; in fact, in most of them, the lower jaw is without teeth. They inhabit small, slow-flowing, weedy or shaded water basins, and many species get their prey (mainly insects) at the surface film. As an adaptation to this feeding habit, they swim in an oblique position, with the head upwards.

In the retina of the tube-mouthed pencilfish, *Nannostomus equus* (Plate 50.1, 2), as in the retina of the characids and anostomids, the double and single cones form regular square mosaics in which the single cones are in the "accessory" position only, i.e. at the corners of the squares (Plate 50.3,4). Two horizontal cell layers were reported.

Reference: WAGNER (1972)

Visual Pigments: A_1 max 503; A_2 max 527 (ALI and WAGNER, 1975b)

Plate 50.1-4. Transverse (1) and (2) and tangential (3) and (4) sections of the retina of the characin, *N. nannostomus* × 560

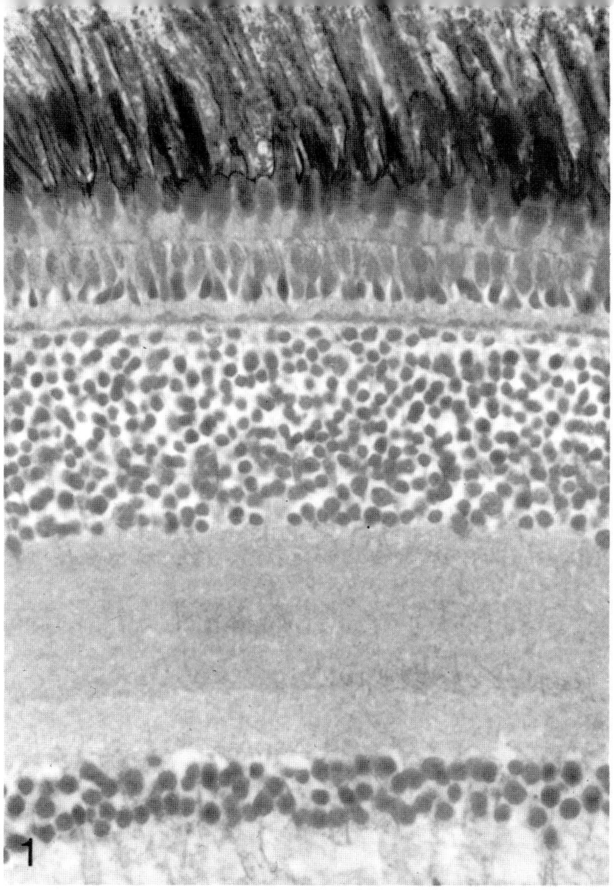
1

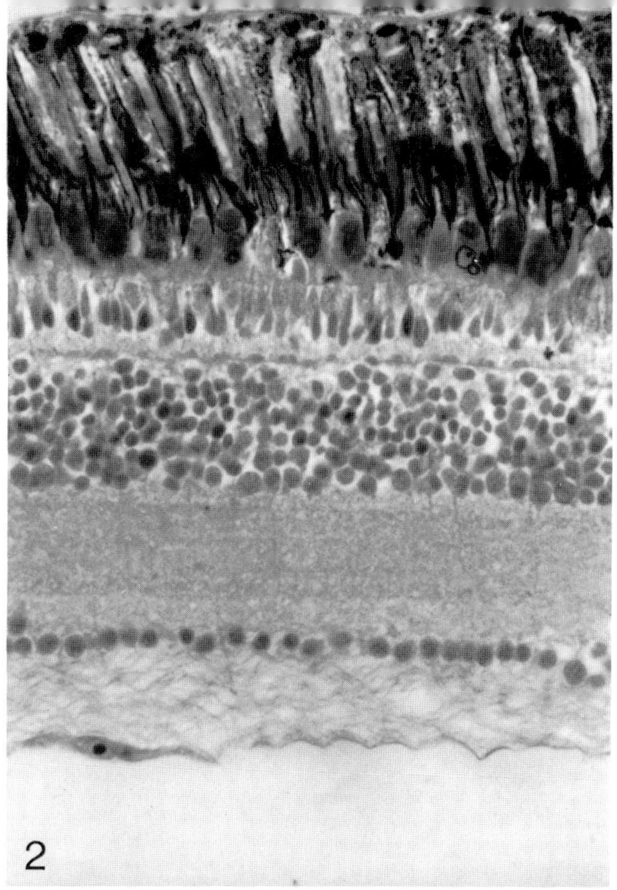
2

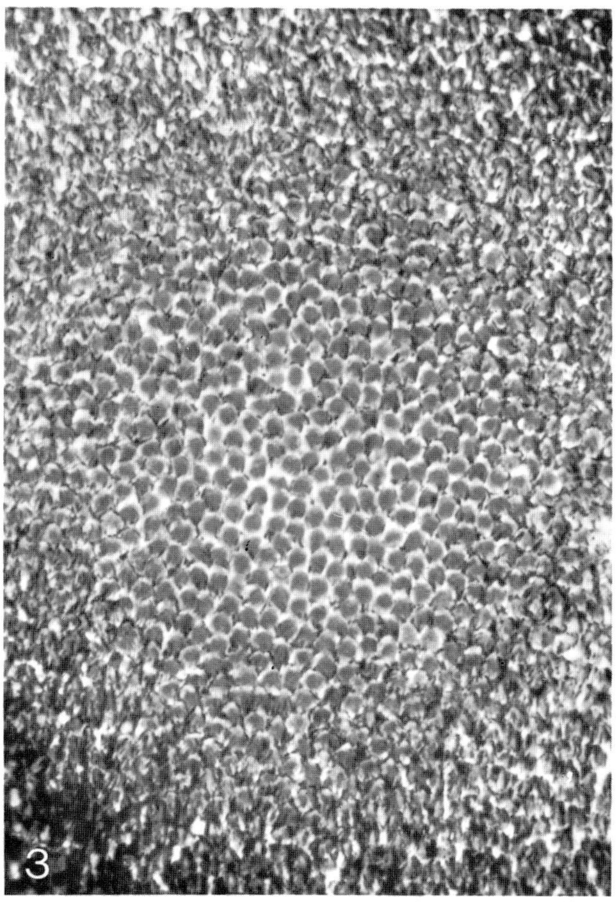
3

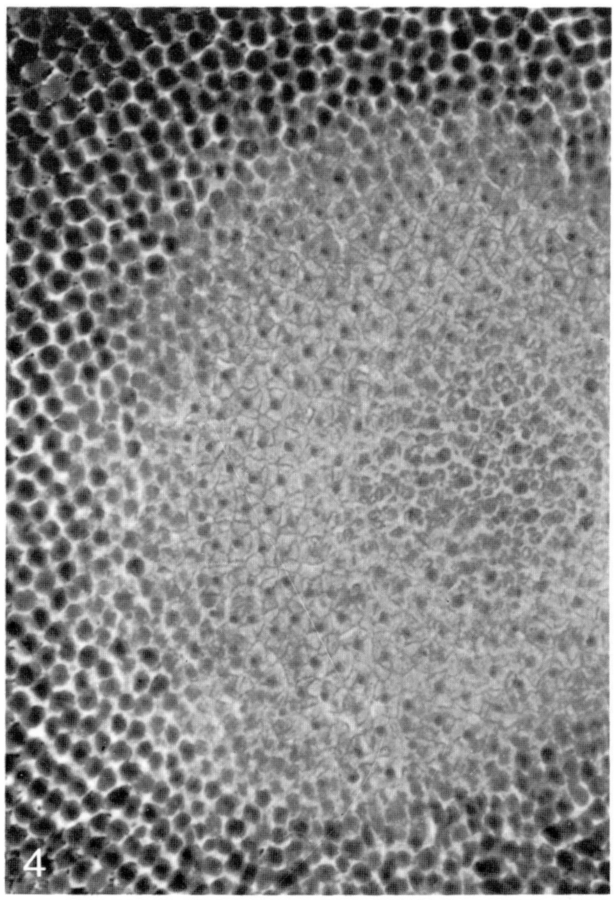
4

Anostomidae (*Headstanders*)

Ostariophysi, Cypriniformes, Characoidei

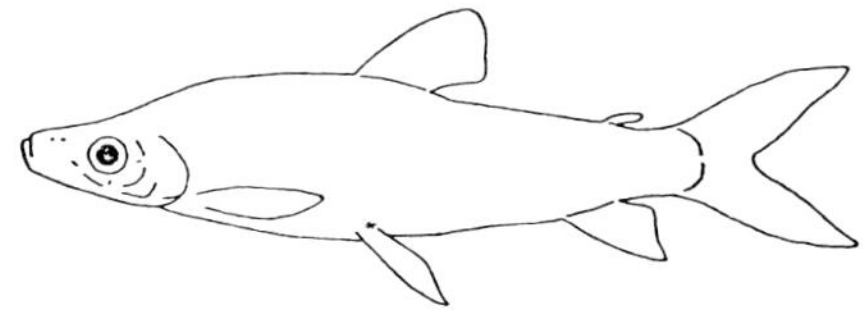

Anostomids are characoids which are distributed throughout Central and South America and characterised by their narrow mouth. The genus *Anostomus* comprises freshwater fishes of elongate, strongly compressed forms in which the upper part of the head is often somewhat flattened. They are inhabitants of standing or sluggish weedy waters in which they swim in an oblique position, with their heads down, continuously seeking their food over the bottom.

The retina of the headstander is characterised by the presence of some reflecting pigment in the epithelial cell processes (Plate 51.1) of the dorsal retina; in the ventral retina only melanin pigment is present and the cone density is higher (Plate 51.2, area retinae). As shown in Plate 51.2,3 (temporal retina), there are three main cone types: the long single cone, the short single cone, and the unequal double cone, more or less regularly arranged in square mosaics. A few triple cones may be seen occasionally (Plate 51.2, *arrow*). Note the coarse granular texture of the double cone and triple cone ellipsoids.

Leporinus fasciatus

This is a genus of fishes with elongate, torpedo-shaped, slightly compressed bodies. Their mouths are relatively small with few teeth. Most species are herbivores and inhabit gravelly, sluggish streams.

The retina of *L. fasciatus* resembles strikingly that of *Anostomus*. Unequal double cones and single cones are present, but no triple cones were observed. Note in Plate 51.4 (temporal retina) the very small single cones and the large horizontal cells. Plate 51.5 is a tangential section passing through the double cone myoids and single cone ellipsoids; note the unequal components of the double cones and the regular square mosaic composed of double cones and so-called accessory single cones (defined by ENGSTRÖM, 1963b).

Visual Pigments: A_1 max 503; A_2 max 527 (ALI and WAGNER, 1975b)

Plate 51.1-5. Transverse (1) and (4) and tangential (2), (3) and (5) sections of the retina of *A. anostomus* and *L. fasciatus*. (1), (3)–(5) ×560; (2) ×720

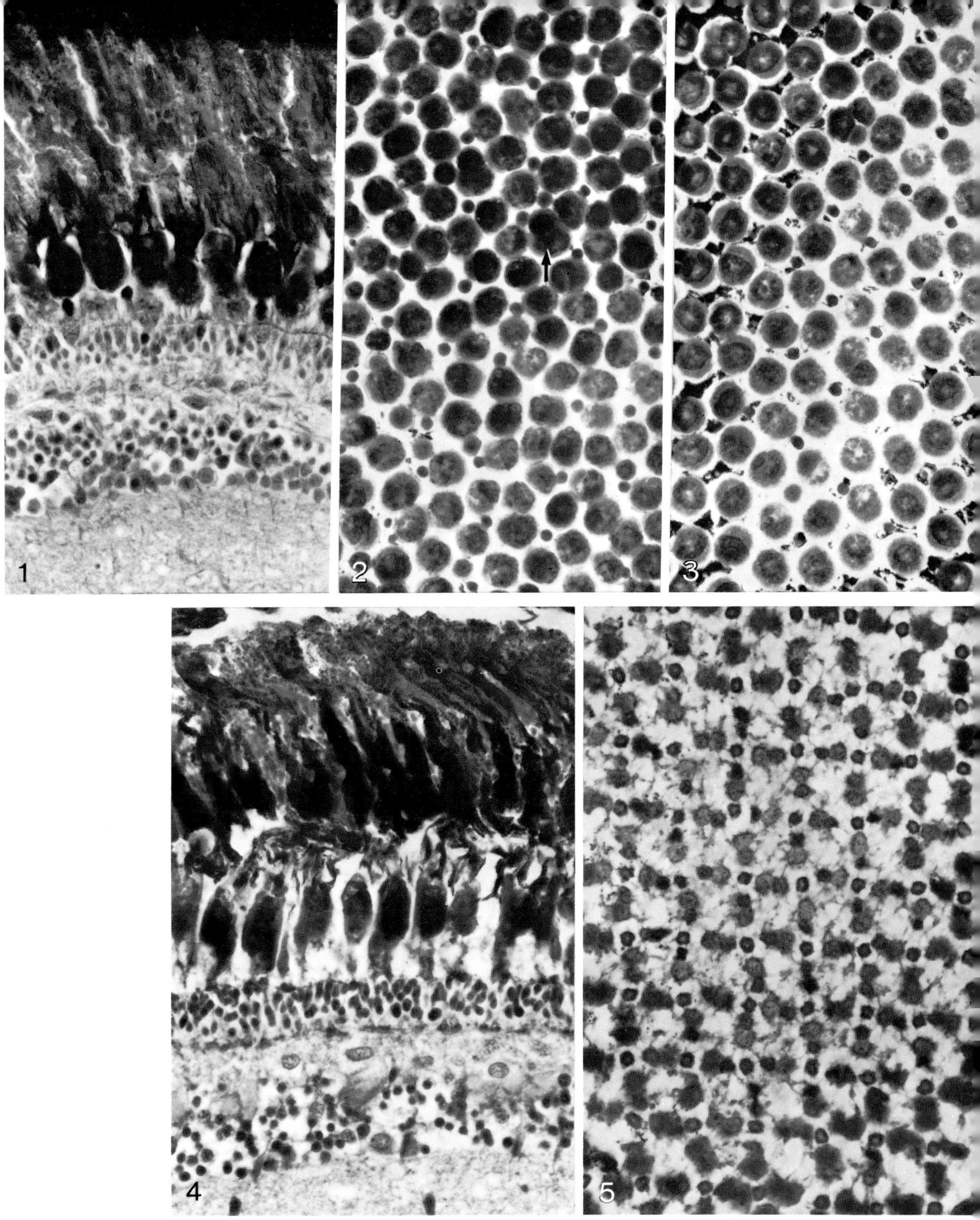

Cyprinidae (*Minnows and Carps*)

Ostariophysi, Cypriniformes, Cyprinoidei

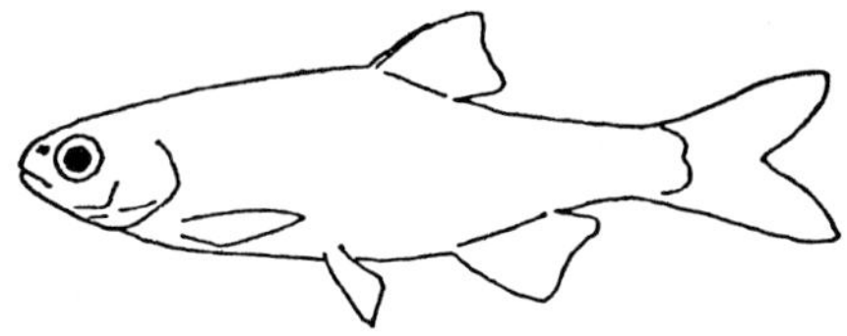

Cyprinoids constitute a large and successful group of freshwater teleosts, having worldwide distribution. Among them, the cyprinids are now considered as the most primitive. They are characterised, as are all of the Ostariophysi, by the presence of a distinctive kind of otophysic connexion, the Weberian apparatus. The cyprinids have soft-rayed fins, toothless jaws and pharyngeal teeth to grind their food in or on the bottom ooze.

The retina of the goldfish (*C. auratus*) conforms to the current morphological type commonly found in the teleosts. The epithelium pigment is abundant, the rods and cones are numerous (Plate 52.1) and undergo extensive retinomotor movements. As shown by Engström (1960), single and unequal double cones (Plate 52.2) are present and arranged in alternate rows. There are two or three layers of tiny horizontal cells. Plate 52.1, 2, and the internal nuclear layers are well developed.

References: Wunder (1925); Verrier (1928a); Walls (1942); Quaguebeur (1955); Lyall (1957b); Marshall and Thinès (1958); Engström (1960); Baburina (1961a); Engström and Rosstorp (1963); Ali (1964a, b); Stell (1965, 1967); Peters and Peters (1965); Bimes et al. (1966a, b); Peyraud (1966); Witkovsky et al. (1974); Scholes (1975); Ali and Wagner (1975a); Stell and Lightfoot (1975); Stell (1975); Stell and Hárosi (1976)

Visual Pigments: Rods A_1 max 467–510; A_2 max 522–536; Cones A_2 max 455–625 (Ali and Wagner, 1975b)

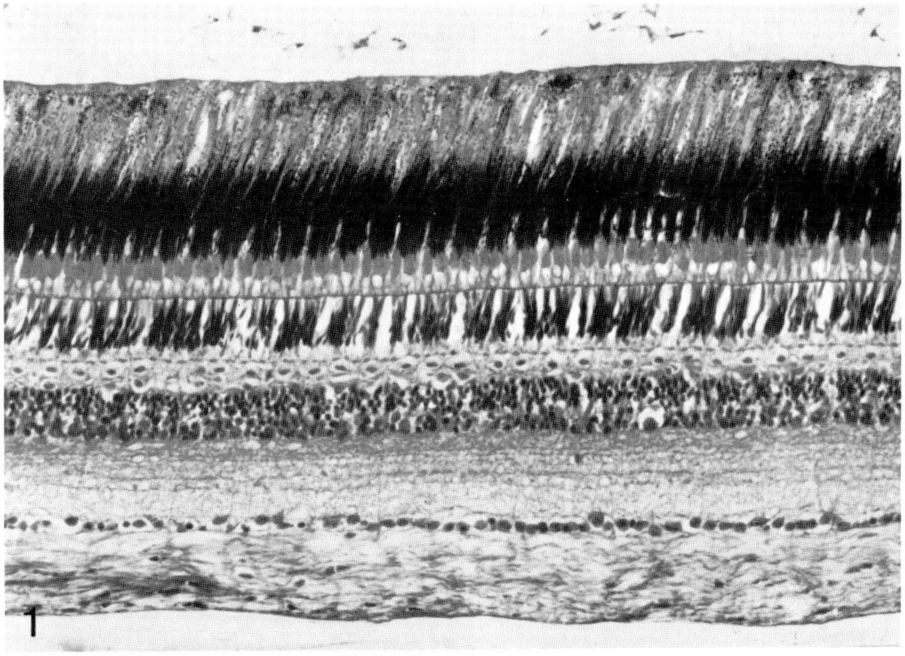

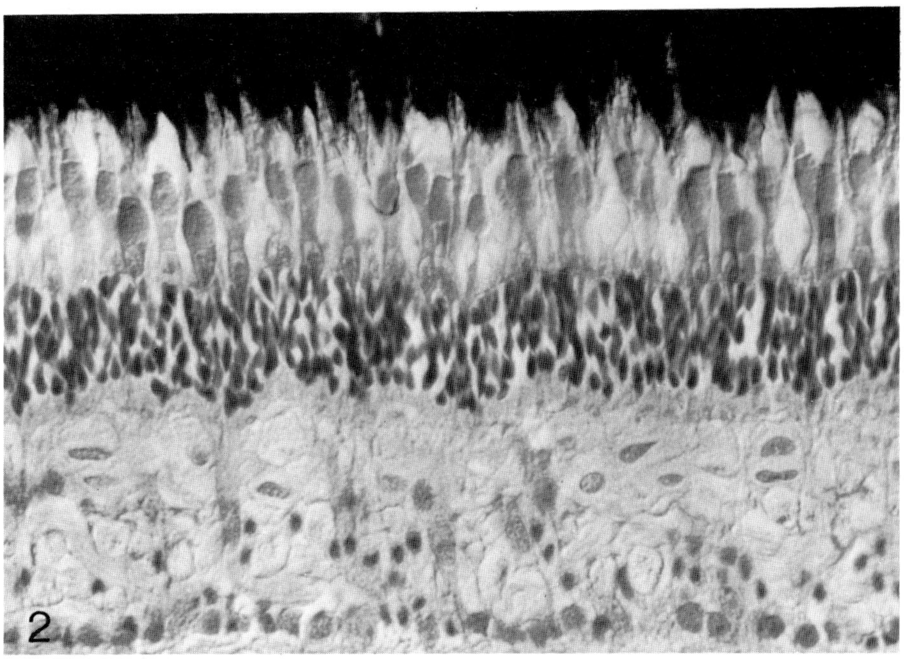

Plate 52.1, 2. Transverse sections of the retina of the goldfish, *C. auratus*. (1) ×235; (2) ×560. (See also Plate 5 in Introduction)

Catostomidae (*Suckers*)

Ostariophysi, Cypriniformes, Cyprinoidei

Suckers are freshwater cyprinoids from North America and Eastern Asia. They are close relatives of the minnows from which they are probably derived; like the minnows, the first four vertebrae are modified, the fins are soft-rayed, and the scales cycloid. The mouth is inferior in position and modified as a sucking device. The longnose sucker frequents cool and relatively deep waters where it feeds on the bottom.

The retina of the longnose sucker differs from that of the goldfish. The rods are comparatively few as shown by the external nuclear layer (Plate 53.1). The cones are large and are of three types: the short single cone, the long single cone, and the unequal double cone (Plate 53.1, 2). Their centre of density (area retinae) is found along the temporo-nasal arc, and as seen in Plate 53.2 the cone types are not distributed in a mosaic. Plate 53.3, 4 illustrates the external and internal horizontal cell layers respectively; they are sectioned tangentially to the retinal surface.

Visual Pigments: A_2 max 524 (ALI and WAGNER, 1975b)

Cobitidae (*Loaches*)

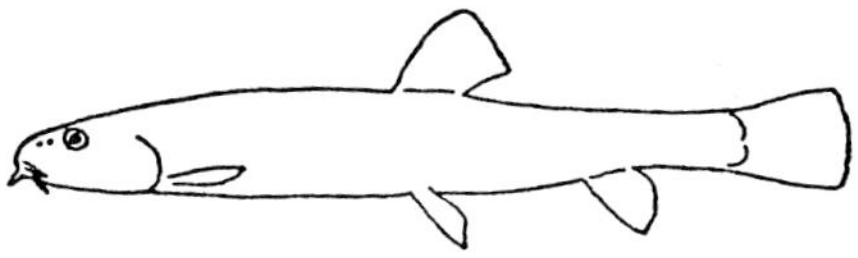

Ostariophysi, Cypriniformes, Cyprinoidei

Loaches are typical Old World cyprinoid fishes. They have a worm-like appearance or a flattened lower surface, both being adaptations for bottom-living. They also have a ventral mouth without teeth and three or more pairs of barbels. Many loaches (*Cobitis, Misgurnus*) burrow into the sand or mud to protect themselves from predators and the winter. Their food consists of worms and insect larvae.

Only single cones and rods have been reported in the retina of *C. barbatula, C. taenia* and *M. fossilis*. Rods and undefined cones have also been reported in the retina of *M. anguillicaudatus*. In all these species both the rods and cones, as well as the epithelial pigment, are capable of retinomotor responses.

References: WUNDER (1925); KOBAYASHI (1957)

Plate 53.1-4. Transverse (1) and tangential (2)–(4) sections of the retina of the longnose sucker, *Catostomus catostomus*. × 560 ▷

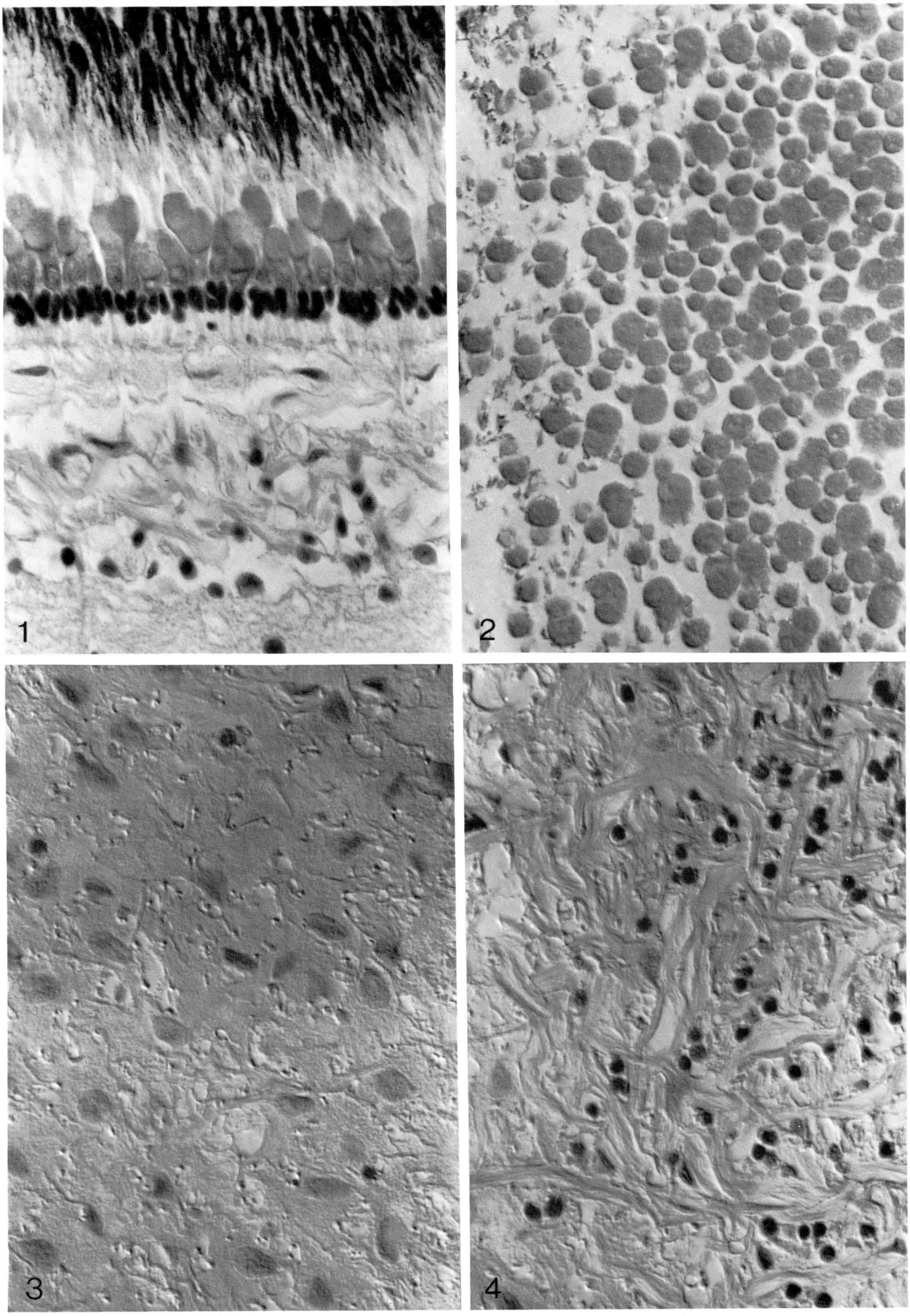

Ictaluridae (*Bullheads, Freshwater Catfishes*)

Ostariophysi, Siluriformes

Siluriformes are Ostariophysi without parietal, symplectic, subopercular, first and second pharyngobranchial, epipleural, and epineural bones. True scales are absent and the body is naked or covered with bony plates. They constitute an unusually well-defined order of fishes, mostly confined to freshwaters of South America and Africa. Within this order, the Loricariidae and Callichthyidae, among other families, are regarded as representing one or more distinct lines of evolution within the catfishes.

The North-American catfishes (Ictaluridae) are characterised by a more or less tadpole-shaped body with a large broad head, and by the presence of sharp spines and eight whiskers. The brown bullhead is a fish of quiet, weedy, muddy lakes and ponds, or of slowly flowing rivers. It is omnivorous and nocturnal. The retina of the bullhead is characterised by a very thick epithelial cell layer, whose processes are filled with some reflecting material of granular texture (Plate 54.1 *dark field,* interference contrast). The large rods are immersed in this granular tapetum lucidum (Plate 54.1). The cones are small, and only single cones are found. The myoid region of the cones (Plate 54.1, 2) has a darkly stained core bordered by a loose membrane (see Plate 54.2). The internal layers are thin and poorly developed. Note the two kinds of horizontal cells seen in tangential section in Plate 54.3, one kind with a large cell body and the other with slender profiles and filamentous processes. Their retina is also characterised by the presence of an optic nerve that branches into five optic papillae as it penetrates the fundus.

References: Arey (1916); Wunder (1925); Verrier (1927); Welsh and Osborn (1937); Arey and Mundt (1941); Polyak (1957); Ali (1964a); Witkovsky et al. (1974); Ali and Wagner (1975a)

Visual Pigments: A_2 max 534 (Ali and Wagner, 1975b)

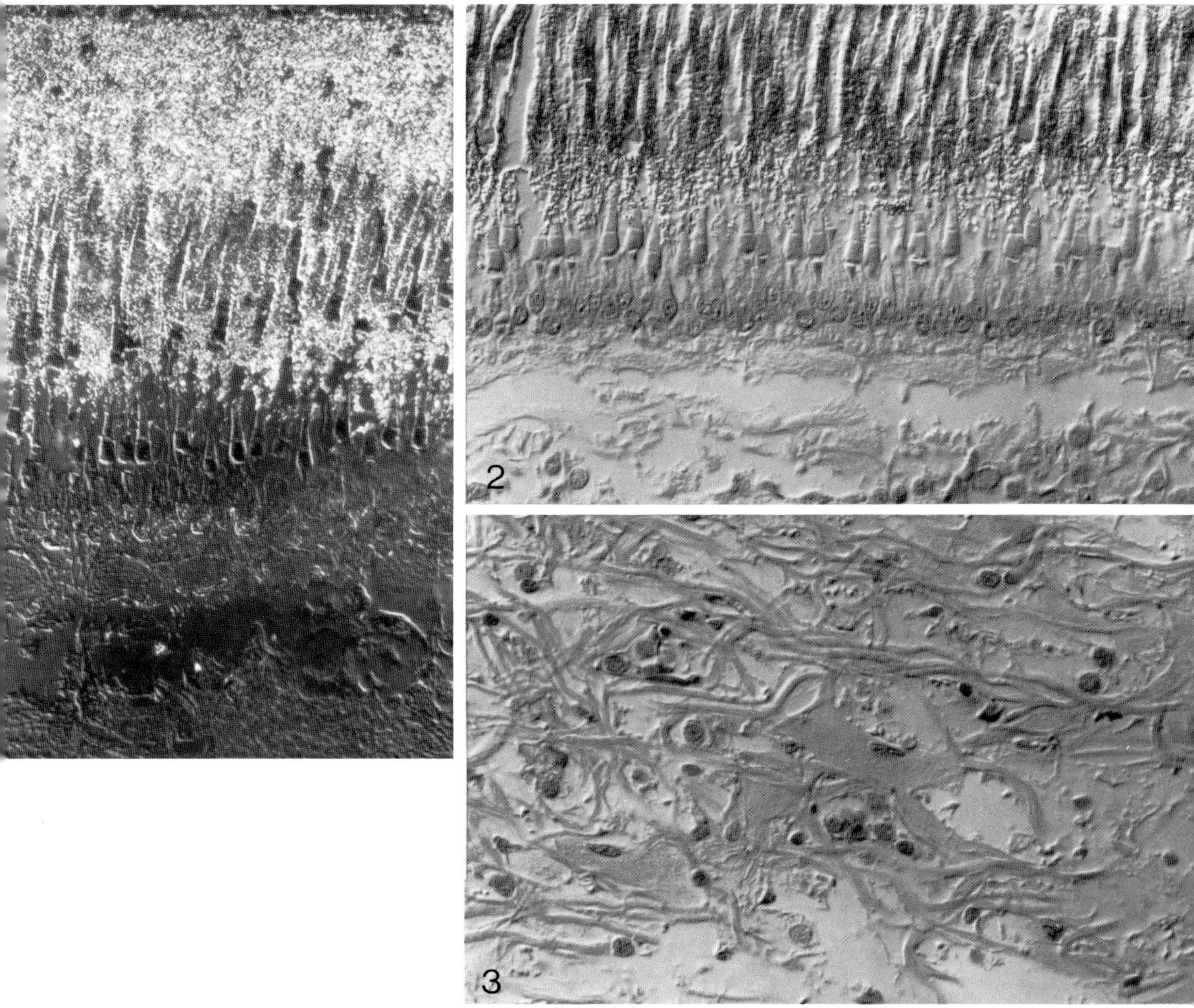

Plate 54.1-3. Transverse (1) and (2) and tangential (3) sections of ventral retina of the brown bullhead, *Ictalurus nebulosus*. $\times 560$

Sisoridae (*Catfishes*)

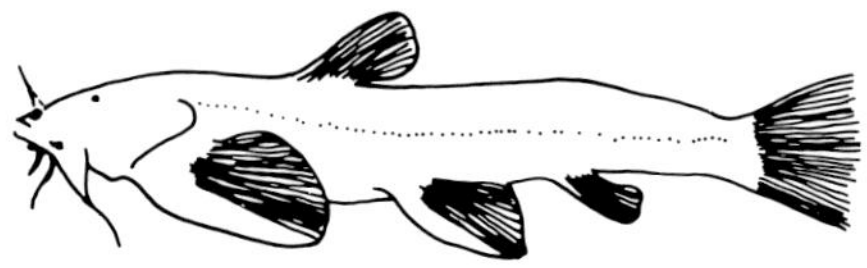

Ostariophysi, Siluriformes

These are Old World catfishes with an entirely naked skin. The eyes are usually covered by skin.

The retina of *Euchiloglanis* sp. has been studied, but no description is given.

Reference: McEwan (1938)

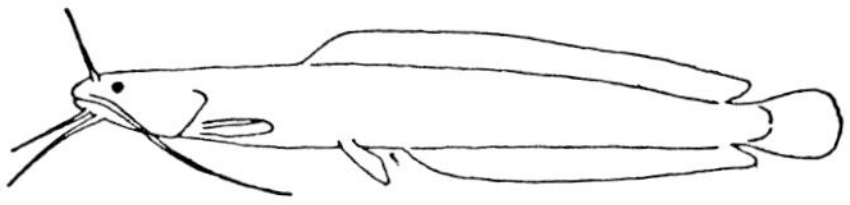

Clariidae (*Airbreathing Catfishes*)

Ostariophysi, Siluriformes

Clariids are elongate, sometimes eel-shaped catfishes with a broad head, a transverse mouth, and four pairs of barbels, which are often very long. Tubular or arborescent accessory air-breathing organs are found posterior to the gill chamber. These organs assist the catfish in very oxygen-deficient waters or when walking out of the water. All these catfishes are voracious predators and feed by night. They burrow in the mud during the dry season.

Only rods have been reported in the retina of *Clarias batrachus*. Stout rods with short myoids were described, and the ellipsoids appear to form two superimposed layers in the illustration. This is an indication that the cones, presumably placed in front of the rods in the histological sections, were overlooked. The internal nuclear layer is similar to that of the brown bullhead.

Reference: VERRIER (1927, 1928a)

Mochokidae (*Upside-Down Catfishes*)

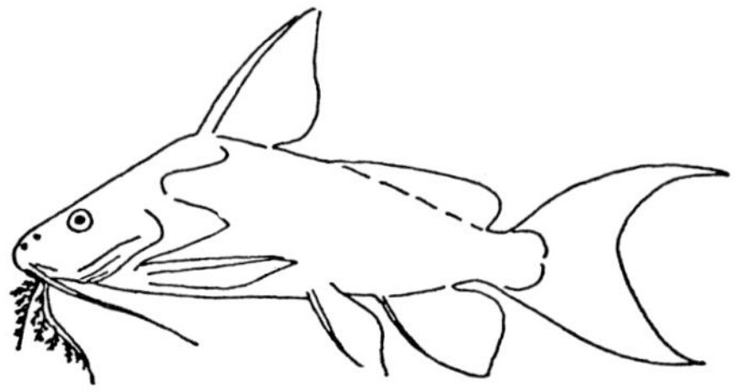

Ostariophysi, Siluriformes

Mochokids are entirely naked catfishes from the African continent. They possess three pairs of barbels. These crepuscular fishes inhabit slow-flowing waters and lagoons, gathering in large shoals during the day, hiding in suitably protected areas. They have the habit of swimming belly-up close to the surface of the water, and in this position they feed on the underneath of aquatic leaves.

The retinas of *Synodontis nigriventris* and *S. notatus* have large rods and small single cones which are numerically represented in the ratio 2:1 (Plate 55.1). The retinomotor movements involve the epithelial pigment and the rods, but not the cones (Plate 55.4, 5). The epithelial cells are packed with some, as yet undefined, reflecting material. The optic nerve fibres are arranged in fascicles between the ganglion cells. A remarkable feature of these retinas is the presence of multiple optic papillae Plate 55.2, 3 radiating from the point of emergence of the optic nerve into the retina (Plate 55.6). Twenty-five to twenty-seven such papilla are present in the upside-down carfish, while only five have been reported in the brown bullhead.

References: Wagner (1970); Ali and Wagner (1975a)

Visual Pigments: A_2 max 533 (Ali and Wagner, 1975b)

Plate 55.1-6. (1) Transverse section of the central region of a dark-adapted retina of *S. nigriventris*. Note that cones are sparse and small whereas rods are more numerous and fairly big. The inner layers are very poor. 5 µm paraffine section; Azan; ×400. (2) Transverse section of the retina of *S. nigriventris* in the region of the multiple optical papilla showing two sites of entry of the optical nerve; light adapted. 5 µm paraffine section; Azan; ×200. (3) Tangential section of the retina of *S. nigriventris* in the region of the multiple optical papilla showing four paired entries of the optical nerve. 5 µm paraffine section; Azan; ×56. (4) and (5) Transverse sections of the light-adapted retina of *S. notatus*; (4) from the ventral and (5) from the central part of the retina. Note the striking difference in the distribution of the retinal pigment epithelium. 5 µm paraffine sections; Azan; ×710. (6) Transverse section of the ventral retina of *S. notatus* showing the course of the optical nerve behind the retina, and six different sites of entry into the retina. 5 µm paraffine section; Azan; ×110

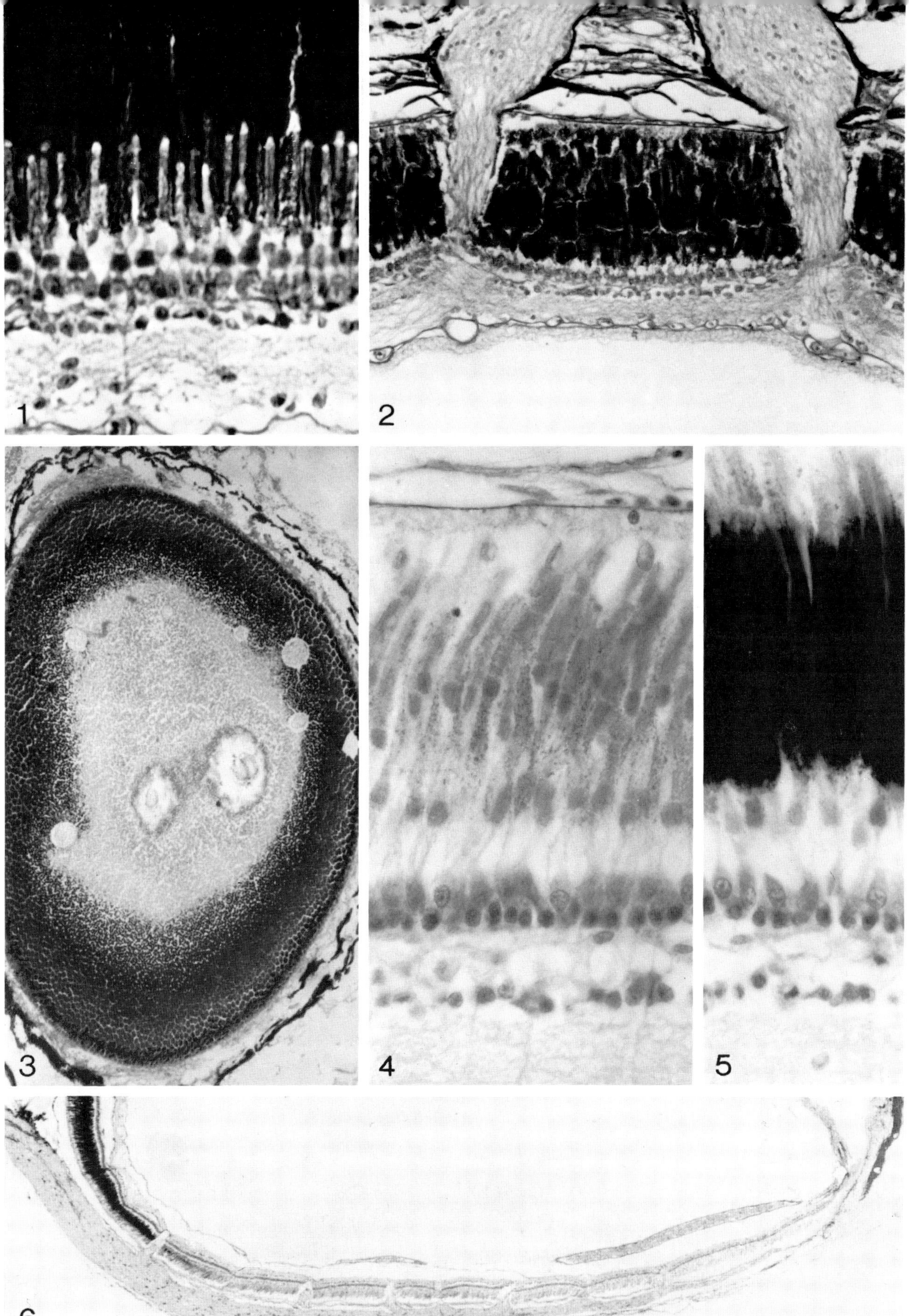
1
2
3
4
5
6

Pimelodidae (*Catfishes*)

Ostariophysi, Siluriformes

Pimelodid catfishes are elongate in varied degrees. They are entirely devoid of scales and bony plates, and are usually provided with long barbels or whiskers extending forwards from the thin, flattened head. These New World catfishes are closely related to the Bagridae of the Old World.

The retina of this catfish is strikingly similar in morphology to that of the ictalurid catfishes. The epithelial cell processes are extremely long, especially in the dorsal retina (Plate 56.1), and contain a granular material which presumably constitutes reflecting tapetum lucidum. This reflecting system is less developed in the ventral retina (Plate 56.2). The single cones compare to those of the bullhead, with long and loose myoid segments in the light-adapted state (Plate 56.1, 2). Note the tangentially sectioned cone ellipsoids in Plate 56.3, surrounded by the tapetal granules. The internal layers are poorly developed, and the horizontal cells are large and intermeshed when observed in tangential sections (Plate 56.4). In the cave catfish, *Pimelodella kronsi* (PAVAN, 1946) the retina has degenerated.

Visual Pigments: A_2 max 527 (ALI and WAGNER, 1975b)

Plate 56.1-4. Transverse (1) and (2) and tangential (3) and (4) sections of the dorsal (1) and ventral (2)–(4) retina of the catfish, *Rhamdia sebae*. × 560 ▷

1

2
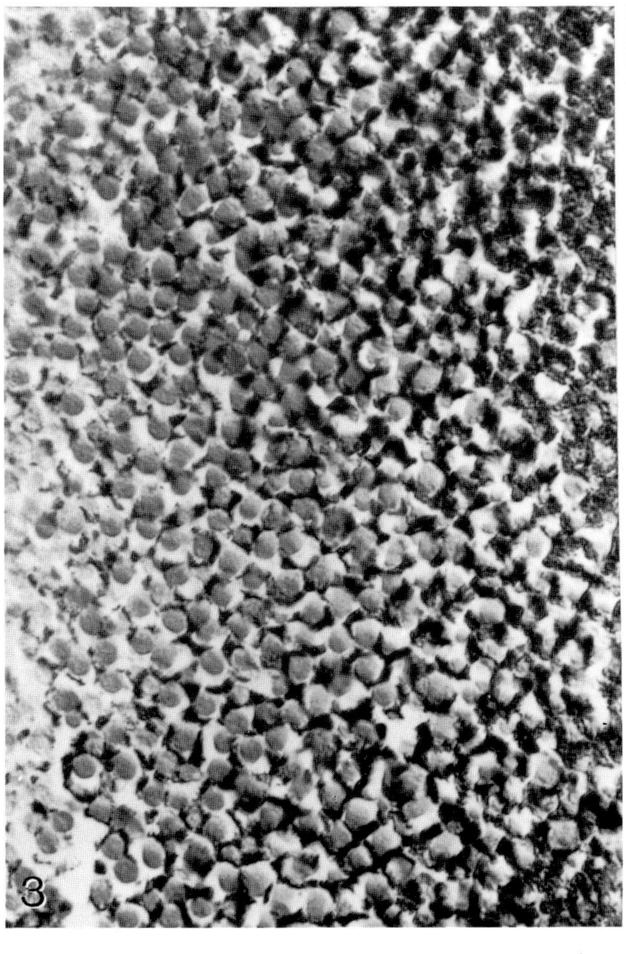
3
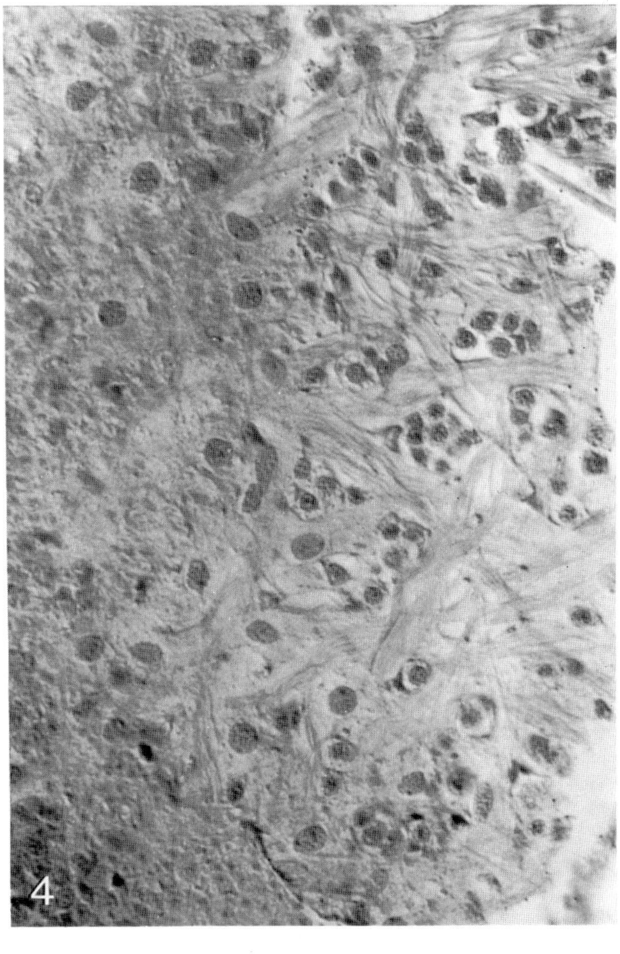
4

Callichthyidae (*Armoured Catfishes*)

Ostariophysi, Siluriformes

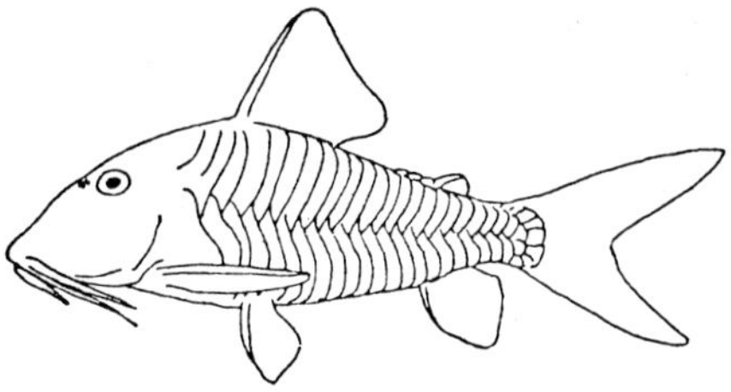

Armoured catfishes are small freshwater fishes characterised by bony plates arranged in two overlapping series on the flanks. They have a small terminal mouth, one or two pairs of maxillary barbels and moveable eyes. They inhabit slowly flowing waters where they search for food and occasionally act as scavengers in mud and sand banks.

The retina of *Corydoras* sp. resembles that of the brown bullhead in its main features. The rods are large and few, forming a single layer of rod nuclei (Plate 57.1). As in the bullhead the epithelial cells and their processes are filled with some granular pigment, presumably a reflecting system (Plate 57.1). The cones, which are exclusively of the single type, are larger than in the bullhead's retina (Plate 57.1, 2). Note the clearly defined cone synaptic pieces, just sclerad to the horizontal cell layer in Plate 57.1 *(between arrows)*, and note their appearance in tangential section (Plate 57.3, *between arrows*). Plate 57.4 shows the retina of *Corydoras julii*.

Plate 57.1-4. Transverse (1) and (4) and tangential (2) and (3) sections of the retina of the corydoras, *Corydoras* sp. and *C. julii*. (1), (3) and (4) × 560; (2) × 720 ▷

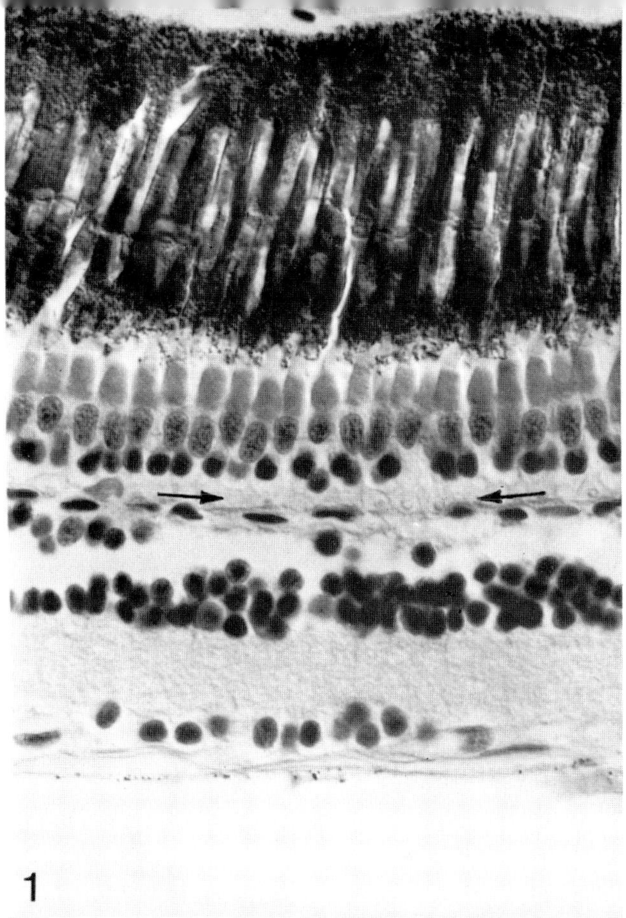
1

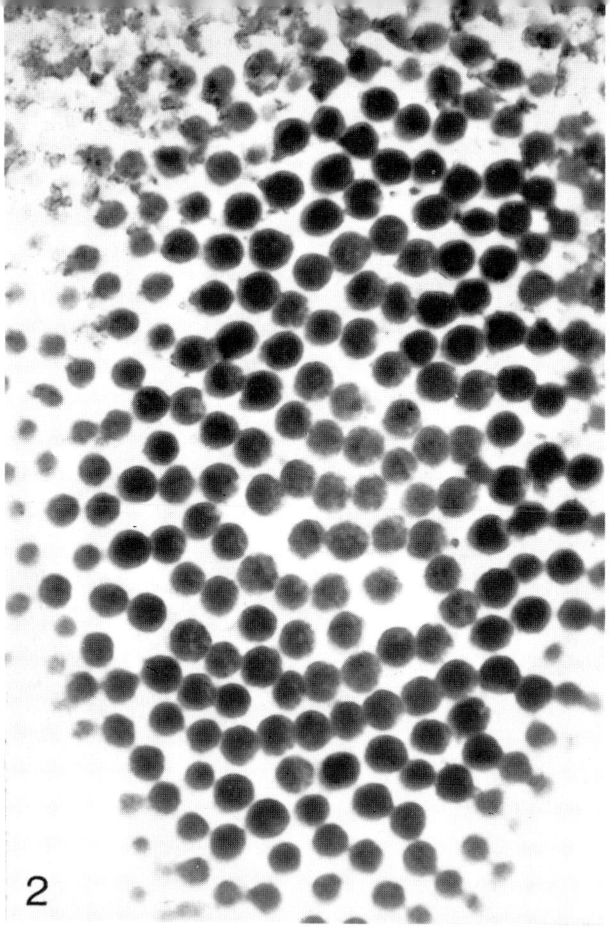
2

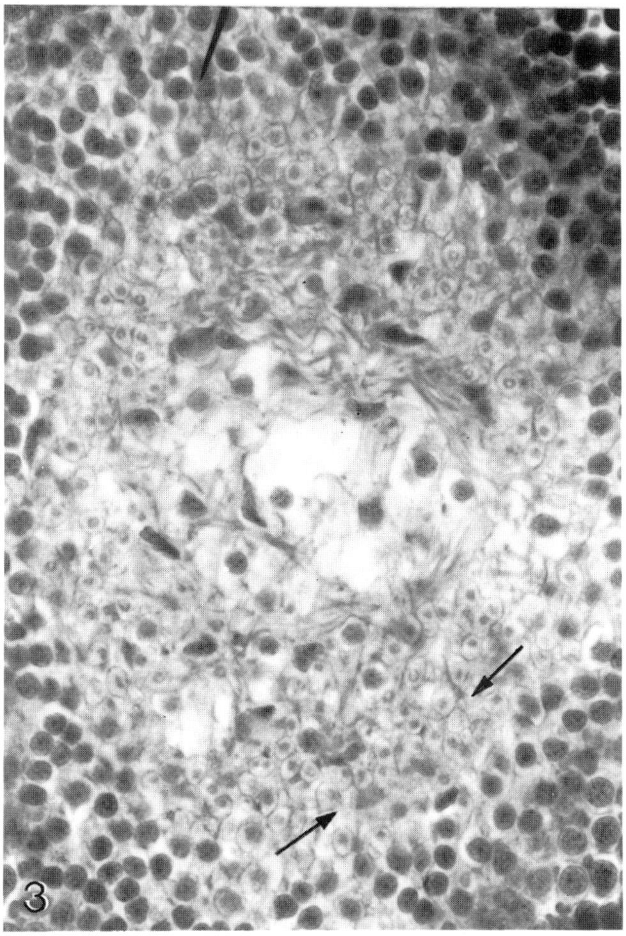
3

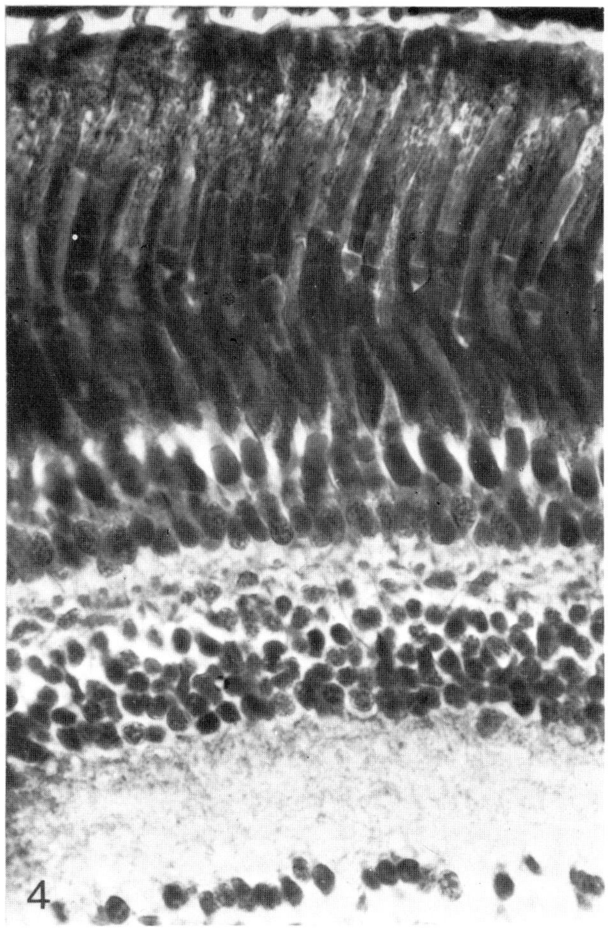
4

Amblyopsidae (*Cave, Spring and Swamp Fishes*)

Paracanthopterygii, Salmopercomorpha, Percopsiformes, Amblyopsoidei

This small family of freshwater teleosts constitutes one suborder of the Percopsiformes, an order of primitive paracanthopterygian fishes. With the exception of *Chologaster,* these fishes have adapted themselves to live in caves; in fact, their distribution is closely related with limestone formations. Adaptive features to the cave habitat include small, rudimentary eyes, and the presence of sensory papillae or tactile organs in prominent rows over the head, body, and tail.

Single as well as double cones have been reported in the retina of *C. cornutus* and *C. papilliferus,* whereas only "irregularly elongate" double cones have been found in *C. agassizi.* In the retinas of the cave fishes *Amblyopsis spelaea, A. rosae,* and *Typhlichthys subterraneus* the visual cells (rods and cones) have degenerated to varied extents.

Reference: EIGENMANN (1899, 1909)

Batrachoididae (*Toadfishes*)

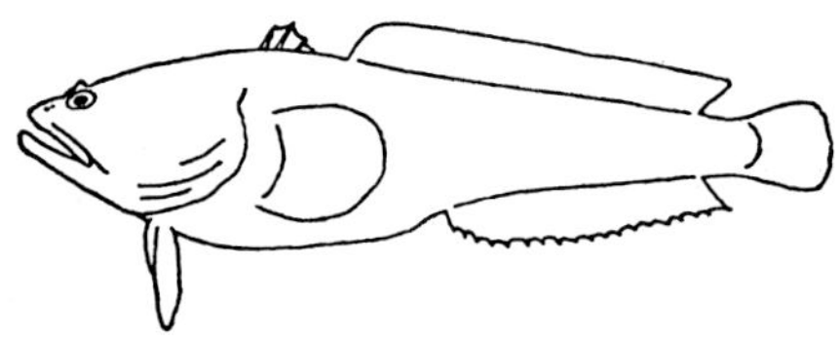

Paracanthopterygii, Salmopercomorpha, Batrachoidiformes

Toadfishes are slow-moving, bottom-dwelling fishes with large mouths and sharp teeth. They have a broad head, which tapers posteriorly to a slender tail. Toadfishes are found in shallow as well as deep shelf waters. The midshipmen (genus *Porichthys*) are equipped with ventro-lateral series of luminescent organs. As in the goosefishes (Lophiidae), the epithelial cells are heavily pigmented and their processes are moderately elongate (Plate 58.1). The rods are short and numerous, while the cones are very large and few. The cone outer segments are very small. Most of the cone population consist of equal double cones (Plate 58.2), but a few single cones are also present. There is one layer of horizontal cells and the internal nuclear layer is poorly developed.

Visual Pigments: A_1 max 496–499 (BEATTY, 1973; ALI and HEUMANN, 1970, 1972)

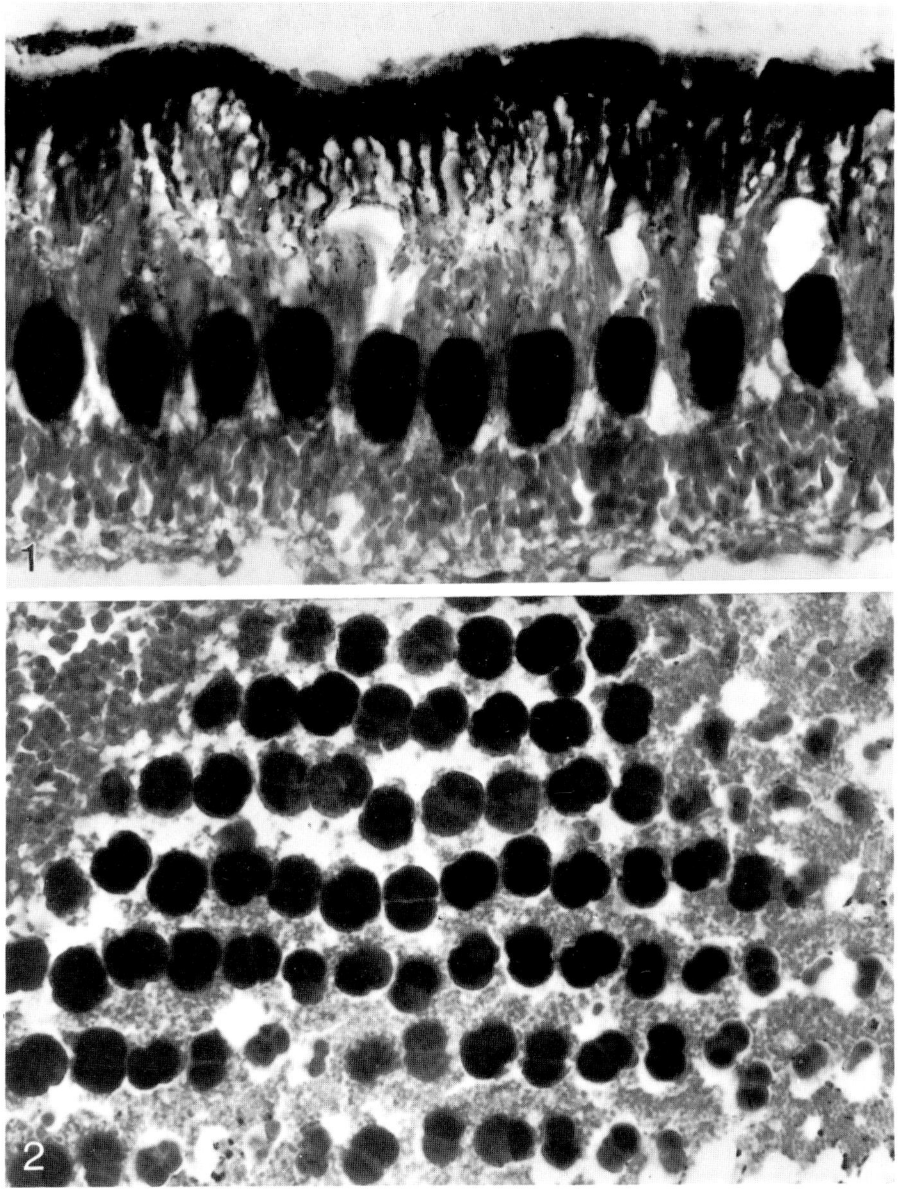

Plate 58.1, 2. Transverse (1) and tangential (2) sections of the retina of the midshipman, *P. porosissimus*. $\times 560$

Gobiesocidae (*Clingfishes*)

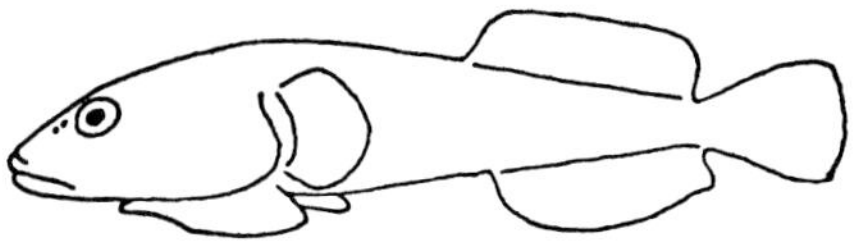

Paracanthopterygii, Salmopercomorpha, Gobiesociformes

The retina of *Gobiesox strumosus* is characterised by a well-developed visual cell layer, two rows of horizontal cells, and a large number of bipolar, amacrine, and ganglion cells (Plate 59.1, 2). Rods and cones are present in relatively large numbers. The single cones are about half the length of the equal double cones (Plate 59.2) and they form together orderly square arrangements (Plate 59.3, 4). Note the long outer segments and the large, pale nuclei of the cones straddling the external limiting membrane (Plate 59.2). Multiple droplets are also very clearly seen in the cone ellipsoids. The retina of clingfishes is known to possess a fovea.

References: VRABEC (1969); WAGNER et al. (1976)

Plate 59.1-4. (1)–(4) Transverse and tangential sections of *G. strumosus*. (1) Dorsal retina (× 880). (2) Part of the ventral retina (× 1,380). (3) Tangential section of the dorsal retina (× 210). (4) Ventral retina (× 1,380). Note the bulk and sparseness of the rods in (1) and (2) ▷

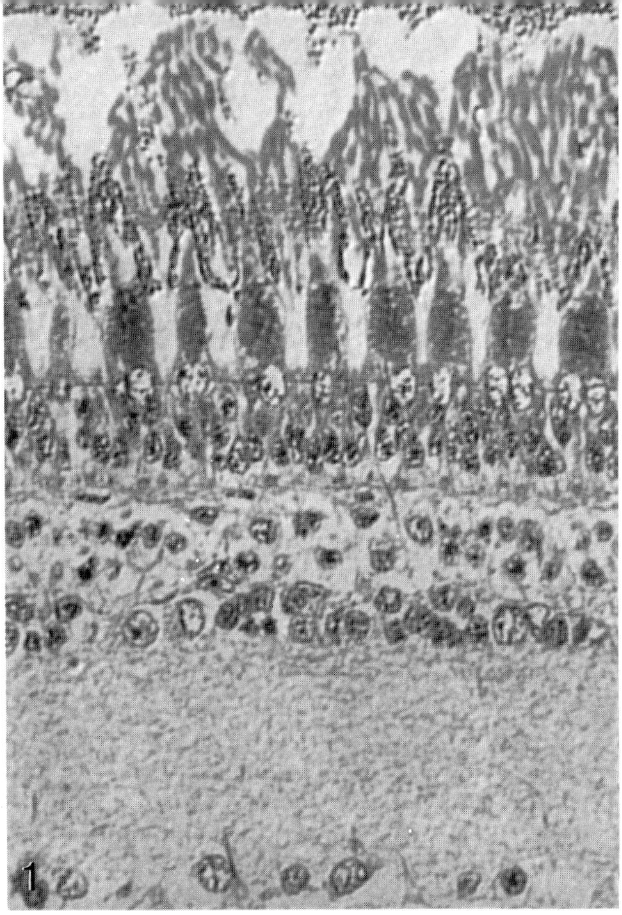
1

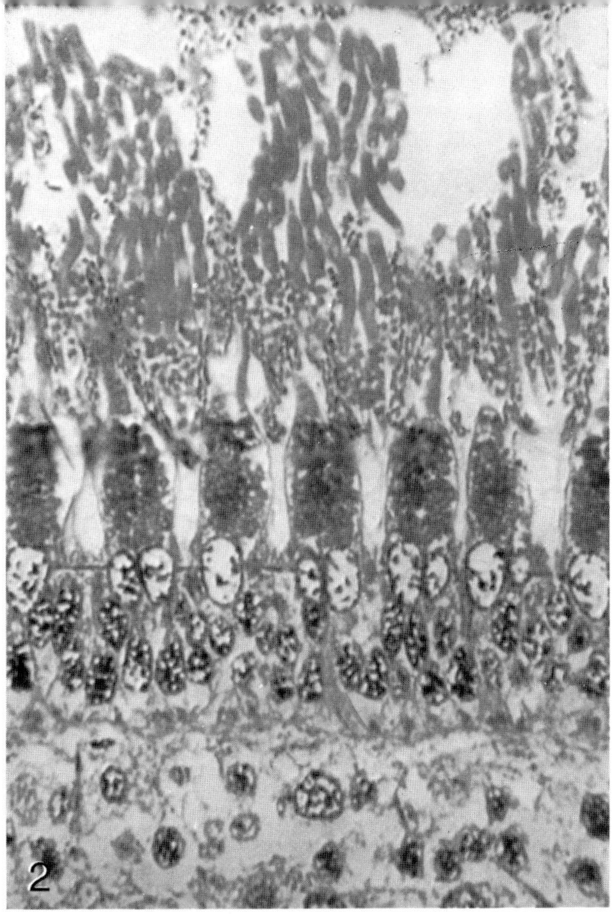
2

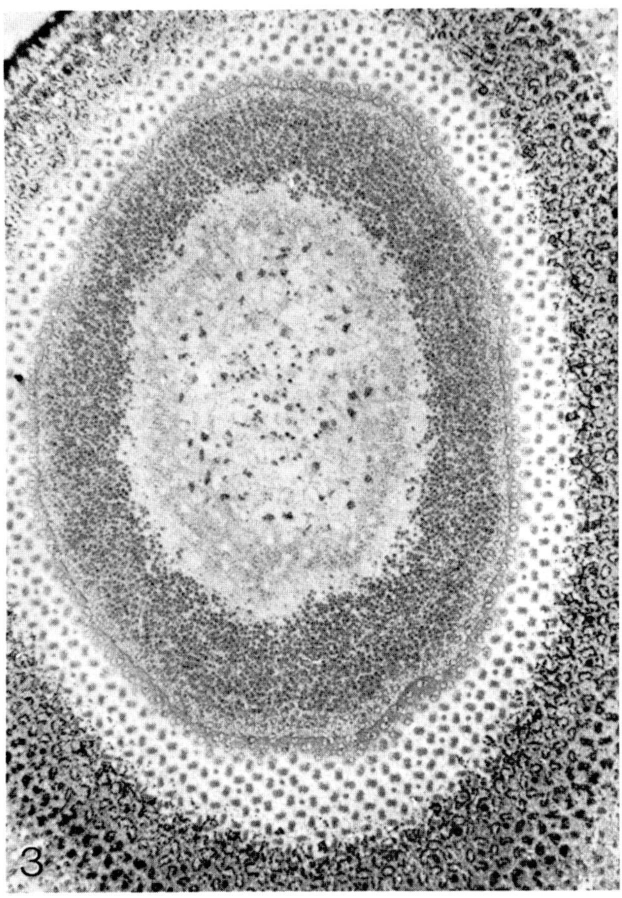
3

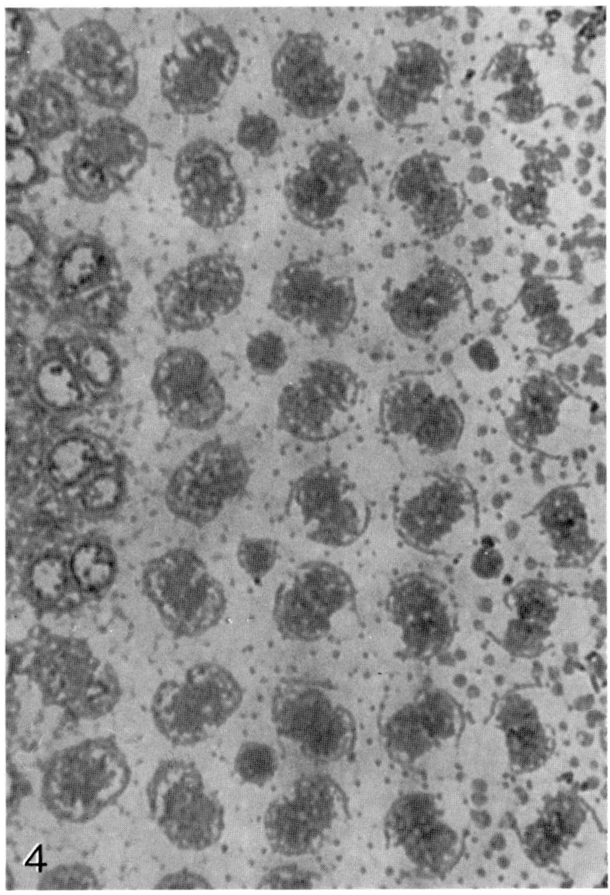
4

Lophiidae (*Goosefishes*)

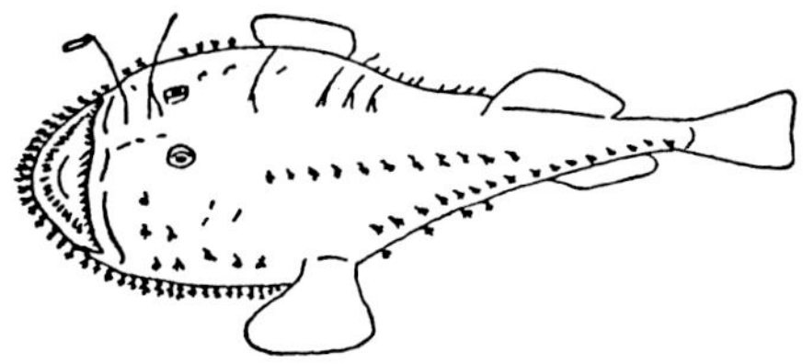

Paracanthopterygii, Salmopercomorpha, Lophiiformes, Lophioidei

Goosefishes are relatively specialised pediculate fishes in which the base of the pectoral fin takes the form of an arm. They have very large and flattened heads and the eyes are directed upwards. Goosefishes are typically benthic fishes of the continental shelf, distributed from the tide line down to hundreds of metres; they are sluggish and attract their prey by means of a tentacular lure hanging over the extremely large mouth.

The retina of the American goosefish (*Lophius americanus*) is poorly developed. The epithelial pigment is abundant and well distributed in fingerlike processes (Plate 60.1, 2). Rods and cones are present, the latter are slender (Plate 60.1) at the temporal periphery of the retina and very bulky (Plate 60.2) in the fundus region. Single, double, triple (Plate 60.3), and some quadruple cones could be distinguished. The internal layers are extremely poor in bipolars and ganglion cells.

References: VERRIER (1928a); ENGSTRÖM (1963b); ANCTIL (1969)

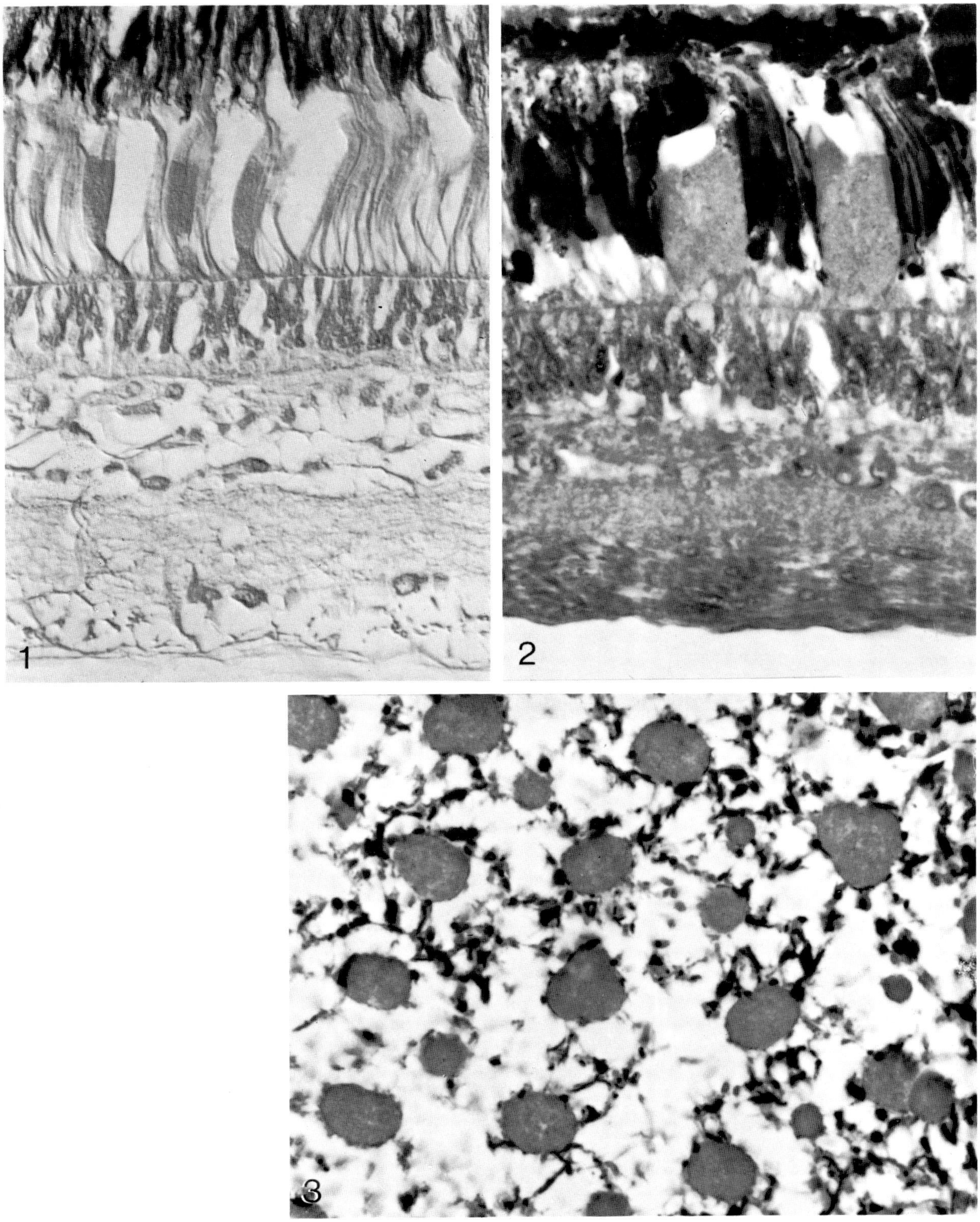

Plate 60.1-3. Transverse (1) and (2) and tangential (3) sections of the retina of the goosefish, *L. americanus*. (1) $\times 560$; (2) and (3) $\times 700$

Ogcocephalidae (*Batfishes*)

Paracanthopterygii, Salmopercomorpha, Lophiiformes, Antennarioidei

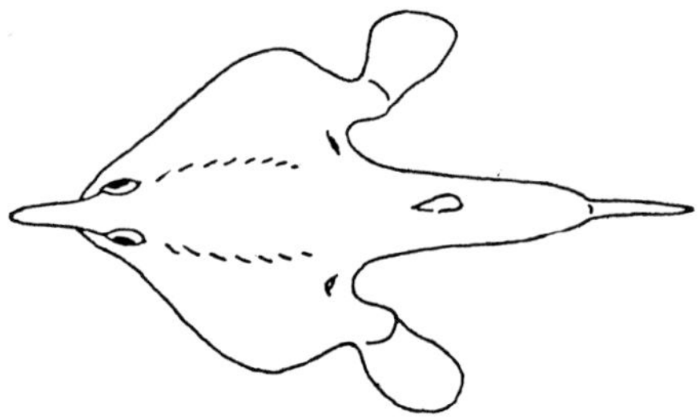

Batfishes are a teleost family somewhat related to the anglerfishes. They are characterised by a flat body, a large head, and the presence of a rostral flap and a fishing pole (escum). Some are deep water forms. Like the anglerfishes, they are bottom dwellers of the continental shelf. The batfishes wander about on the bottom by moving alternately their pelvic and pectoral fins. They are reported to be luminescent, but no specialised light organ has been found on their body.

The retina of *Ogcocephalus vespertilio,* from the Brazilian coastal waters, is fairly well developed. The pigment epithelium is well endowed with pigment and processes. The visual cell layer contains a large number of cones and, to a certain extent, rods. Long and short single cones as well as double cones are present (Plate 61.1, 2), and arranged in square patterns (Plate 61.3, 4) at different levels of the cells. Two layers of horizontal cells can be readily identified (Plate 61.1) and a relatively large number of bipolar and amacrine cells are present. The ganglion cells usually form a single layer in transverse section.

The retina of *Halicmetus ruber* is characterised by densely packed epithelial pigment. Only rods are present as visual cells, and they are longer in the ventral retina.

Reference: BRAUER (1908)

Visual Pigments: A_1 (ALI and HEUMANN, 1970)

Plate 61.1-4. Transverse (×330) sections of dorsal (1) and ventral (2) retinas of *O. vespertitilio*. Tangential sections (×525) of dorsal (3) and ventral (4) retinas are also shown ▷

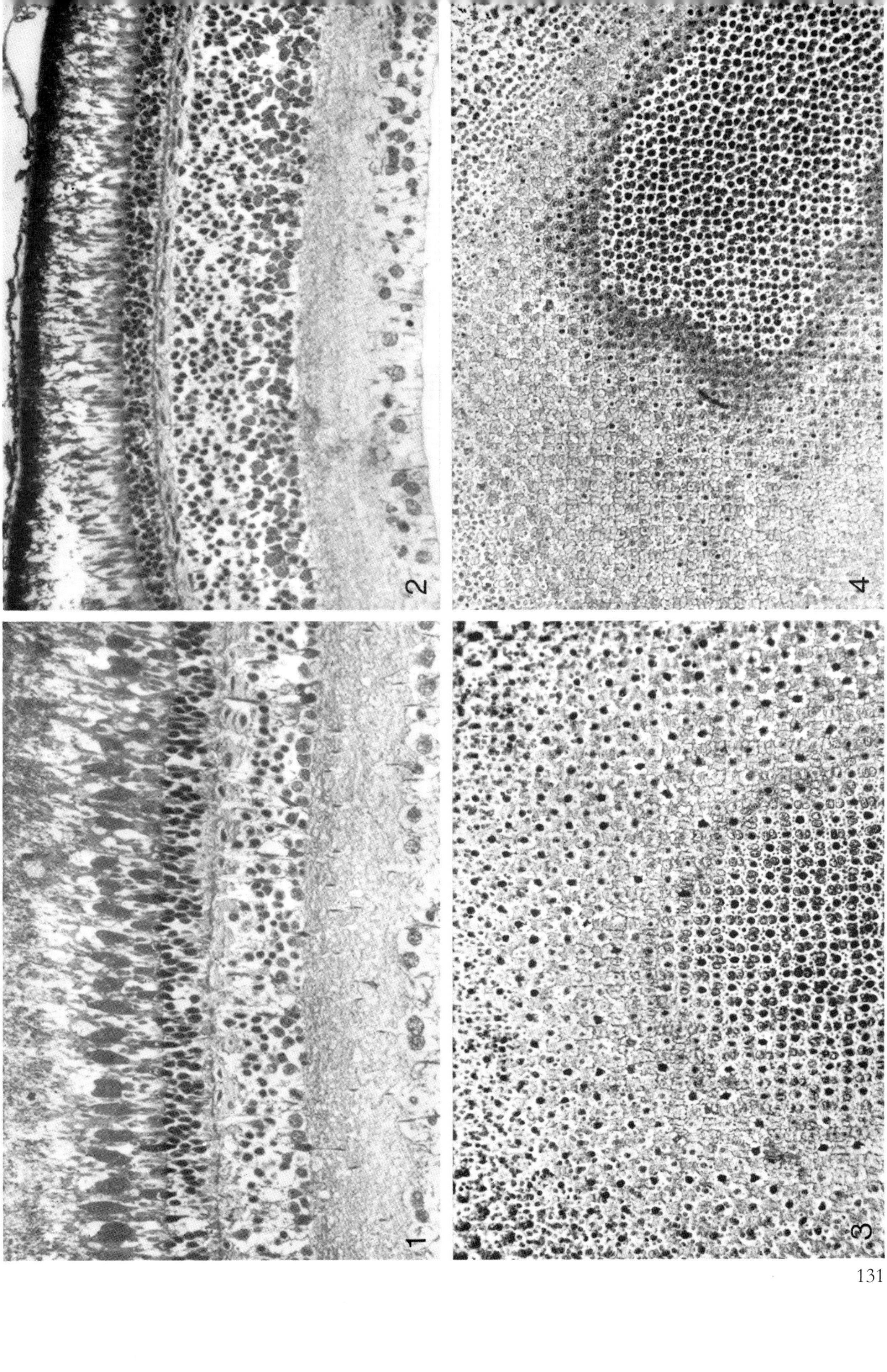
1
2
3
4

Oneirodidae (*Deep-Sea Anglers*)

Paracanthopterygii, Salmopercomorpha, Lophiiformes, Ceratioidei

Ceratioids differ from the lophioid and antennarioid anglers by the absence of pelvic fins. These are mesopelagic and bathypelagic and are highly specialised. They possess an escal lure in the rostral region, which is equipped with a luminous organ. Some families produce parasitic males reduced to the status of a sperm-producing appendage.

The oneirodids are small, usually black anglers with luminescent esca and small eyes.

The retina of *Oneirodes niger* appears normally developed. The epithelial pigment forms a thick layer. It is a pure-rod retina, and the rod layer constitutes one third of the thickness of the retina. The bipolars and amacrines are few and sparse. Brauer noted some cone-like elements in the visual cell layer, but interpreted them as artifacts.

Reference: Brauer (1908)

Gigantactinidae (*Deep-Sea Anglers*)

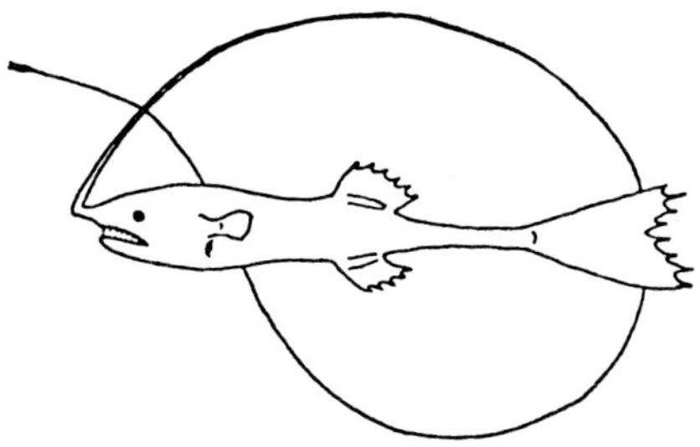

Paracanthopterygii, Salmopercomorpha, Lophiiformes, Ceratioidei
Bathypelagic fishes (cf. Oneirodidae)

The retina of *Gigantactis vanhoeffeni* resembles that of *O. niger,* except that the epithelial pigment is less abundant and the rods are shorter.

Reference: Brauer (1908)

Ceratiidae (*Seadevils*)

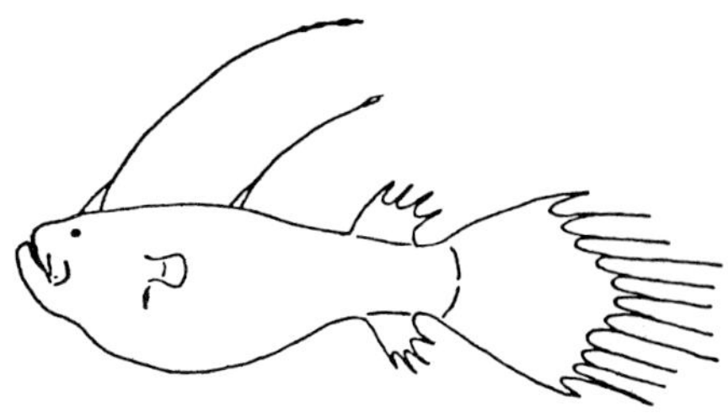

Paracanthopterygii, Salmopercomorpha, Lophiiformes, Ceratioidei

In this family there are parasitic males attached to the females during the adult stage (cf. Oneirodidae).

Ceratias holboelli and *Cryptosaras cousei* (warted seadevil) have pure-rod retinas (Plate 62); the rod ellipsoid is very short and the outer segment relatively long. The ventral retina is thicker than the dorsal retina. In the retina of the females, the rods are strikingly more numerous, and the internal nuclear layers better developed than in the retina of metamorphosed free-living males.

References: Brauer (1908); Munk (1964b, 1966a)

Linophrynidae (*Deep-Sea Anglers*)

Paracanthopterygii, Salmopercomorpha, Lophiiformes, Ceratioidei
There are parasitic males, as in the Ceratiidae (cf. Oneirodidae).

The retinas of *Aceratias* (= *Linophryne macrorhinus*) and *L. arborifera* are essentially similar to those of other ceratioid fishes. However, they have tubular eyes pointing forwards.

References: BRAUER (1908); MUNK (1964b, 1966a)

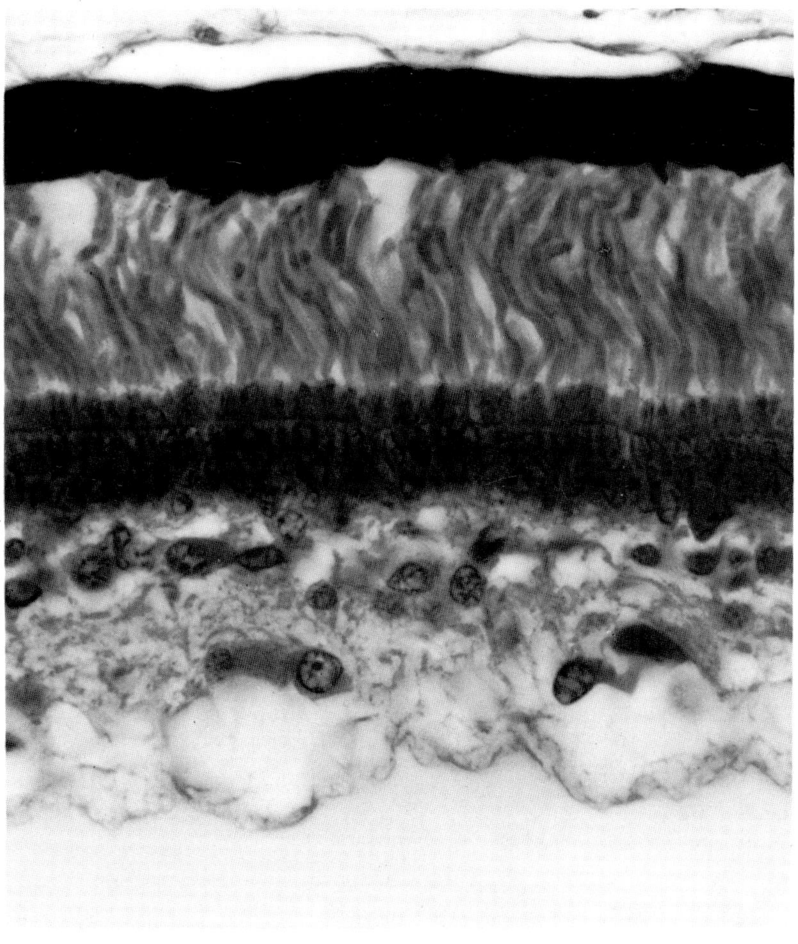

Plate 62. Photomicrograph of the transversely cut retina of a female warted seadevil *C. cousei.* × 945

Gadidae (*Codfishes*)

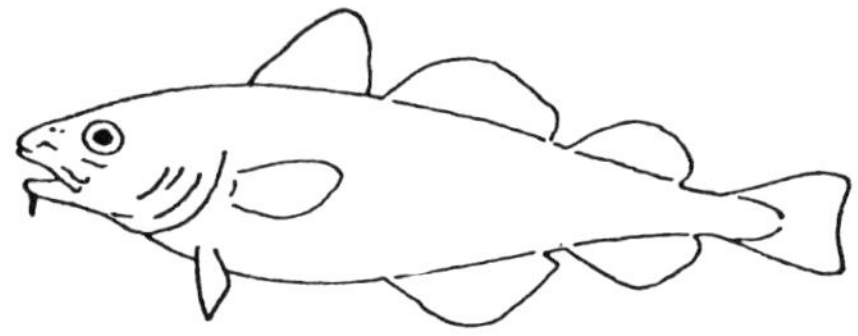

Paracanthopterygii, Salmopercomorpha, Gadiformes, Gadoidei

Atlantic codfishes are soft-finned, primarily marine shelf water fishes. The large pelvic fins are situated under or in front of the pectorals, and the codfishes are also characterised by the presence of a mental barbel. The Atlantic cod is distributed from the surface down to 500 m. It is a predaceous fish feeding mostly on other fishes and invertebrates.

The pigment in the retinal epithelial cell processes is very abundant, the rods are long, slender, densely packed, and the single and double cones are numerous. Note the difference in cone positions between the peripheral (Plate 63.1) and fundus (Plate 63.2) retina; in the former single cones are close to the limiting membrane, while the double cones are partly extended sclerally.

References: Cobbold (1862); Wunder (1925); Walls (1942); Lyall (1957b); Engström (1961; Ali and Hanyu (1963); Ali et al. (1968); Wagner (1972)

Greenland Cod

The Greenland cod (*Gadus ogac*) is found in islets and close to shore. It is rarely found offshore in deeper water.

The retinal epithelial pigment is less abundant than in the Atlantic cod, and the internal nuclear layers are poorly developed (Plate 63.3).

References: Ali et al. (1968); Ali and Wagner (1975a)

Tomcod

The tomcod (*Microgadus tomcod*) is limited in distribution to coastal waters and estuaries, feeding mainly on amphipods, annelids, shrimps, etc.

The retina is similar to that of the Atlantic cod; the cones are present in large number and the internal nuclear layer is well developed (Plate 64.1).

Reference: Ali et al. (1968)

Pollock

The pollock (*Pollachius virens*) is found chiefly offshore on the continental shelf, and is piscivorous.

Plate 63.1-4, Plate 64.1-4. Transverse sections of the retinas of various codfish species (see text). Plate 63.1,2 and Plate 64.4, ×440; Plate 63.3, ×330; Plate 63.4 and Plate 64.1, ×355; Plate 64.2, ×305; Plate 64.3, ×255 ▷

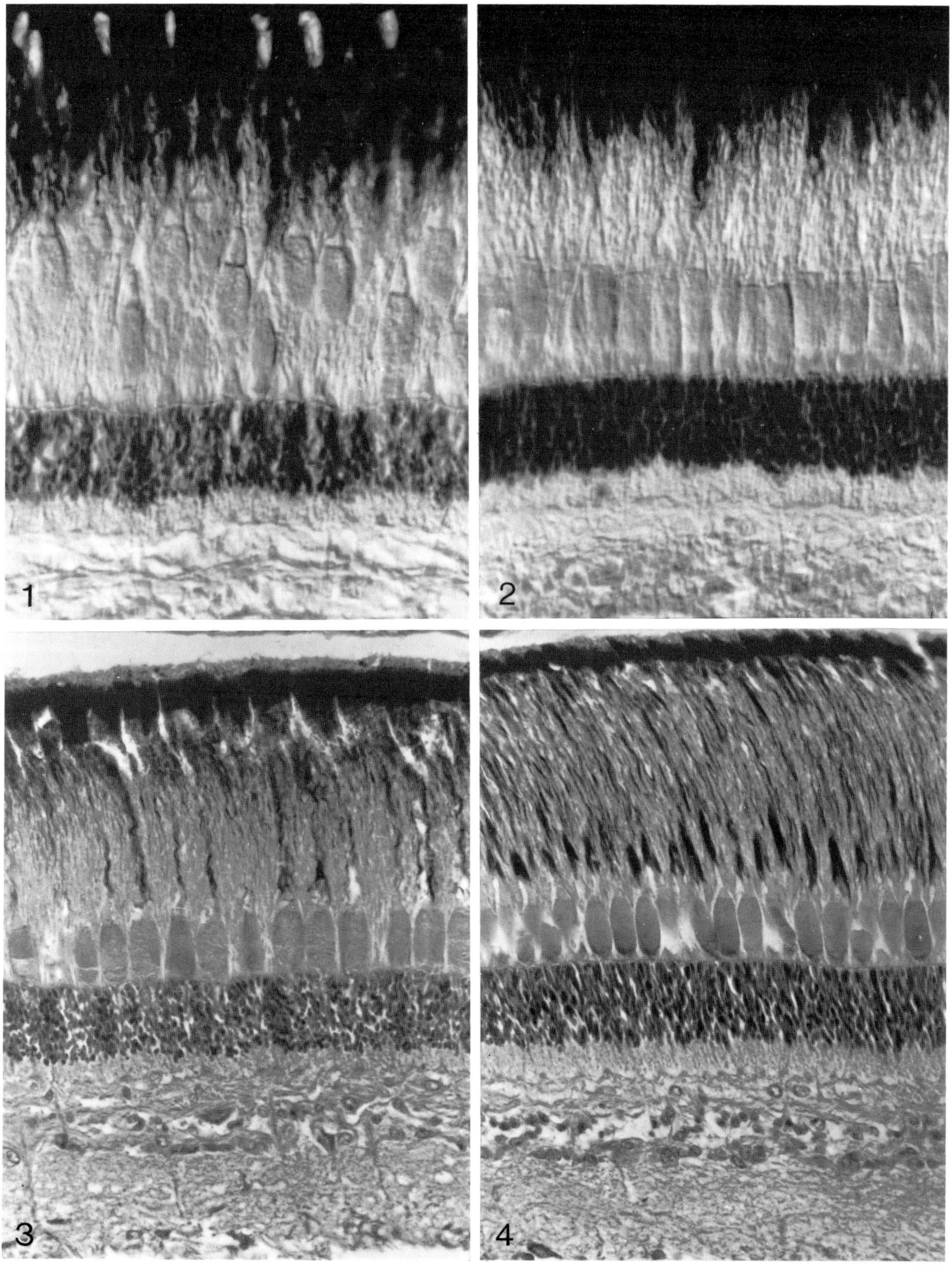
1
2
3
4

The retina is poorly developed (Plate 64.2). The pigment is less abundant than in other codfishes and is in the dark-adapted state. The rods are relatively short and the cones less numerous.

Reference: ALI et al. (1968)

White Hake

The white hake (*Urophycis tenuis*) is a bottom-fish ranging from shallow water to over 1,000 m. It feeds on crustaceans and small fish.

The retina is very poorly developed; the rods are particularly long and numerous while the cones, which are of the single and double types, are sparsely distributed (Plate 64.3). The internal layers are more or less ill-defined.

Reference: ALI et al. (1968)

Haddock

The haddock (*Melanogrammus aeglefinus*) lives mainly on the bottom. A small subterminal mouth limits the haddock to small food articles (crustaceans, molluscs, etc.).

The retina resembles that of the Greenland cod (Plate 63.4); the rods are numerous and very long.

Reference: ALI et al. (1968)

Blue Hake

The blue hake (*Antimora rostrata*) has a mental barbel and the second ray of the first dorsal fin is modified into a filament. It inhabits deep shelf or oceanic waters.

The retina of the blue hake represents an extreme specialisation within the codfish family. It is a pure-rod retina (Plate 64.4) with little epithelial pigment which is fully retracted sclerally, and with barely visible internal nuclear layers.

Reference: ALI and HANYU (1963)

Visual Pigments: A_1 max 485–523; A_2 max 527 (ALI and WAGNER, 1975b)

Plate 64.1-4. (Legend see page 134) ▷

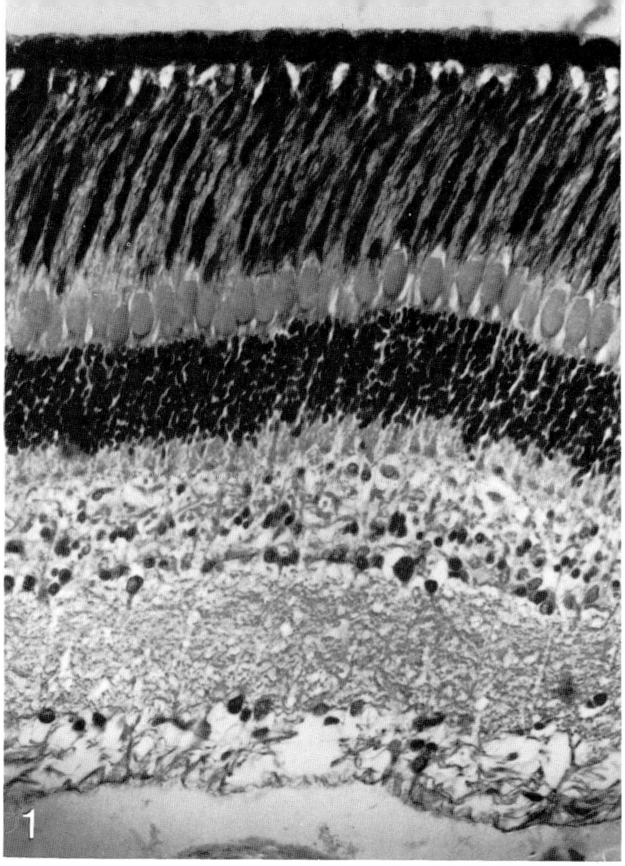

1

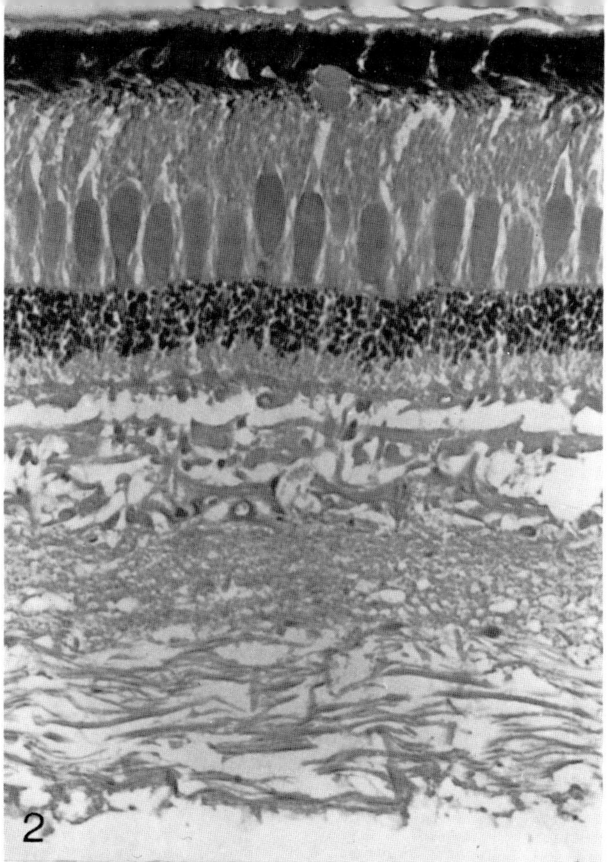

2

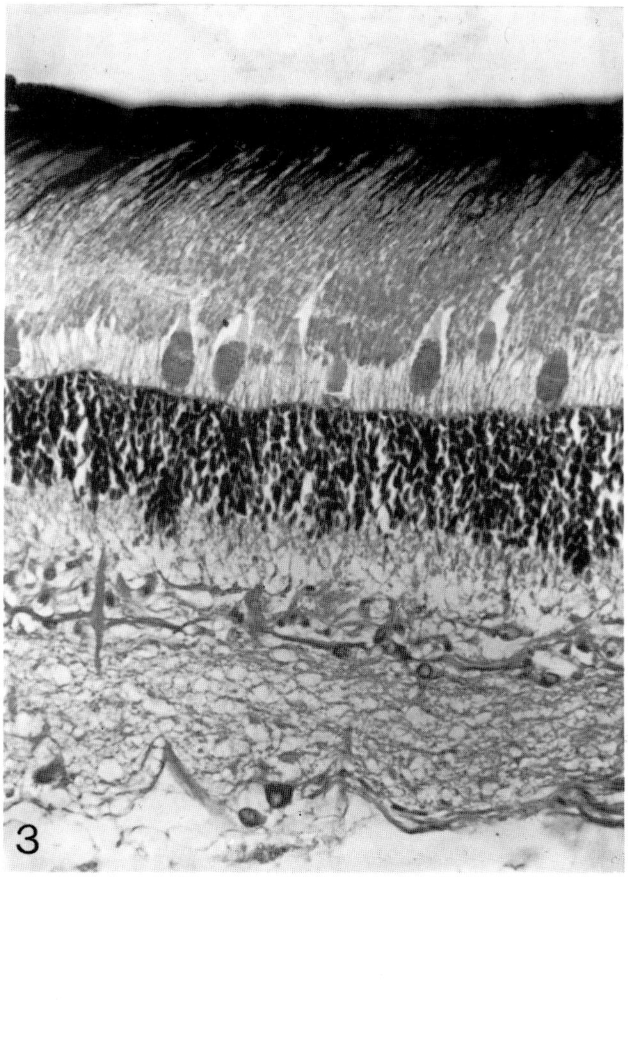

3

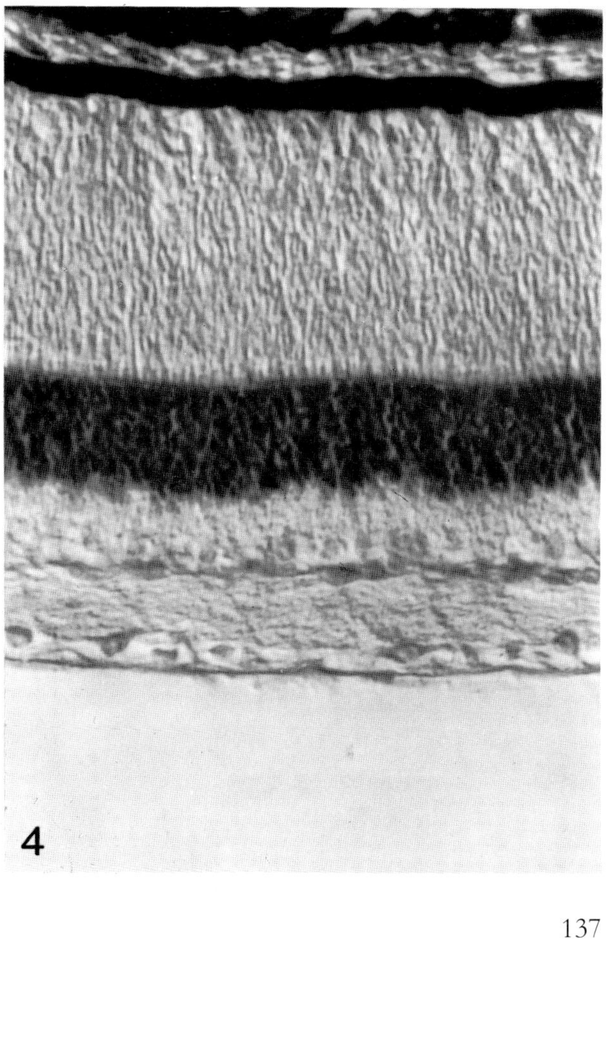

4

Merlucciidae (*Hakes*)

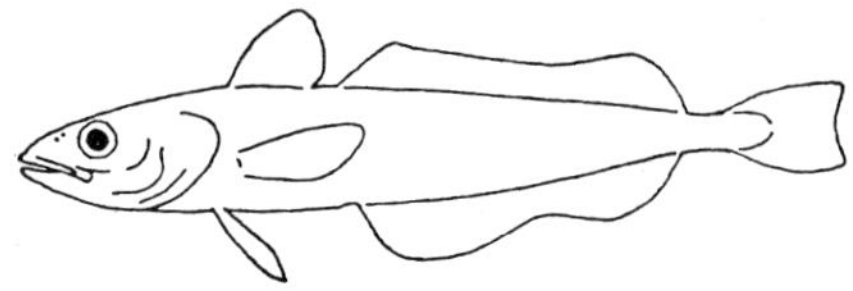

Paracanthopterygii, Salmopercomorpha, Gadiformes, Gadoidei

Hakes are very closely related to gadids. The silver hake, (*Merluccius bilinearis*) is benthic, or nearly benthic, and is found at various depths; it pursues its food in a pelagic fashion, coming near the surface chiefly at night. It eats young fish, squids, euphausiids, and decapods.

The retina is strikingly similar to that of most gadid fishes, especially that of the pollock (*Pollachius virens*). The epithelial pigment is present in short processes, the rods are very abundant, the cones are relatively few, and the internal layers are only moderately developed (Plate 65).

Reference: Ali et al. (1968)

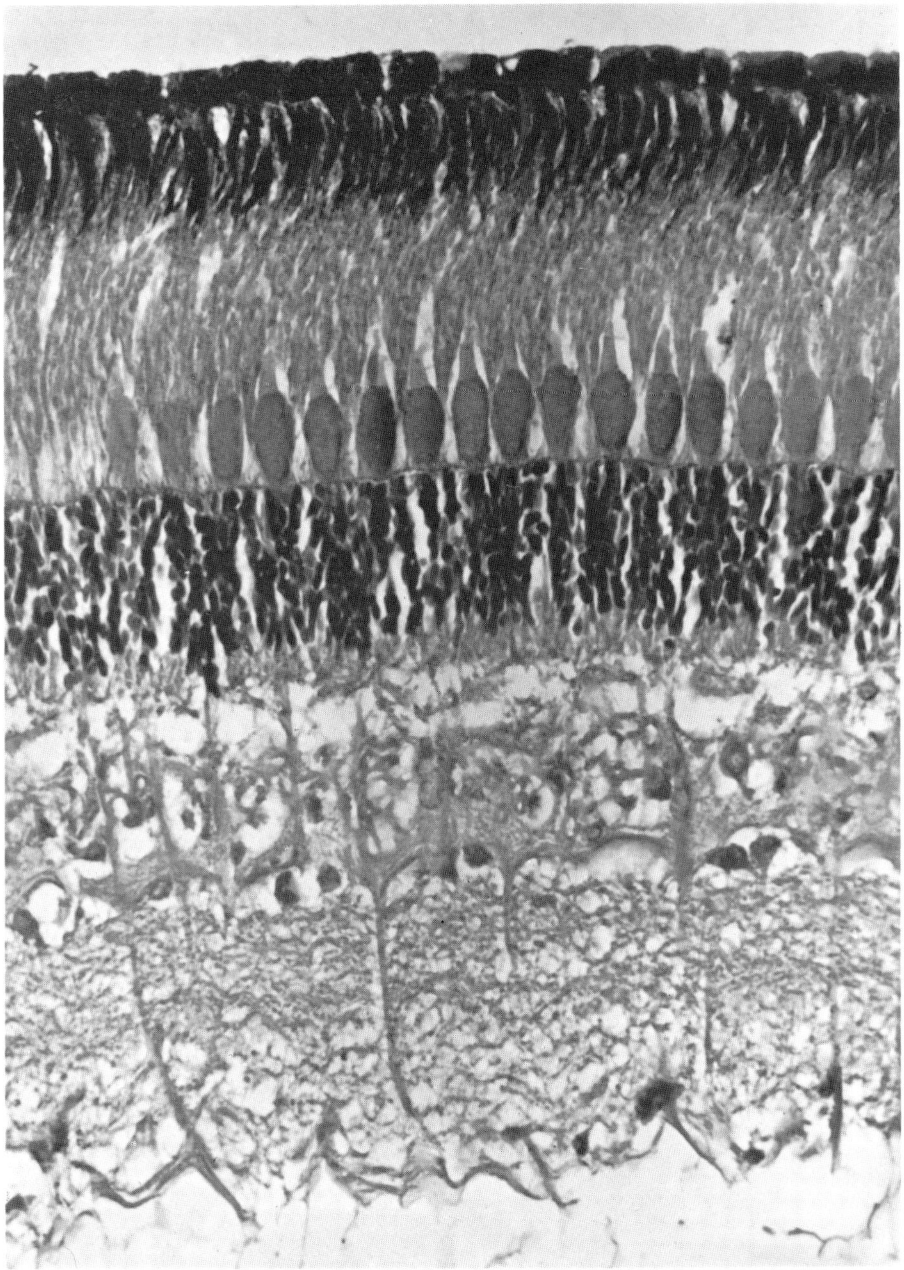

Plate 65. Transverse section of the ventral fundus-retina of the silver hake, *M. bilinearis*. ×500.

Ophidiidae (*Cusk-Eels and Brotulas*)

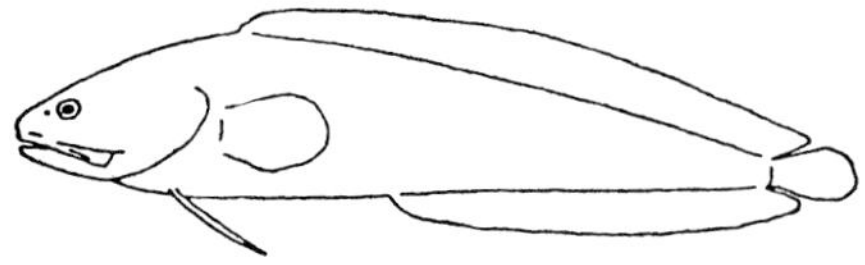

Paracanthopterygii, Salmopercomorpha, Gadiformes, Ophidioidei

Cusk-eels are usually elongated forms with the pelvic fins reduced in size, modified into chin whiskers, or entirely lacking. This is a family of widely differentiated species, ranging in habitat from the greatest depths of the ocean, to freshwater caves. Both habitats may have some blind species. Some species are occasionally caught in shallow water (tide pools and estuaries). The cusk-eels are essentially benthic or bathybenthic fishes.

The retinas of the deep-sea *Neobythites nigripinnis, Barathronus affinis, B. erikssoni, Bassogigas profundissimus* and *Sciadonus kullenbergi* show various degrees of degeneration: a simple pure-rod retina with short rods or fragmented rod outer segments surrounded by macrophages, or a retina reduced to a mass of heavily pigmented cells (*Sciadonus*). More or less degenerate retinas have been described in the cave-dwellers *Lucifuga subterraneus, Stygicola dentatus, Typhleotris madagascariensis, Thyphliasina pearsi* and *Dipulus caecus*. Grouped photoreceptors (rod bundles?) have been reported in the retina of the larval *Cataetyx memorabilis*.

In a specimen of *Ophidion welshi* caught in shallow coastal water, the retina has epithelial cells with bodies that are heavily filled with melanin pigment; and the short blunt processes contain lipid tapetal material which is missing in histological preparations (Plate 66.1).

It is an almost pure-rod retina; the rods are very numerous, having a short outer segment, and appear to be arranged in several tiers (Plate 66.1). A few small single cones are present at the sclerad margin of the visual cell layer, adjacent to the epithelial cell processes (Plate 66.1, 2).

References: BRAUER (1908); EIGENMANN (1909); THINÈS (1960); MUNK (1964a, 1966a); MEYER-ROCHOW (1972)
See also NICOL et al. (1973), on tapetum lucidum in cusk-eels.

Visual Pigments: A_1 (ALI and HEUMANN, 1972)

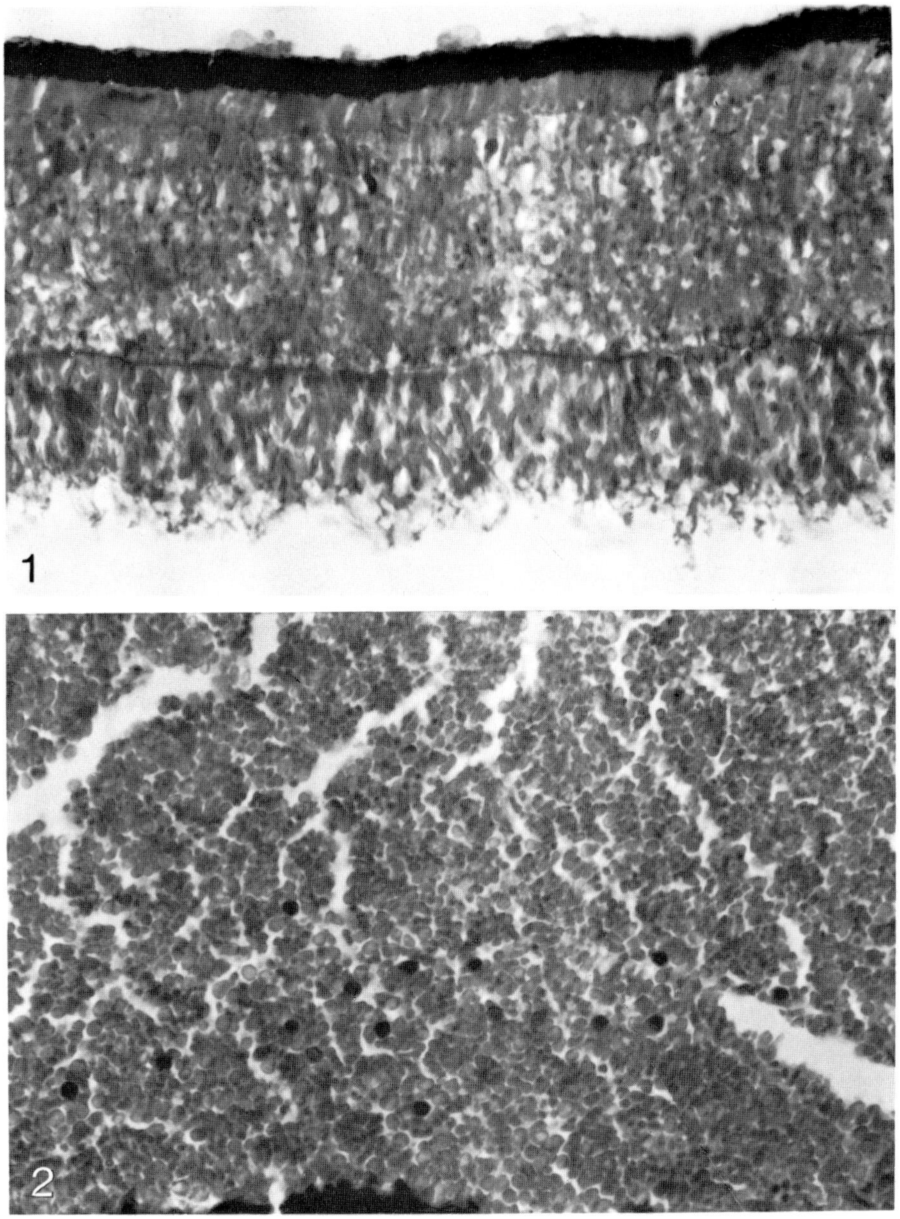

Plate 66.1, 2. Transverse (1) and tangential (2) nasal retina of the cusk-eel, *Genipterus blacodes*. × 560

Zoarcidae (*Eelpouts*)

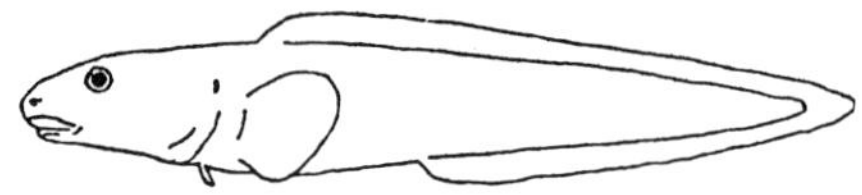

Paracanthopterygii, Salmopercomorpha, Gadiformes, Zoarcoidei

Eelpouts are slender, eel-like benthic fishes with the anal fin continuous with the caudal fin. They are sluggish in general and spend most of their time hiding among seaweeds and stones. They are found at varying depths on the continental shelf. They feed on shelled molluscs, crustaceans and echinoderms.

The retina of the oceanpout, *Macrozoarces americanus,* has a well-developed epithelium with abundant pigment, relatively few rods, and large bulky cones (Plate 67.1). The internal nuclear layers are ill-defined.

The retina of the eelpouts varies from a well-developed condition in *L. turneri* (Plate 67.2) with numerous rods and cones and thick internal nuclear layers, to a poor condition in *L. reticulatus* (Plate 67.3) with few cones, many rods and ill-defined internal layers. An intermediate situation is found in *L. lavalei* (Plate 67.4). The epithelial pigment is densely packed sclerally. The cones are of the single and (chiefly) double types (Plate 67.5, 6).

References: Brauer (1908); Ali and Hanyu (1963); Engström (1963b); Ali et al. (1968); Anctil (1969); Wagner (1972); Ali and Wagner (1975a)

Plate 67.1-6. Transverse (1)–(4) and tangential (5) and (6) sections of the retinas of various eelpouts. (1) ×580; (2) ×500; (3) ×465; (4) and (5) ×365; (6) ×765 ▷

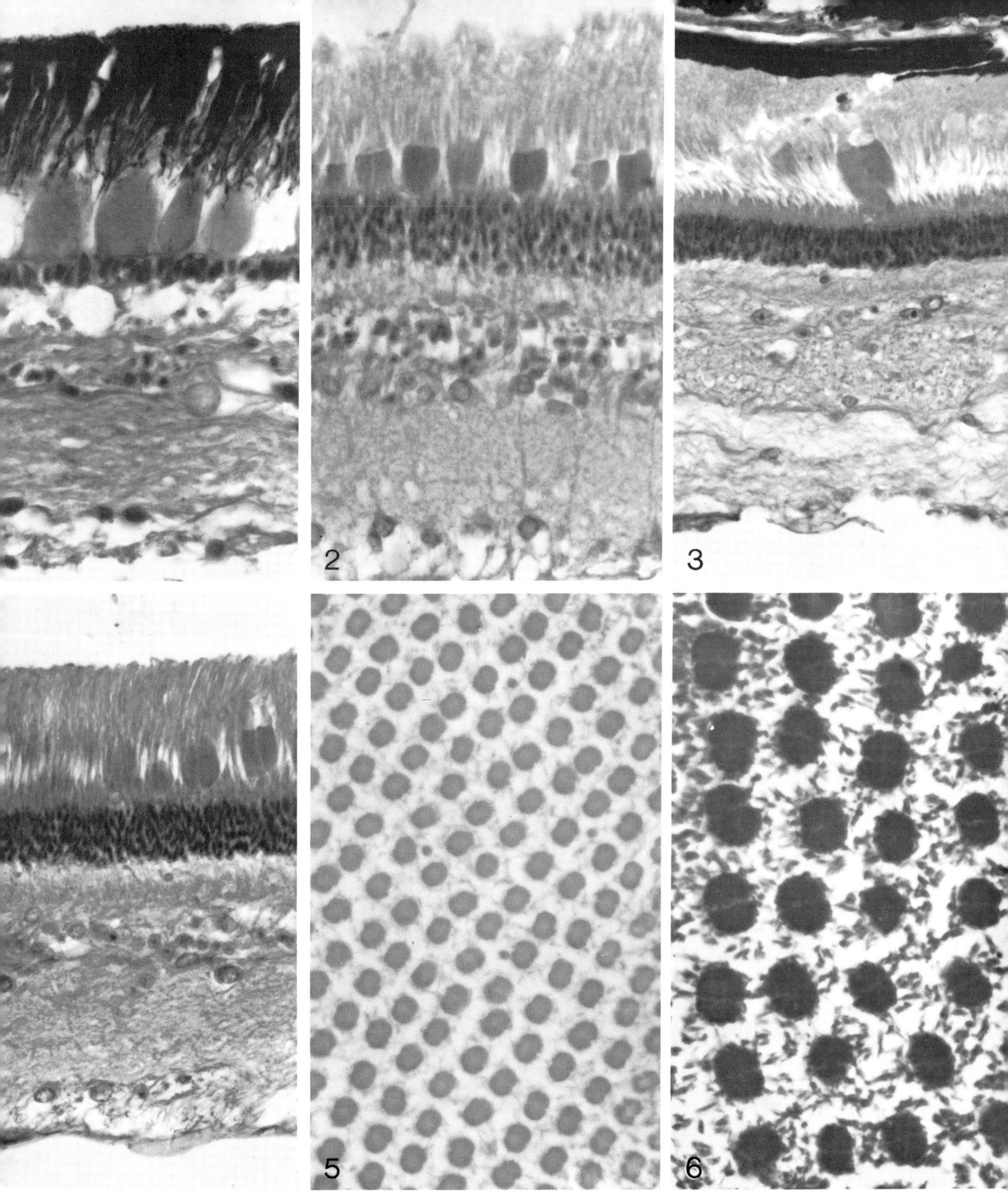
2
3
5
6

Macrouridae (*Grenadiers*)

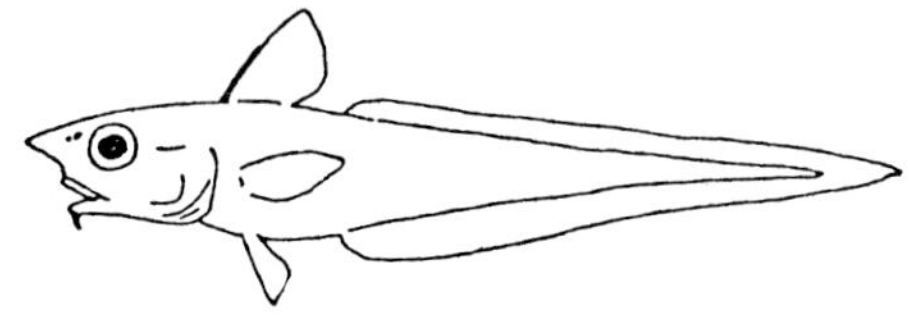

Paracanthopterygii, Salmopercomorpha, Gadiformes, Macrouroidei

Grenadiers are characterised externally by having large heads, projecting snouts, and slender bodies that taper to whiplike tails. The eyes are large. They are closely allied to the cod family but differ in having one stout spine in the first dorsal fin. These fishes are benthic in habits and live in deep waters from the continental shelf to the oceanic abyssal bottoms. They are loose in texture, weak swimmers, and feed on bottom invertebrates.

The retina of the grenadier (*Coryphaenoides rupestris*) is very simply built and poorly developed. Note in Plate 68.1 the densely packed epithelial pigment sclerally, the numerous rods whose length takes nearly half of the retinal thickness, and the very sparsely distributed bipolar and ganglion cells vitreally.

References: Brauer (1908); Ali and Hanyu (1963); Munk (1964a, 1966a)

Nezumia

Nezumia (*Nezumia bairdii*) is a bottom fish usually found on soft mud in depths ranging from 100 to 2,400 m.

It has a pure-rod retina as in *C. rupestris,* with essentially the same characters, except that the internal layers are almost absent (Plate 68.2).

Reference: Ali and Hanyu (1963)

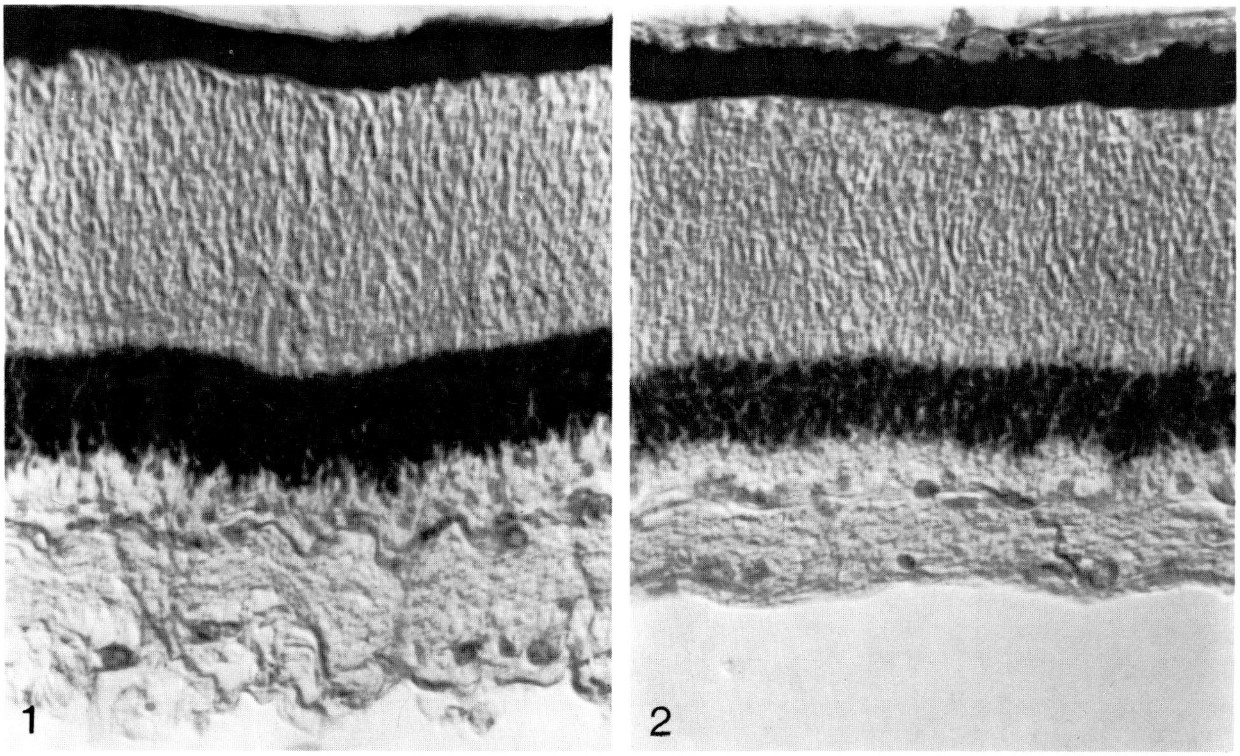

Plate 68.1, 2. Transverse sections of the retinas of the grenadiers, *C. rupestris* (1) and *N. bairdii* (2). ×440

Exocoetidae (*Flyingfishes and Halfbeaks*)

Acanthopterygii, Atherinomorpha, Atheriniformes, Exocoetoidei

Typical flyingfishes are marine, epipelagic teleosts with long, stiff pectoral fins with which they plane several feet through the air, chiefly in attempts to escape their predators (such as dolphins). Some exocoetids have a pyramid-shaped cornea with flattened surfaces to permit emmetropic vision in air as well as the typical hypermetropic (farsighted vision), of epipelagic teleosts in water.

The retina of *E. volitans* shown in Plate 69.1 is in a dark-adapted state, with the abundant pigment retracted sclerally, the numerous cones fully extended, and the very numerous and slender rods. Single cones are present, and there are unequal (Plate 69.2) as well as equal (Plate 69.3) double cones in this retina. The area retinae is ventro-temporal (Plate 69.2). The nuclear layers are moderately well developed and three layers of small horizontal cells are present in Plate 69.1.

The retina of *Fodiator acutus* resembles essentially that of *Exocoetus volitans,* with the exception that rods are less abundant and that unequal double cones are found more regularly (Plate 70.2). Note the light-adapted ventral retina and the well-developed bipolar as well as the ganglion cell layers in Plate 70.1.

References: ANCTIL and ALI (1970); ALI and WAGNER (1975a)

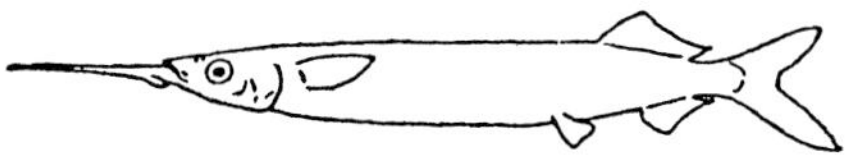

Halfbeaks

Halfbeaks, previously included in a separate family (Hemirhamphidae), are very elongate, coastal, pelagic fishes. Adult halfbeak, *Hyporamphus unifasciatus* have fast and sudden movements. They feed on small crustaceans, molluscs, and plant material.

The retina of the halfbeak is essentially similar to that of the flyingfish, as is evident in Plate 70.3 showing a semi-light-adapted dorsal retina. The cones are very tiny in comparison with the long, slender rods. The horizontal cells are small and disposed in three layers, one of which is shown in Plate 70.4.

Reference: ALI et al. (1973)

Visual Pigments: A_1 max 475–499 (ALI and HEUMANN, 1970; MUNZ and MCFARLAND, 1973)

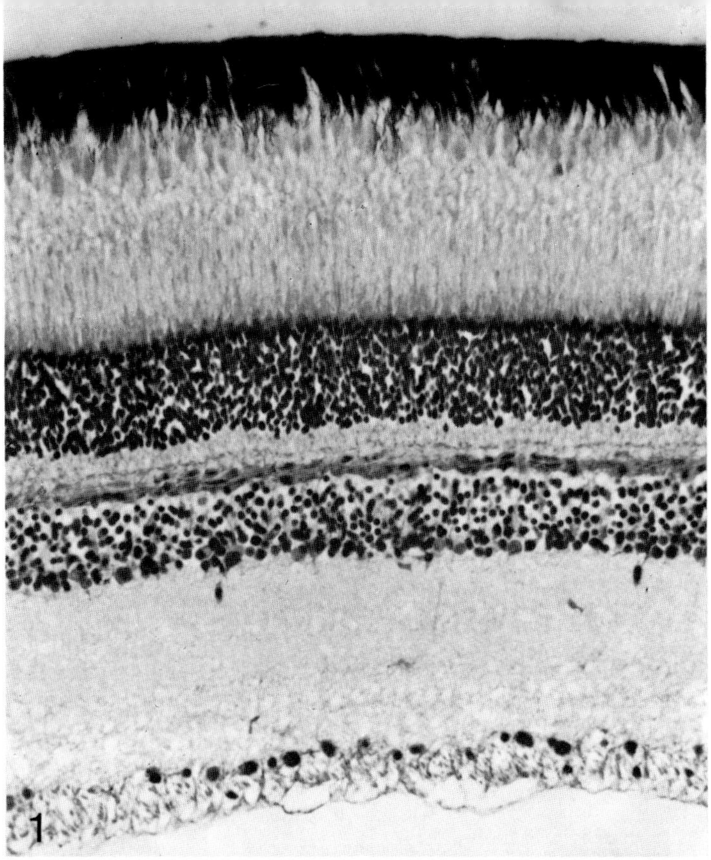

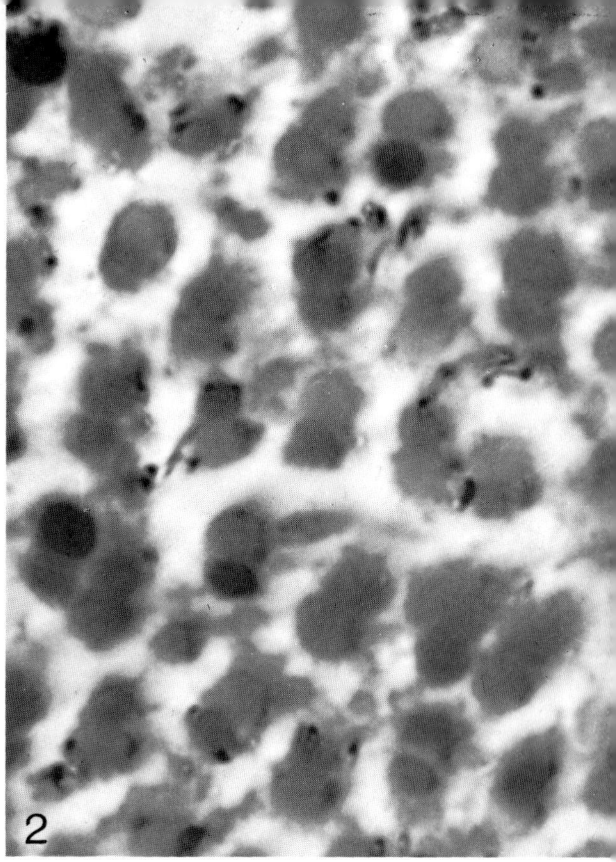

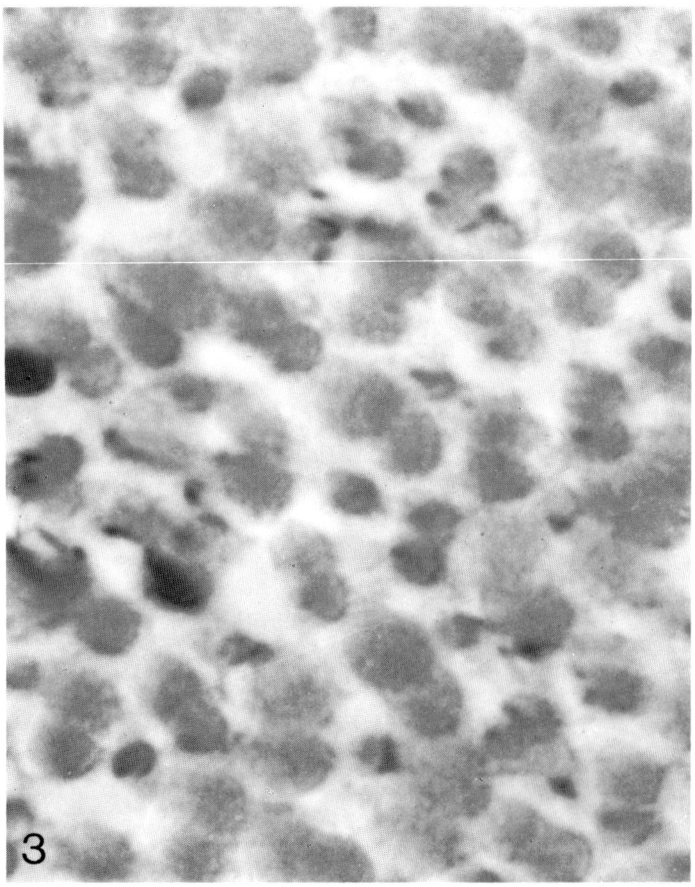

Plate 69.1-3. Transverse (1) and tangential (2) and (3) sections of the retina of the flyingfish *E. volitans*. (1) ×320; (2) and (3) ×1,105

Plate 70.1-4. Transverse (4) and (6) and tangential (5) and (7) sections of the retinas of the flyingfish *F. acutus*, and the halfbeak, *H. unifasciatus*. (4) ×600; (5) ×182; (6) ×225; (7) ×560 ▷

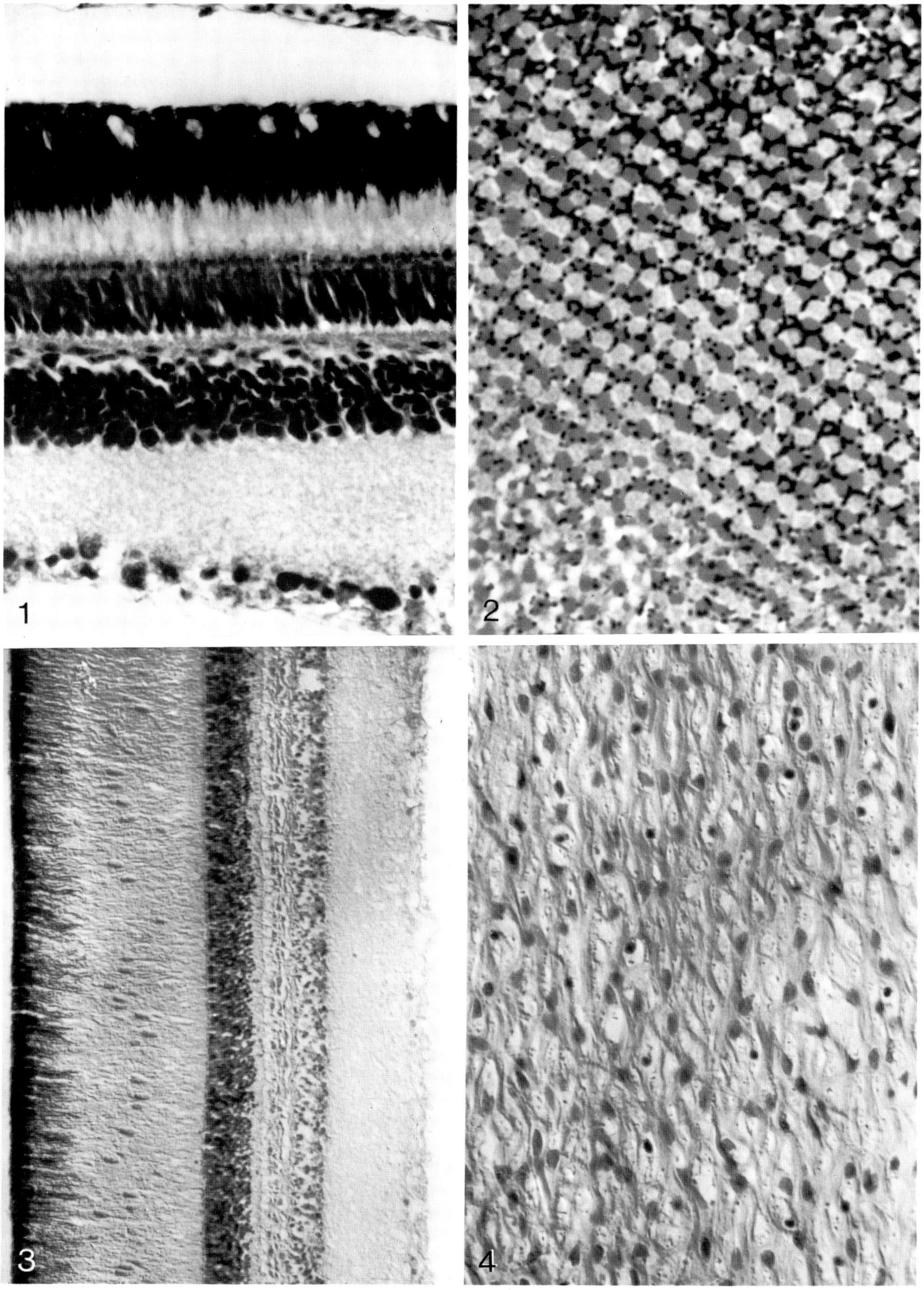
1
2
3
4

Belonidae (*Needlefishes*)

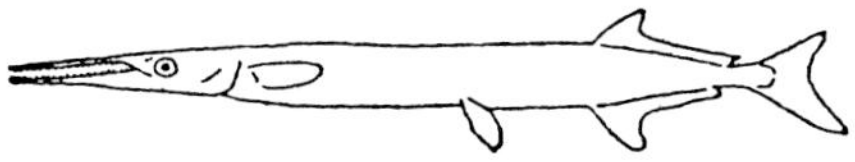

Atherinomorpha, Atheriniformes, Exocoetoidei

Needlefishes or garfishes are closely related to the flyingfish (Exocoetids) family. They have an elongate body and a long jaw armed with many fine teeth, and are fast swimmers. They are essentially coastal, pelagic or epipelagic. Although their body is rather rigid they swim relatively well, and are able to jump vertically out of the water when chased by larger predators.

The retina of *Strongylura timucu* is characterised by a large number of rods and cones, three to four distinct layers of horizontal cells, a moderately well-developed internal nuclear layer, and one layer of ganglion cells (Plate 71.1). Single and double cones are present (Plate 71.2) and arranged in square patterns (Plate 71.3, 4). Cones are characterised by unusually long outer segments and large ellipsoidal mitochondria, giving a coarsely granular texture to the inner segment (Plate 71.2). Some cones appear to possess a globule or droplet at the scleral end of their ellipsoid.

References: HANNOVER (1843); FRIIS (1879)

Visual Pigments: A_1 max 501 (MUNZ and MCFARLAND, 1973)

Plate 71.1-4. (1) Transverse section of the dorsal retina of *S. timucu*. Note the large, conspicuous Müller's fibres *MF* ×330. (2) Transverse section of the dorsal retina of *S. timucu* ×1,320. (3) Tangential section of the dorsal retina of *S. timucu* showing the cone arrangement at the ellipsoidal level (×330). (4) Tangential section of the ventral retina (×525)

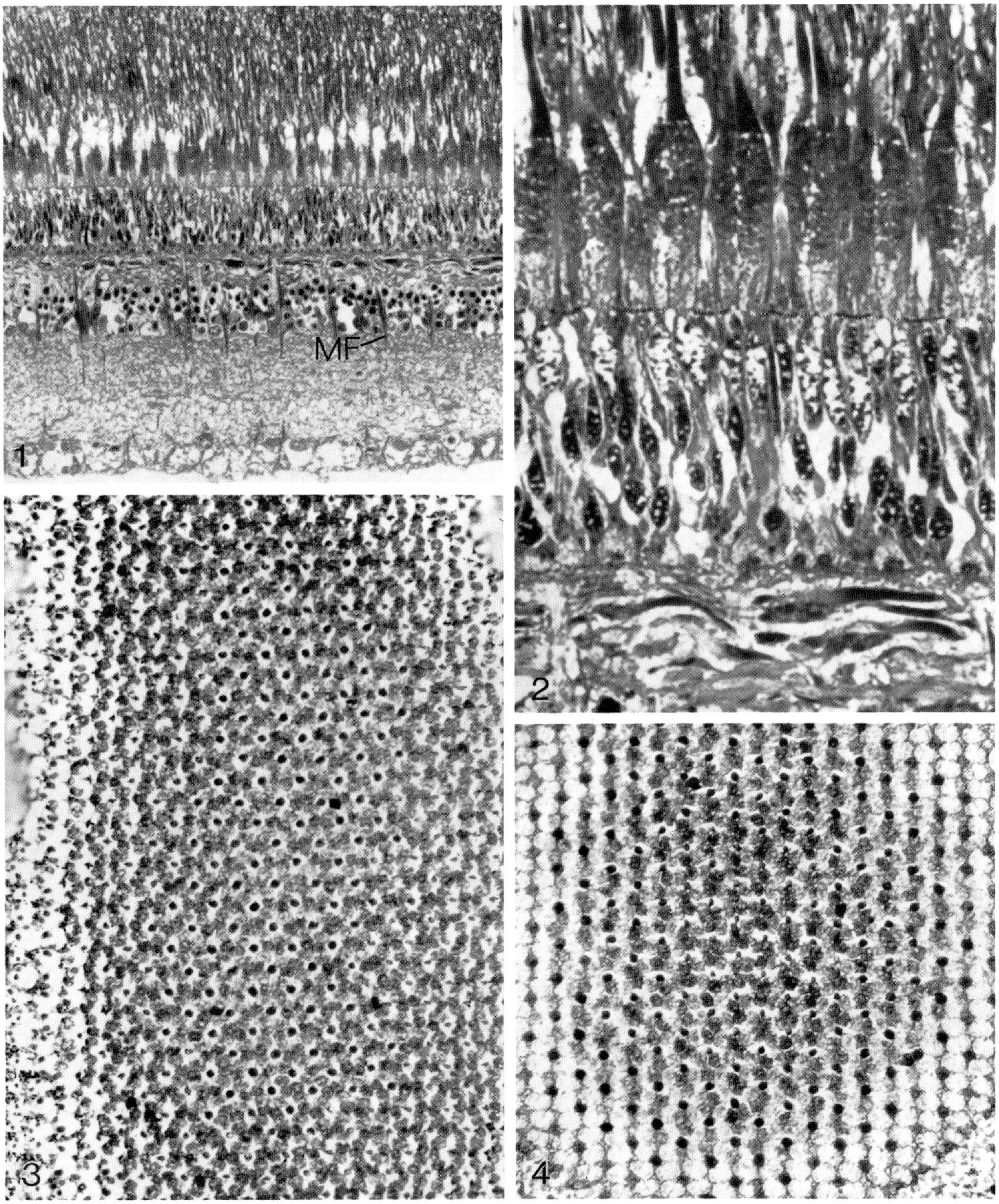

Plate 71.1-4. (Legend see opposite page)

Cyprinodontidae (*Killifishes*)

Acanthopterygii, Atherinomorpha, Atheriniformes, Cyprinodontoidei

Cyprinodontoids are small, common fishes, with the characteristic appearance of fishes that spend much of their time capturing food near the water surface: the mouth is in a forward position at the upper tip of the head. The lateral line and adipose fin are absent, however an air bladder is present. The killifishes (Cyprinodontidae) are a large family of voracious fishes living in fresh or brackish waters.

Their retina conforms to the well-developed teleostean type. There are many rods and three types of cones: the short single cone, the long single cone (Plate 72.1), and the unequal double cone (Plate 72.2). All the cone types possess a distinct globule at the scleral tip of the ellipsoid (Plate 72.1, 2). The mosaic formed by the cones is viewed tangentially in Plate 72.3. A ventral band-shaped area retinae is present in this species and is characterised by very slender cones. The change in cone dimension takes place abruptly in the area as seen in Plate 72.4. Cone ellipsoids contain oil droplets of mitochondrial origin in several species, notably *F. heteroclitus* (Plate 72.5, 6). See also Plate 4, (2).

References: EIGENMANN (1899); BUTCHER (1938); LYALL (1957b); ENGSTRÖM (1963b); MUNK (1970); WAGNER (1972); ALI and WAGNER (1975a); ANCTIL and ALI (1976)

Visual Pigments: A_1 max 500–506; A_2 max 522 (ALI and WAGNER, 1975b)

Plate 72.1-4. Transverse (1), (2) and (4) and tangential (3) sections of the retina of the killifish, *F. heteroclitus*. (1) and (2) ×1,105; (3) and (4) ×440 ▷

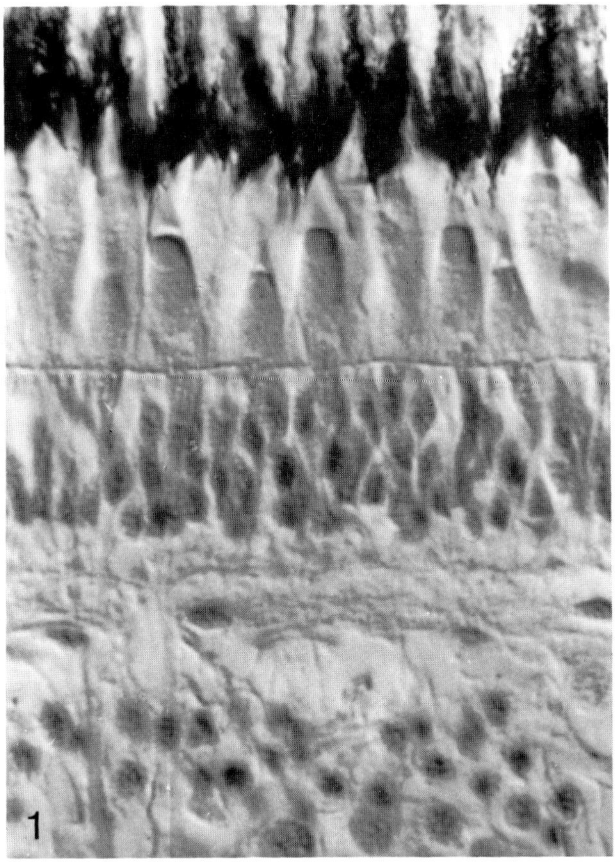

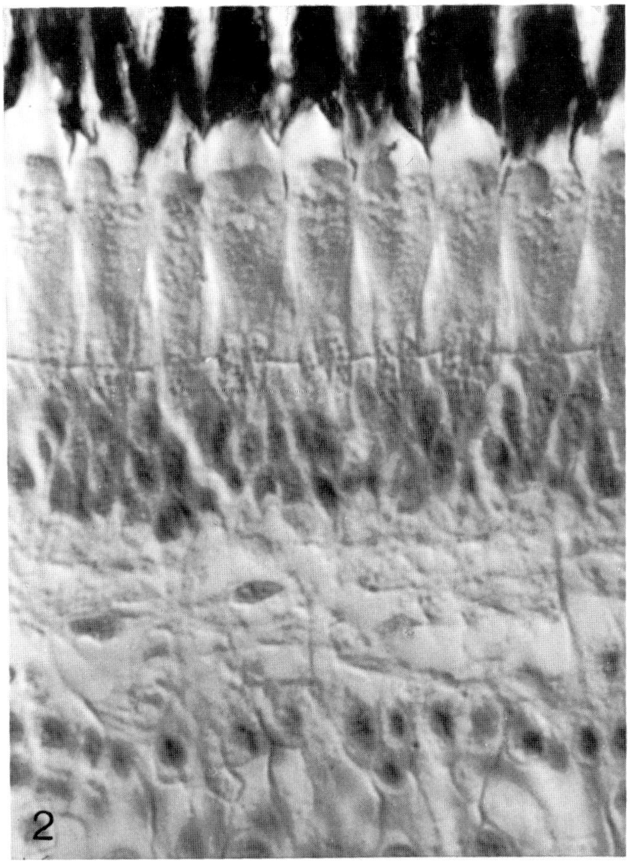

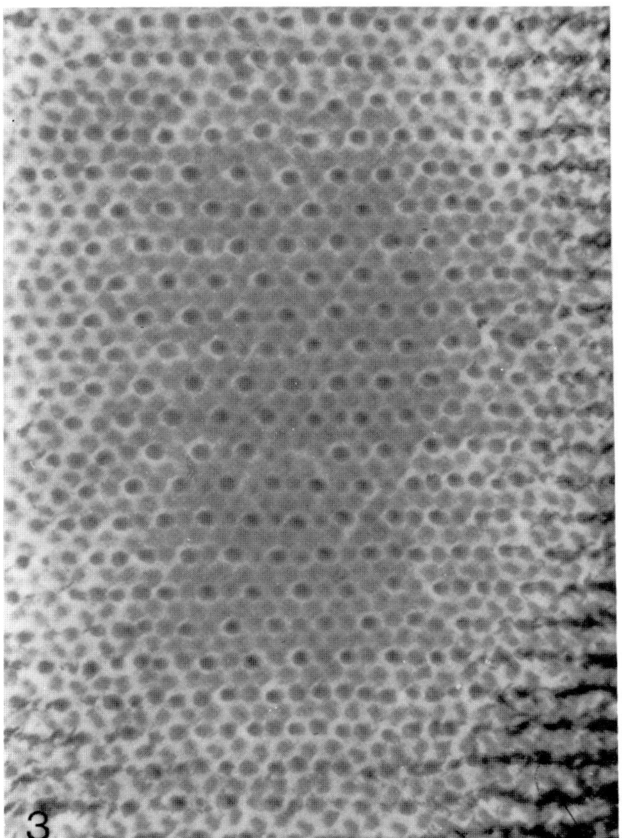

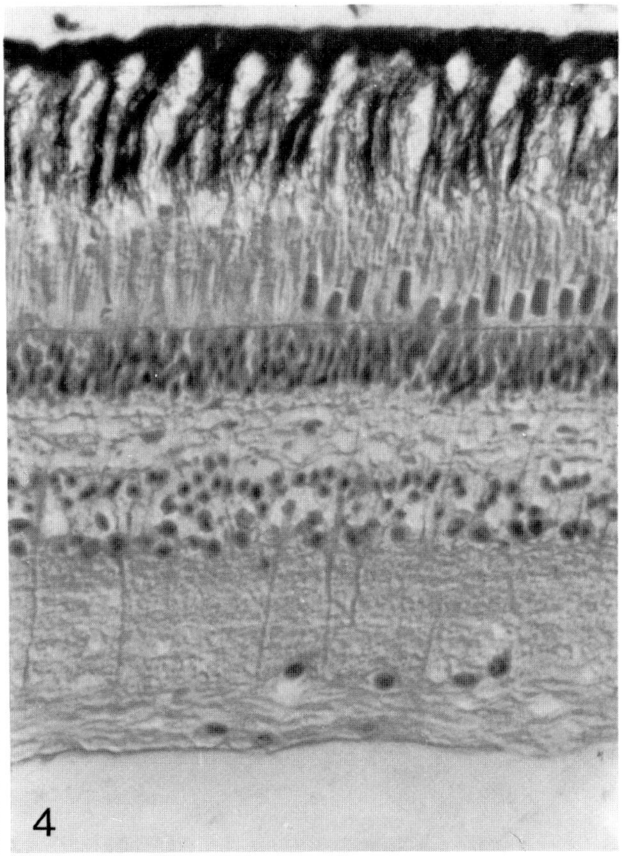

Anablepidae (*Four-Eyed Fishes*)

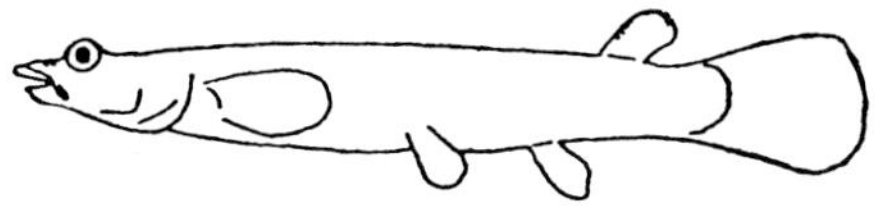

Acanthopterygii, Atherinomorpha, Atheriniformes, Cyprinodontoidei

The four-eyed fish differs from all the other cyprinodontoids by its larger number of vertebrae and peculiar eyes. This freshwater fish spends much of its time cruising just below the surface of the water with the upper half of its eyes protruding above the water. The water line is at the centre of each eye, and at this point the eye is clearly divided by a longitudinal epithelial band. The cornea and the retina are also divided; this appears to be an adaptation for air-and-water vision, enabling the fish to watch for predators or prey both above and below the water surface.

The retina is well-developed. There are significant differences between the upper (aquatic vision: Plate 73.1) and lower (aerial vision: Plate 73.2) portions of the retina of *Anableps microlepis*. In the ventral retina there are about twice as many cones as in the dorsal retina and they are smaller; the internal nuclear layers are better-developed in the ventral retina (aerial vision). In *A. tetrophthalmus,* there are short and long single cones and double cones, which are arranged in square patterns, and there are three layers of horizontal cells. Oil droplets of mitochondrial origin are found in the ellipsoids of several species.

References: SCHNEIDER-ORELLI (1907); SCHWASSMANN and KRUGER (1965); WAGNER (1972); BORWEIN and HOLLENBERG (1973)

Visual Pigments: A_1 max 506 (ALI and WAGNER, 1975b)

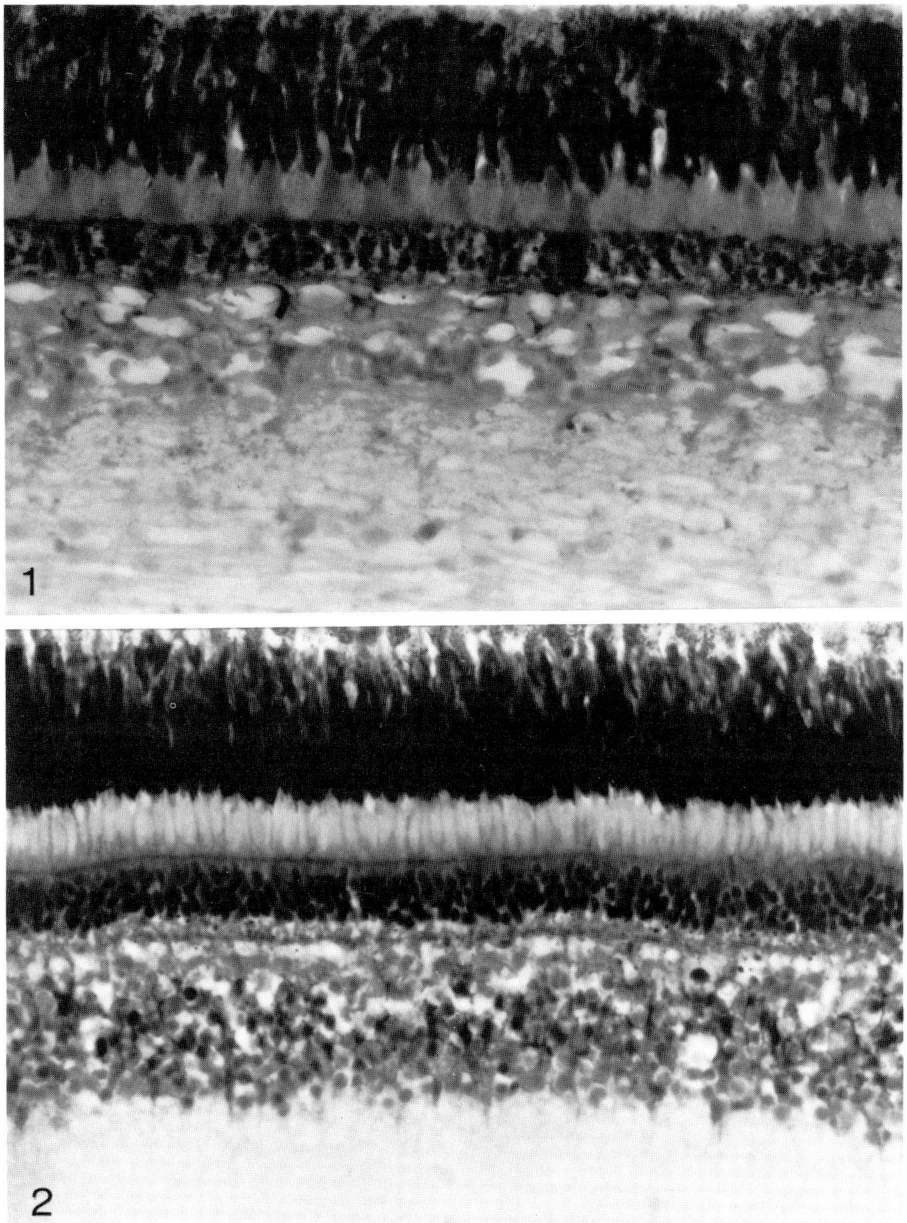

Plate 73.1, 2. Transverse sections of the dorsal (1) and ventral (2) retina of the four-eyed fish, *A. microlepis*. ×560

Poeciliidae (*Livebearers*)

Acanthopterygii, Atherinomorpha, Atheriniformes, Cyprinodontoidei

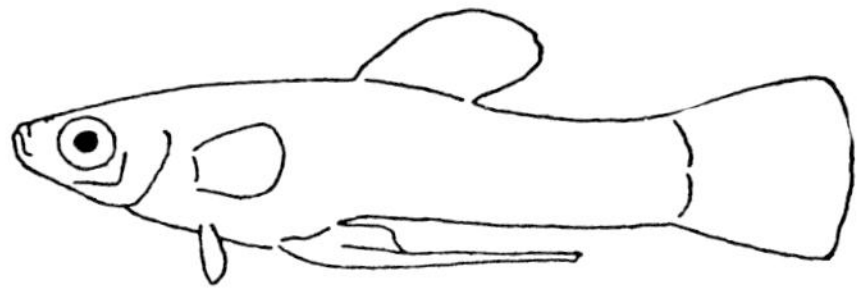

Poecilids differ from other cyprinodontoids by the absence of exoccipital condyles and the presence of a gonopodium in the male, being a modification of the anal fin rays designed for viviparous habits. The livebearers are shoaling fishes that inhabit shallow, stagnant waters. Some frequent brackish waters. They feed mainly on insect larvae and algal materials.

The retina of the guppy *Poecilia reticulata*, (=*Lebistes reticulatus*) resembles closely that of the cyprinodontids. Both rods and cones are fairly numerous. In the light-adapted retina (Plate 74.1), the three types of cones (short and long single cones, unequal double cones) are found close to the limiting membrane, the nucleus being inside in the short single cone, outside in the long single cone, and across the limiting membrane in the double cones. In the dark-adapted retina (Plate 74.2), it is noted that the short single cones show negligible movements, whereas the long single cones move to an intermediate position, and the double cones to a fully sclerad position. These cones are arranged in square mosaics of double cones with the short single cones at the corners, and a long single cone at the centre (Plate 74.3,4). Oil droplets of mitochondrial origin are found in the cone ellipsoids of several species.

References: BUTCHER (1938); MÜLLER (1952); ENGSTRÖM (1963b); LANG (1965); BERGER (1966); WAGNER (1972); ALI and WAGNER (1975a)

Visual Pigments: A_1 max 498–502; A_2 max 521 (ALI and WAGNER, 1975b)

Plate 74.1-4. Transverse (1) and (2) and tangential (3) and (4) sections of the retina of the guppy, *P. reticulata*. ×560

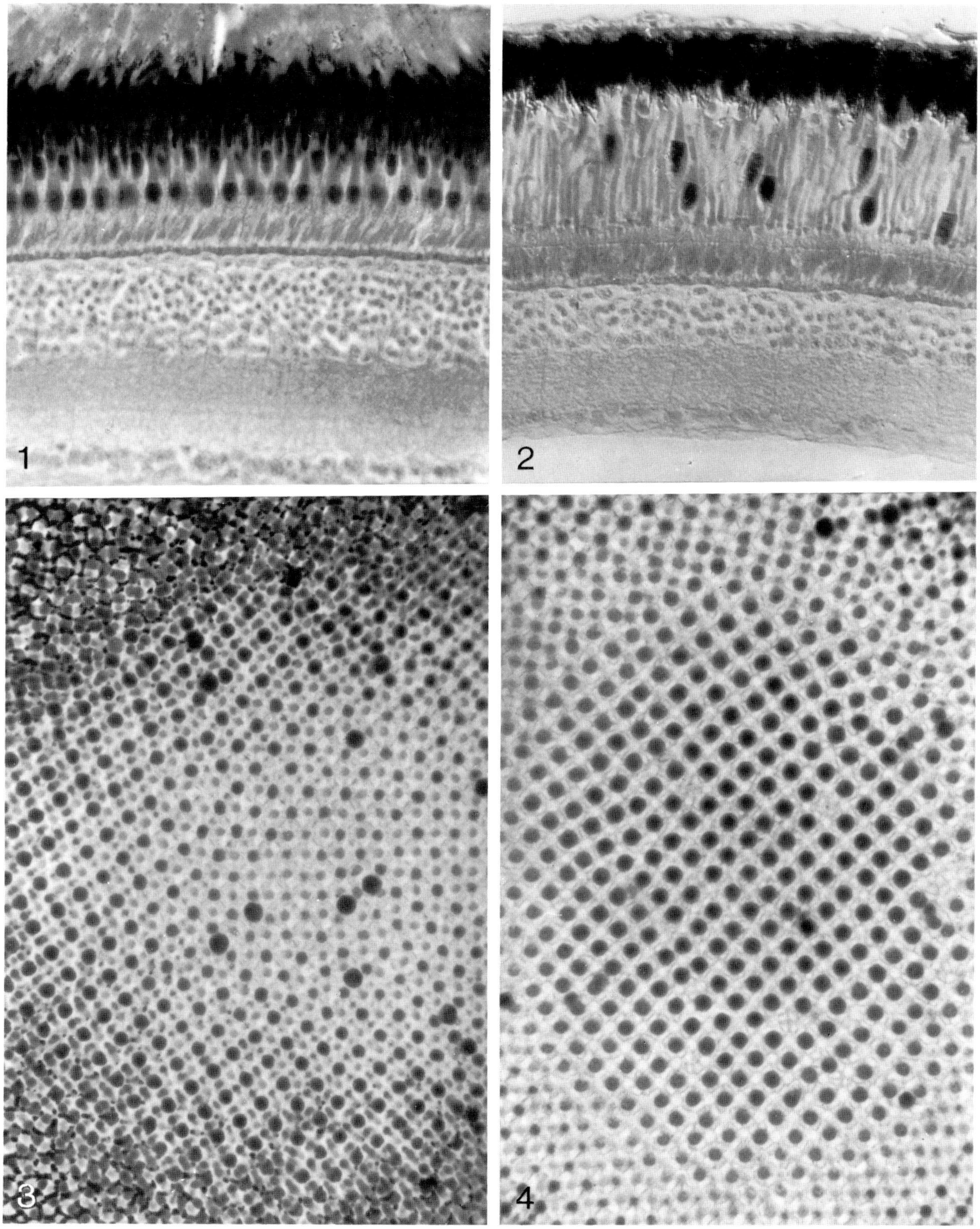

Plate 74.1-4. (Legend see opposite page)

Atherinidae (*Silversides*)

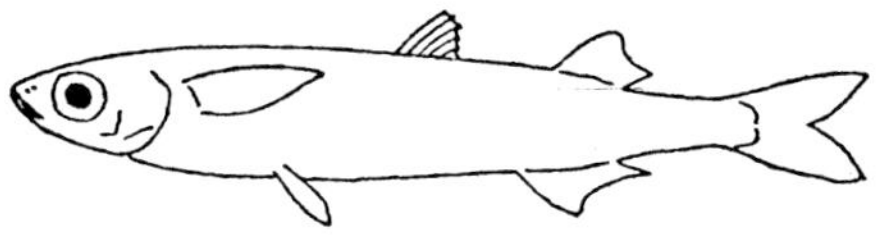

Acanthopterygii, Atherinomorpha, Atheriniformes, Atherinoidei

Silversides are characterised by a broad silvery band along their sides, the presence of two separate dorsal fins, and the absence of a lateral line. They are chiefly coastal marine, schooling fishes.

Rods are present in large numbers in the retina of *Austromenidia* sp., and small single cones as well as unequal and equal double cones have been reported. The retinal epithelial pigment, rods and cones are reported to undergo retinomotor changes in response to light and dark in *Atherina mochonpontica*.

References: SVERDLICK (1940, 1955); PROTASOV et al. (1960)

Visual Pigments: A_1 max 501–508 (ALI and WAGNER, 1975b)

Melamphaeidae (*Bigscales*)

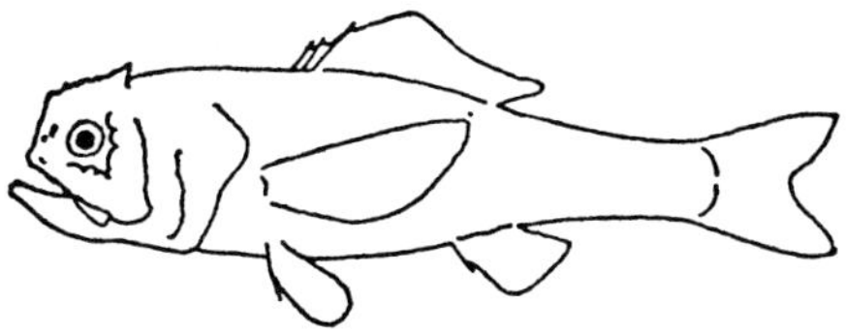

Acanthopterygii, Percomorpha, Beryciformes, Stephanoberycoidei

The Beryciformes include some groups of similar subperciform fishes that are generally considered to be antecedent to the Perciformes.

The melamphaeids are small mesopelagic or bathypelagic fishes with blunt heads covered with spines and crests, truncate snouts, and big eyes. They differ from other stephanoberycoids by the presence of large cycloid scales on their bodies.

In the retina of *Melamphaes buborbitalis* the epithelial pigment is densely packed sclerally and the rods, which are very numerous, are the only visual cells present. In the retina of *Poromitra nigrofulvus* (Plate 75) there are two superimposed layers of rods with short ellipsoids, the sclerad-most rods possessing long and slender myoids running between the vitread-most rods. The internal nuclear layers of both species are poorly developed.

References: BRAUER (1908); LOCKET (1969)

Visual Pigments: A_1 max 488 (ALI and WAGNER, 1975b)

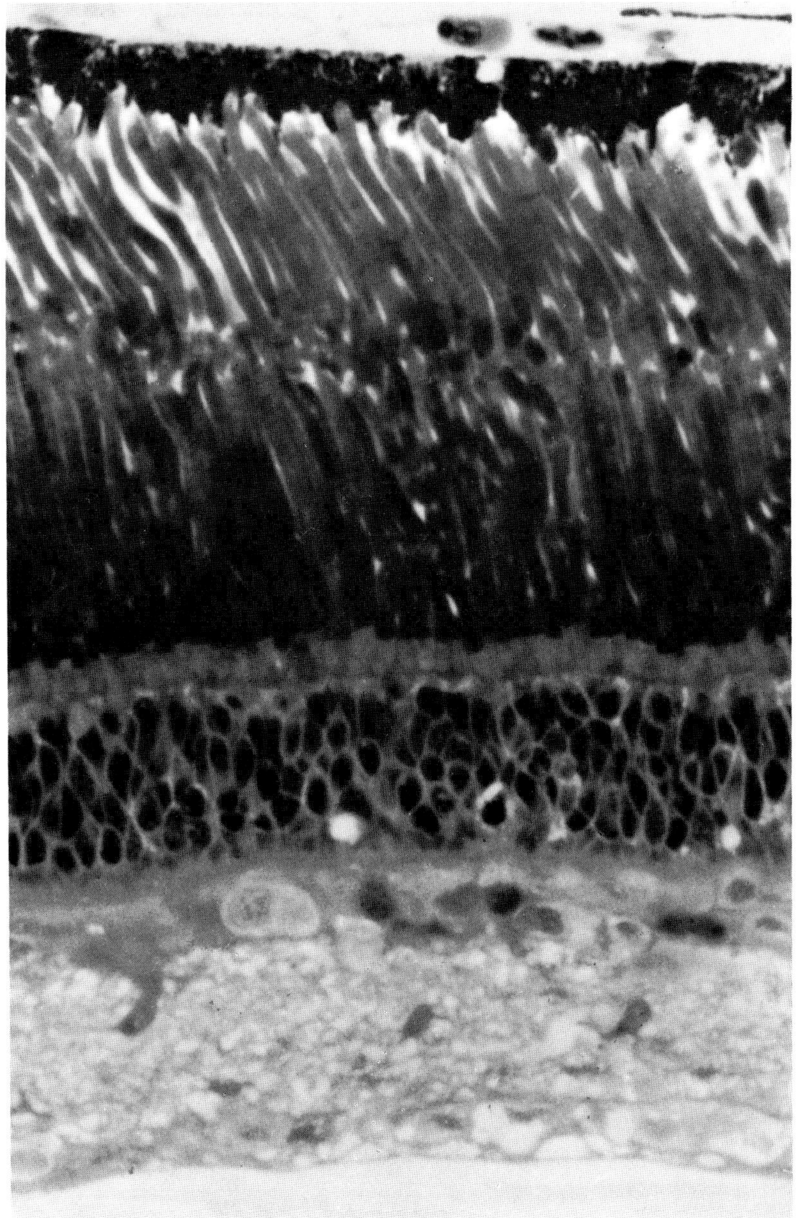

Plate 75. Photomicrograph of a transverse section of the retina of *P. nigrofulvus,* showing two banks of rods

Diretmidae (*Diretmids*)

Acanthopterygii, Percomorpha, Beryciformes, Berycoidei

Diretmids are mesopelagic fishes related to the Berycidae, with bodies that are higher and more compressed. Their eyes are very large.

The retina of *Diretmus argenteus* is well developed and highly specialised. Its main characteristics are (1) the presence of multiple, superimposed layers of rods throughout the retinal regions (especially in the ventro-temporal region; Plate 76), the vitread-most being considerably longer than the more sclerad ones, (2) the presence of small single cones in a restricted area (near the ora terminalis) of the specialised ventro-temporal region (Plate 76), and (3) a corresponding increase in thickness of the internal nuclear layers (bipolar and ganglion cells) in the latter region.

Reference: MUNK (1966b)

Zeidae (*Dories*)

Percomorpha, Zeiformes

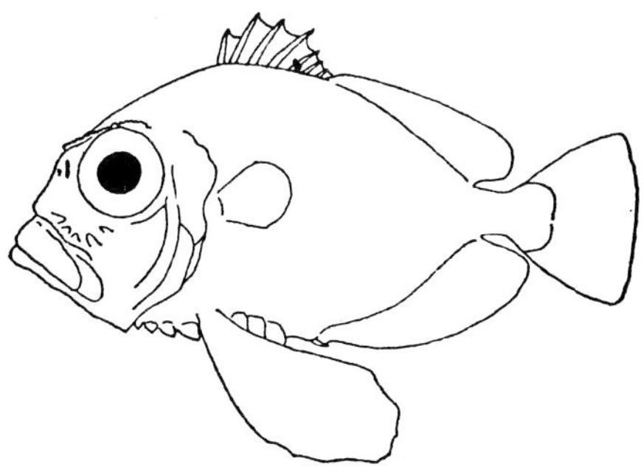

The oceanic Zeiform fishes are a small group of spiny-rayed fishes, usually compressed and deep-bodied. They normally have enormously distensible jaws set at an oblique angle. The John dories inhabit only moderate depths of the seas.

The retina of the Japanese dory *Zenion japonicum,* caught at depths of 100–300 m, is characterised by the presence of double cones forming a sharp area retinae in the ventral region. The single cones are absent. A reflecting tapetum lucidum is reported in this species, but it is not stated whether it is of chorioidal or retinal origin.

Reference: TAMURA (1957)

Visual Pigments: A_1 max 492 (ALI and WAGNER, 1975b)

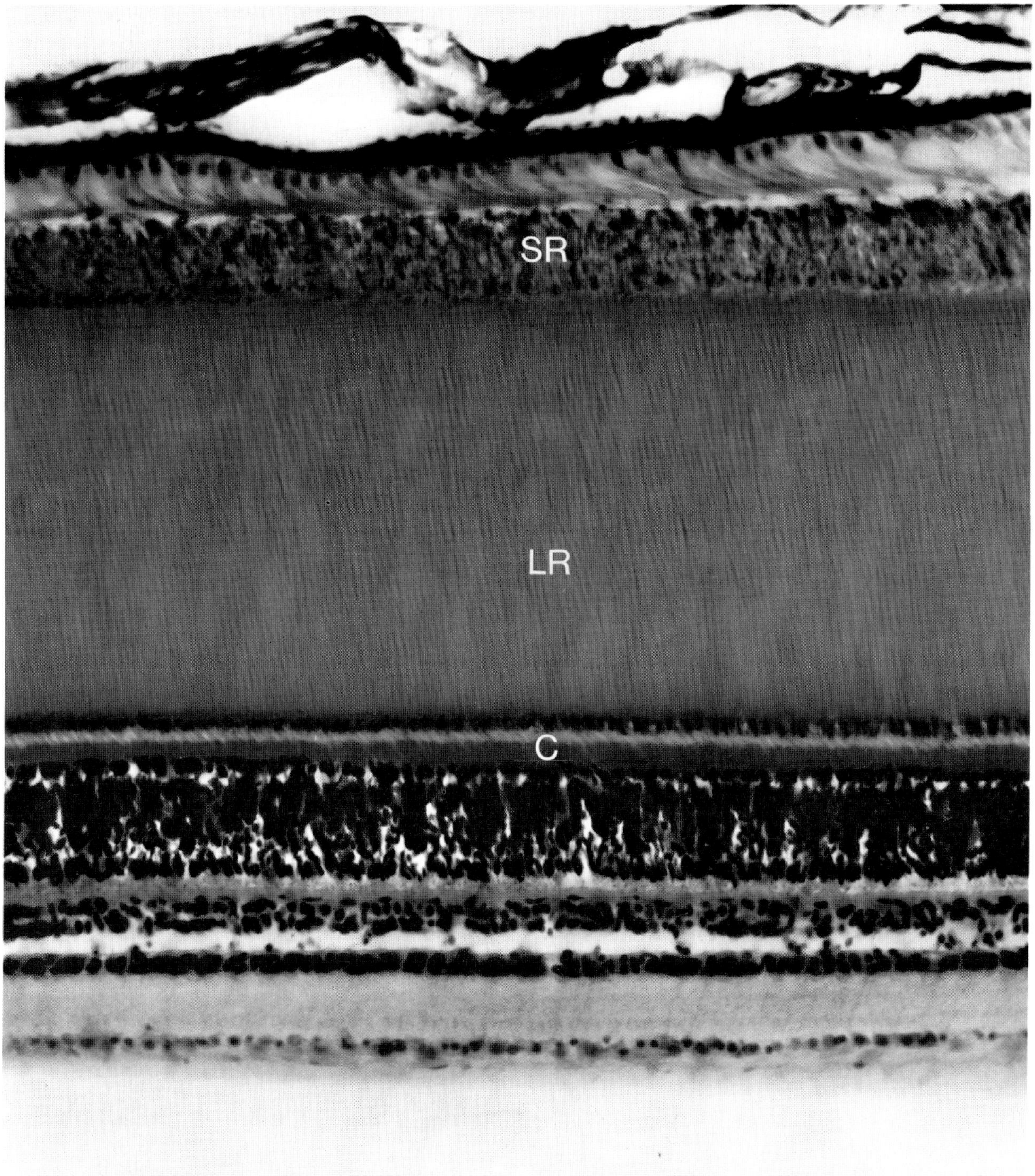

Plate 76. Photomicrograph of the transversely cut ventro-temporal retina of the diretmid fish, *D. argentus*. $\times 385$. *C* cone layer, *LR* layer of long rod acromeres, *SR* multiple layers of short rod acromeres

Stylephoridae (*Tube-Eyes*)

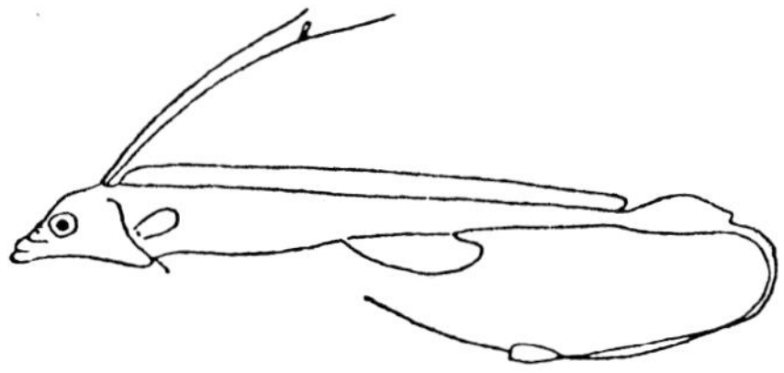

Acanthopterygii, Percomorpha, Lampridiformes, Stylephoroidei

Lampridiform fishes are rare, bathypelagic forms with large eyes and small, more or less protractile mouths. The stylephorids have an elongated body, subcylindrical rather than ribbonlike, a naked skin, a very long lower jaw, and rostrally directed tubular eyes.

The main retina has three superimposed layers of rods; the cones are absent. The bipolar, amacrine, and ganglion cells are numerous and distributed in multiple layers (*Stylephorus chordatus*).

Reference: MUNK (1966a)

Gasterosteidae (*Sticklebacks*)

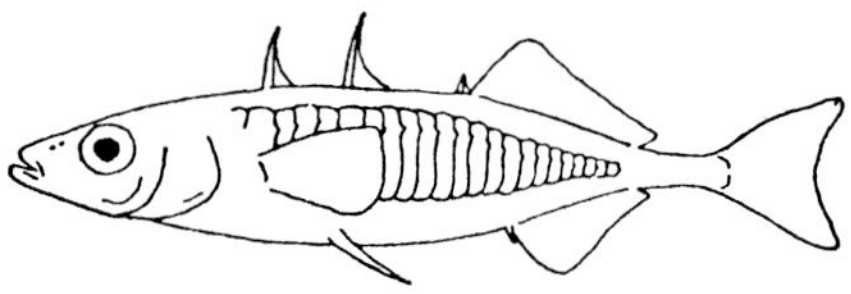

Acanthopterygii, Percomorpha, Gasterosteiformes, Gasterosteoidei

Sticklebacks constitute a group of small, pugnacious, primarily marine fishes. Their dorsal, anal, and pelvic fins are normally equipped with sharp spines. They live in shallow shore waters and feed on copepods, isopods, euphausiids, etc.

The retina is well-developed with a great abundance of epithelial pigment distributed in long processes, a large number of slender cones (equal double cones and single cones), a certain number of rods and well-developed internal nuclear layers (Plate 77).

References: MENNER (1929); WUNDER (1925); VERRIER (1928a); MÜLLER (1952); ENGSTRÖM (1963b); ALI et al. (1968); WAGNER (1972); ALI and WAGNER (1975a)

Visual Pigments: A_1 max 501; A_2 max 522 (ALI and WAGNER (1975b)

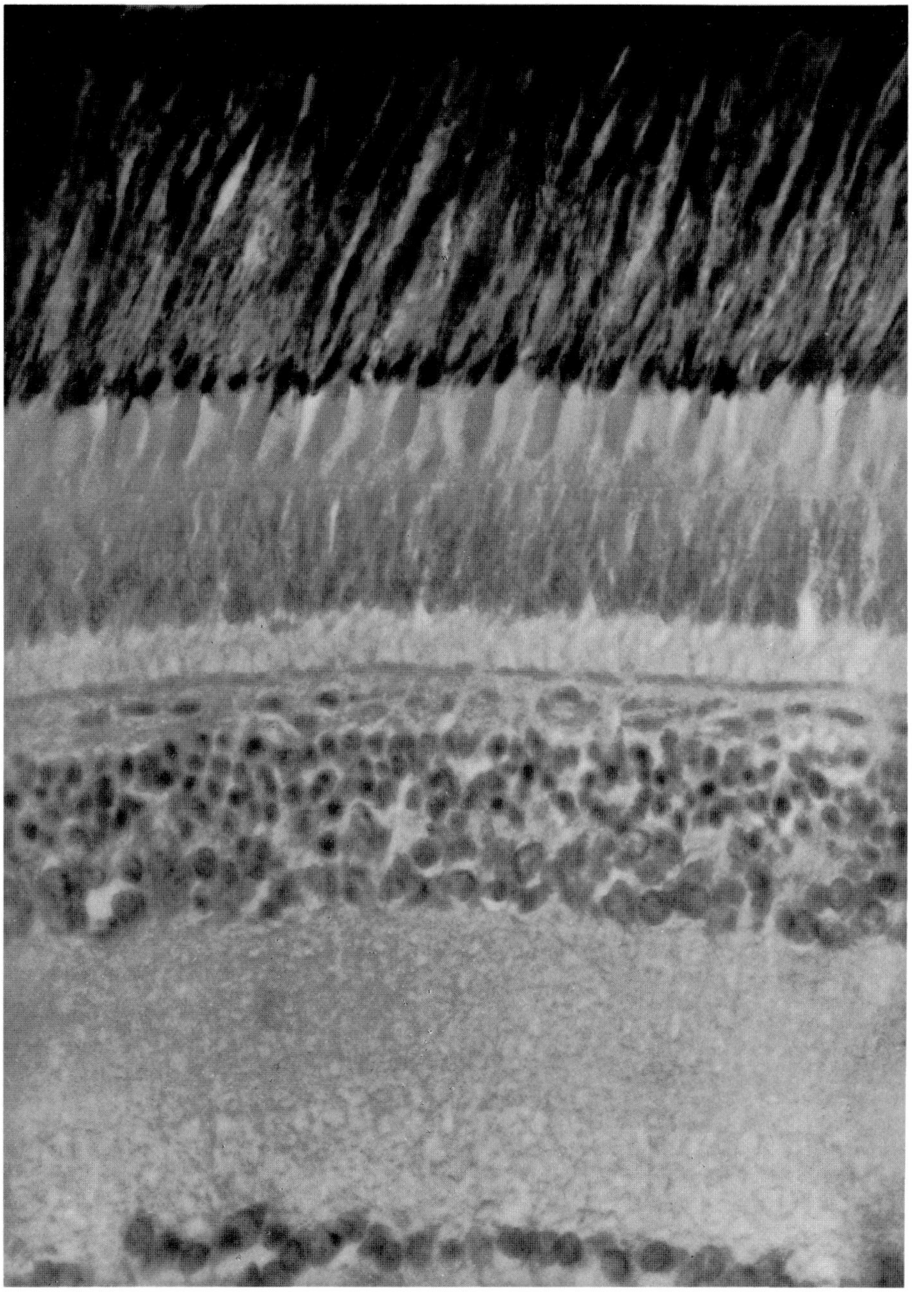

Plate 77. Transverse section of the retina of the three-spine stickleback, *Gasterosteus aculeatus*. × 825

Syngnathidae (*Seahorses and Pipefishes*)

Acanthopterygii, Percomorpha, Gasterosteiformes, Syngnathoidei

Seahorses and pipefishes are specialised marine inshore fishes characterised by segmented bodies and tails encased in jointed bony rings; they have lobate gills, tubular snouts and their heads are cocked at tight angles to the main axis of the body. Seahorses lack the caudal fin and have prehensile tails; they are found in weedy areas and move slowly, feeding principally by sight.

The retina is characterised by abundant epithelial pigment distributed in long processes, by the presence of a few rods and a large number of slender cones, and by the well-developed bipolar and amacrine cell layers (Plate 78.1). Note the foveal region in the ventral retina (Plate 78.2), and the single and equal double cones observed tangentially (Plate 78.3). Note also the filamentous cone cells in the peripheral dorsal retina in Plate 78.4.

References: CARRIÈRE (1885); VERRIER (1928a); KAHMANN (1934, 1936); ENGSTRÖM (1963b); WAGNER (1972)

Plate 78.1-4. Transverse (1), (2) and (4) and tangential (3) sections of the retina of the seahorse, *Hippocampus kuda*. (1) and (4) × 560; (2) × 360; (3) × 1,400

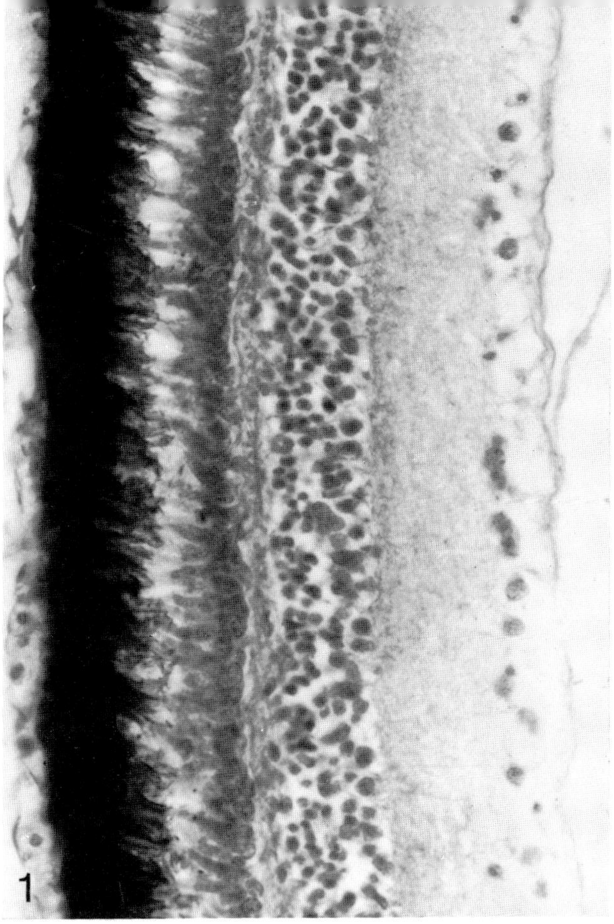

1

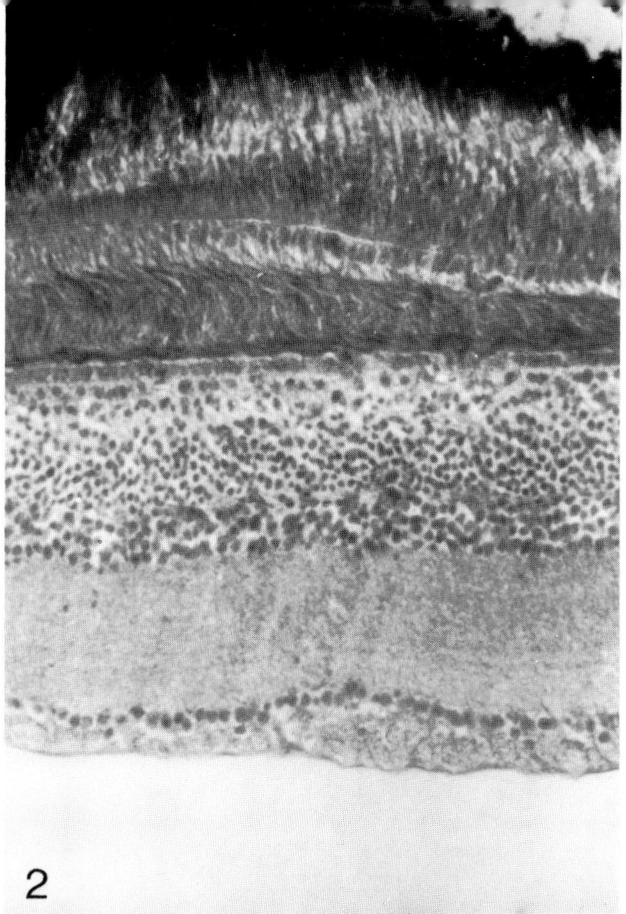

2

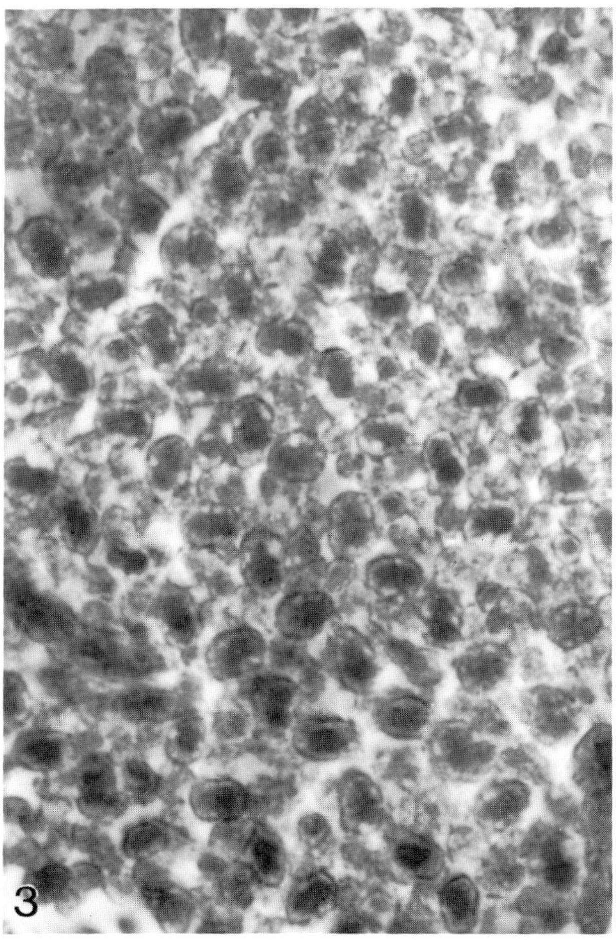

3

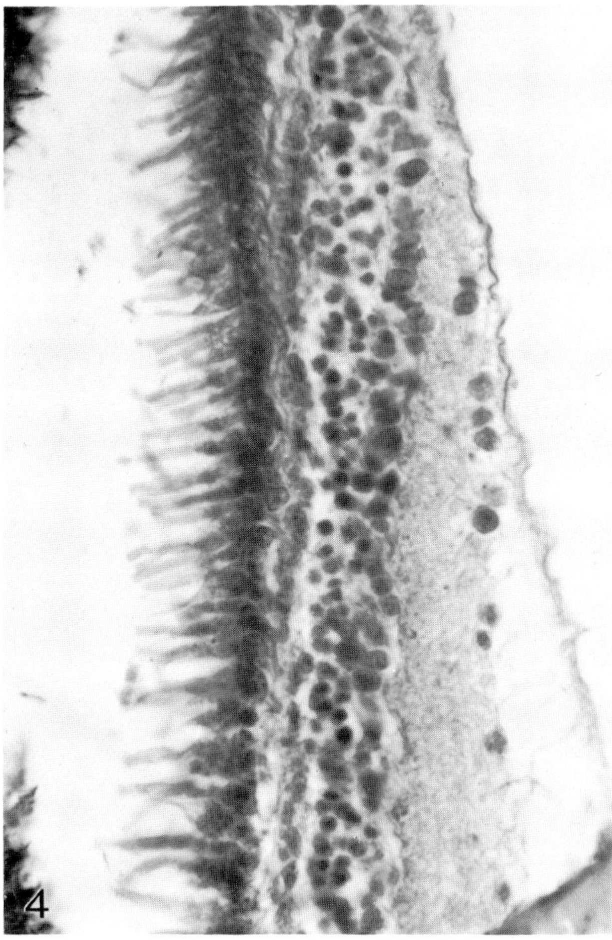

4

Centropomidae (*Snooks*)

Acanthopterygii, Percomorpha, Perciformes, Percoidei

Shovel-headed snooks are primarily marine coastal percoid fishes that are related to the serranids. These carnivorous fishes are found in abundance in mangrove areas, in estuaries, and some even in freshwater.

The retinas of *Centropomus undecimalis* and *C. ensiferus* have the characteristic appearance of a teleostean retina with numerous slender rods, and large single and equal double cones arranged in regular square mosaics. There are four layers of horizontal cells, and the rod horizontal cell layer is intercalated between the medial and inner layers of cone horizontal cells. There are three layers of bipolar cells.

Plate 79.1 illustrates a Golgi-impregnated rod horizontal cell; Plate 79.2, a cone internal horizontal cell; Plate 79.3 a small, pyriform amacrine cell. Note the multiple branches and terminal arborisation of the rod horizontal cell, compared to the terminal structures of the cone horizontal cell.

References: Villegas (1960, 1961); Villegas and Villegas (1963); Stell (1965); Selvin de Testa (1966); Parthe (1972)

Visual Pigments: A_1 max 499–500 nm; A_2 max 520 nm (Ali and Wagner, 1975b)

Plate 79.1-3. Thick, transverse sections of the retina of *C. undecimalis,* impregnated according to the Golgi procedure. Enlargement not available

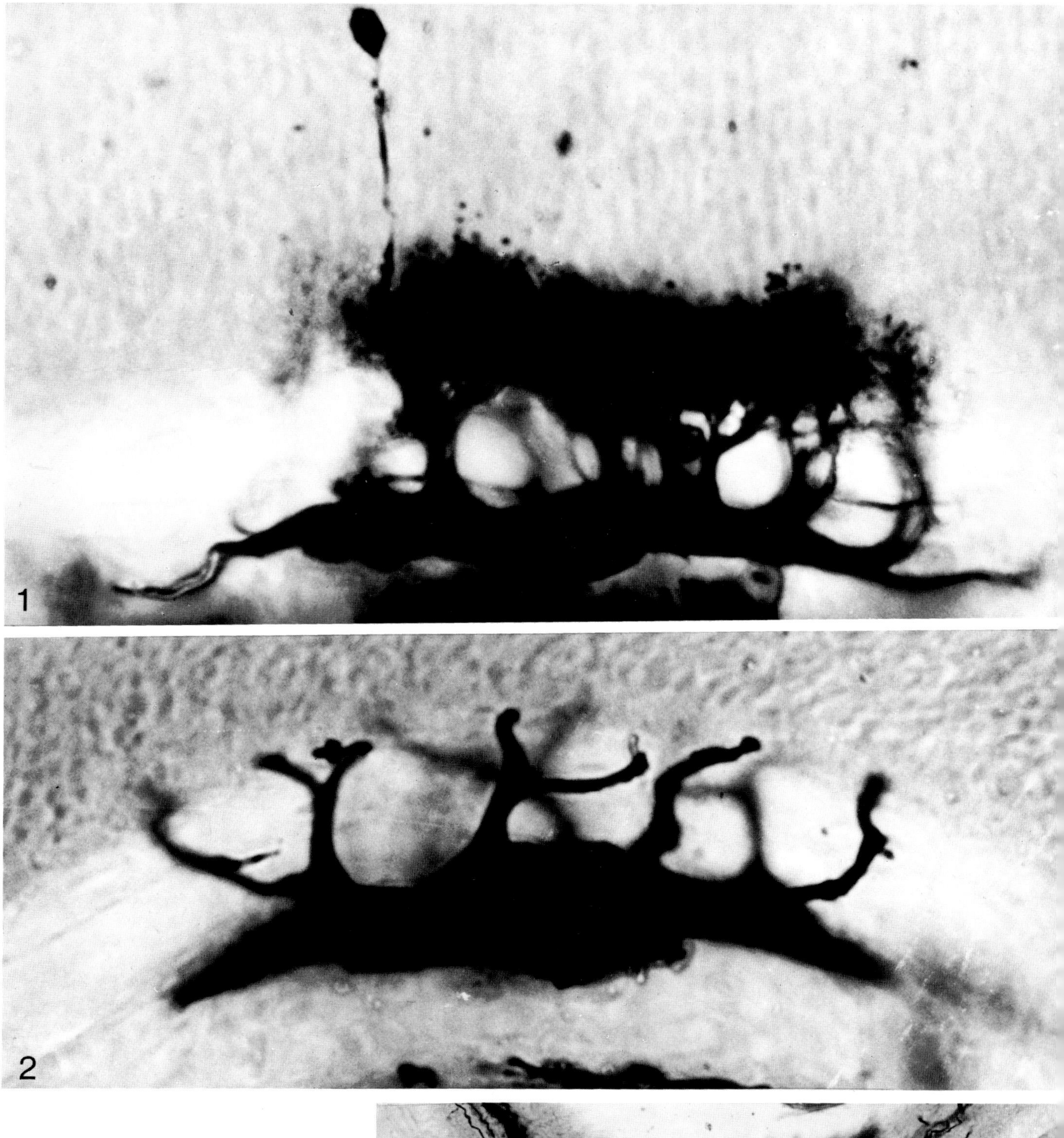

Plate 79.1-3.
(Legend see opposite page)

Serranidae (*Sea Basses and Groupers*)

Acanthopterygii, Percomorpha, Perciformes, Percoidei

This is a large family of marine shore fishes. The major characteristics that identify the serranids are (1) a more or less elongate and compressed body, (2) a large mouth and sharp teeth, and (3) the ctenoid scales on the body. The species of small size and the juveniles of the larger-sized species form schools, while the adults of the larger species tend towards a solitary life. Some groupers are diurnal; others are crepuscular. The soapfish, on the other hand, is strictly nocturnal. The diet of these fishes is composed of other fishes and benthic crustaceans.

The retina of most serranid species described so far is very well-developed. The single cones and equal double cones are numerous, especially in the area retinae temporalis, where in species such as the kelp bass *Paralabrax clathratus* the retina is invaginated to form a foveal depression (Plate 80.1). The cones are arranged in regular square mosaics with central single cones (Plate 80.2). In species from deeper coastal waters (*Acanthistius* sp., *Malakichthys wakiyae*), only double cones are seen and the rods are numerous.

References: VERRIER (1928a, b); KAHMANN (1934, 1936); SVERDLICK (1940, 1955); TAMURA (1957); SCHWASSMANN (1968)

Visual Pigments: A_1 max 492–502; A_2 max 534 (ALI and HEUMANN, 1970, 1972; MUNZ and MCFARLAND, 1973; ALI and WAGNER, 1975b)

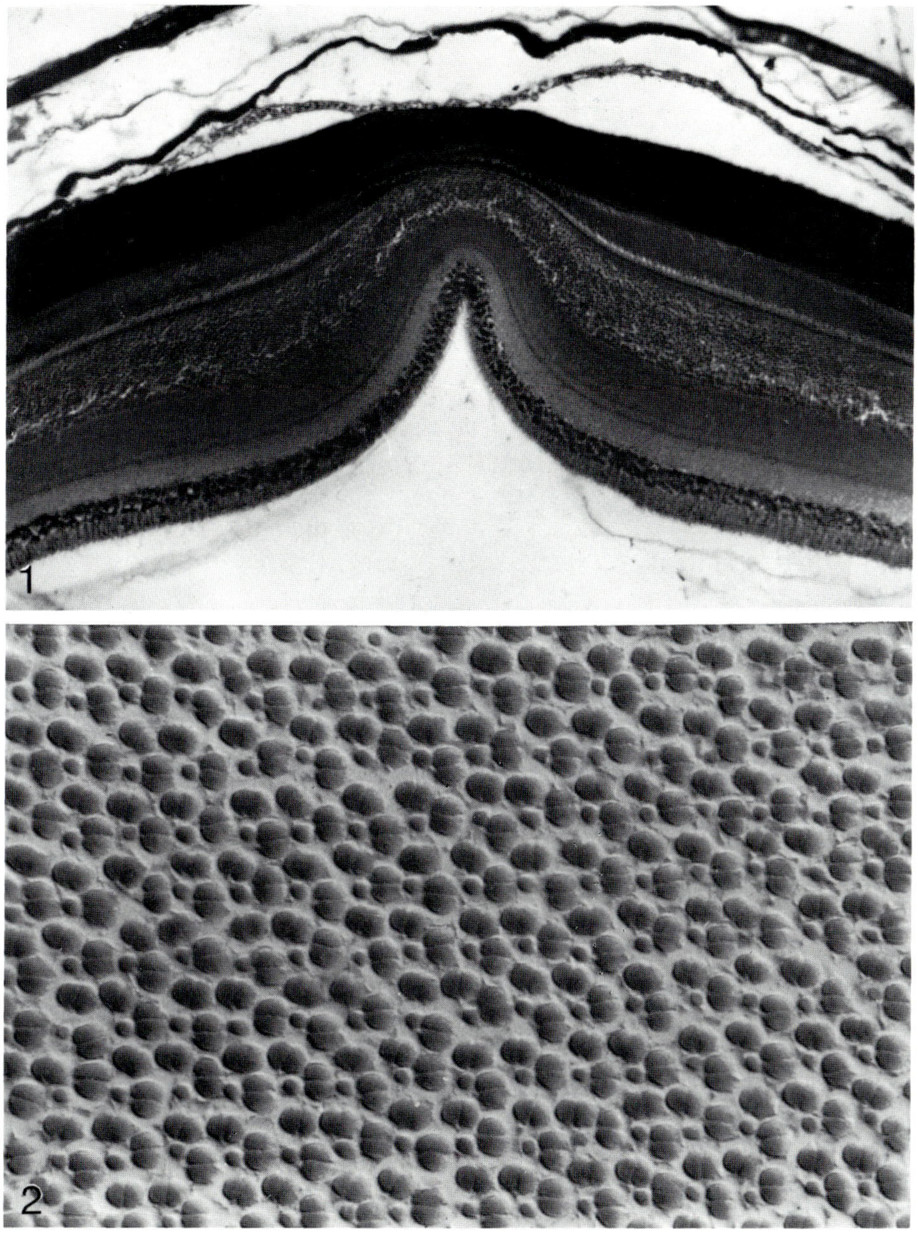

Plate 80.1,2. Transverse (1) and tangential (2) sections of the temporal retina of the kelp bass, *P. chathratus*. (1) ×140; (2) ×560

Centrarchidae (*Sunfishes*)

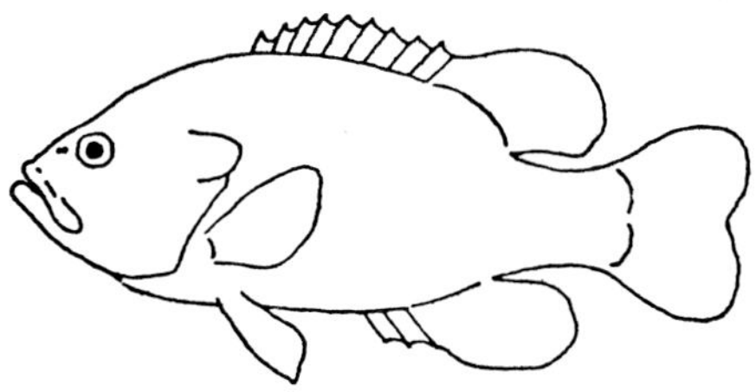

Acanthopterygii, Percomorpha, Perciformes, Percoidei

Bass

This is a relatively small family of strictly fresh-water fishes, native to warm lakes and rivers of NorthAmerica. The large mouth bass (*Micropterus salmoides*) lives in warm, weedy waters and is tolerant of turbid or silted conditions. These predaceous fishes feed on fish, crayfish, and insects.

The retina of the largemouth bass is characterised by a densely pigmented epithelium with long processes, and a large number of densely packed cones (Plate 81.1). The single and equal double cones are arranged in regular square mosaics (Plate 81.2). There are two to three layers of small horizontal cells and the bipolar cells are abundant. The ganglion cells are densely set in one to two layers (Plate 81.1).

References: SHAFER (1900); EIGENMANN and SHAFER (1900); WALLS (1942); POLYAK (1957); ENGSTRÖM (1963b); WANG (1968)

Black Crappie

The black crappie (*Pomoxis nigromaculatus*) inhabits lakes, ponds and slow-moving streams, usually where there is a dense growth of aquatic plants. It is a gregarious species, swimming often in schools. Small fishes and aquatic insects constitute its diet.

Rods are more numerous and cones are fewer and stouter in this species than in the retina of the largemouth bass. As seen in Plate 81.3, the single cones are considerably smaller than the double cones (see also Plate 81.4), the horizontal cells are relatively large, and the bipolar cells are few and sparse. The equal double cones and central single cones are organised in regular square mosaics (Plate 81.4).

Visual Pigments: A_2 max 524–527 (ALI and WAGNER, 1975b)

Plate 81.1-4. Transverse (1) and (3) and tangential (2) and (4) sections of the retinas of the largemouth bass, *M. salmoides,* and the black crappie, *P. nigromaculatus*. (1), (3) and (4) ×440; (2) ×560

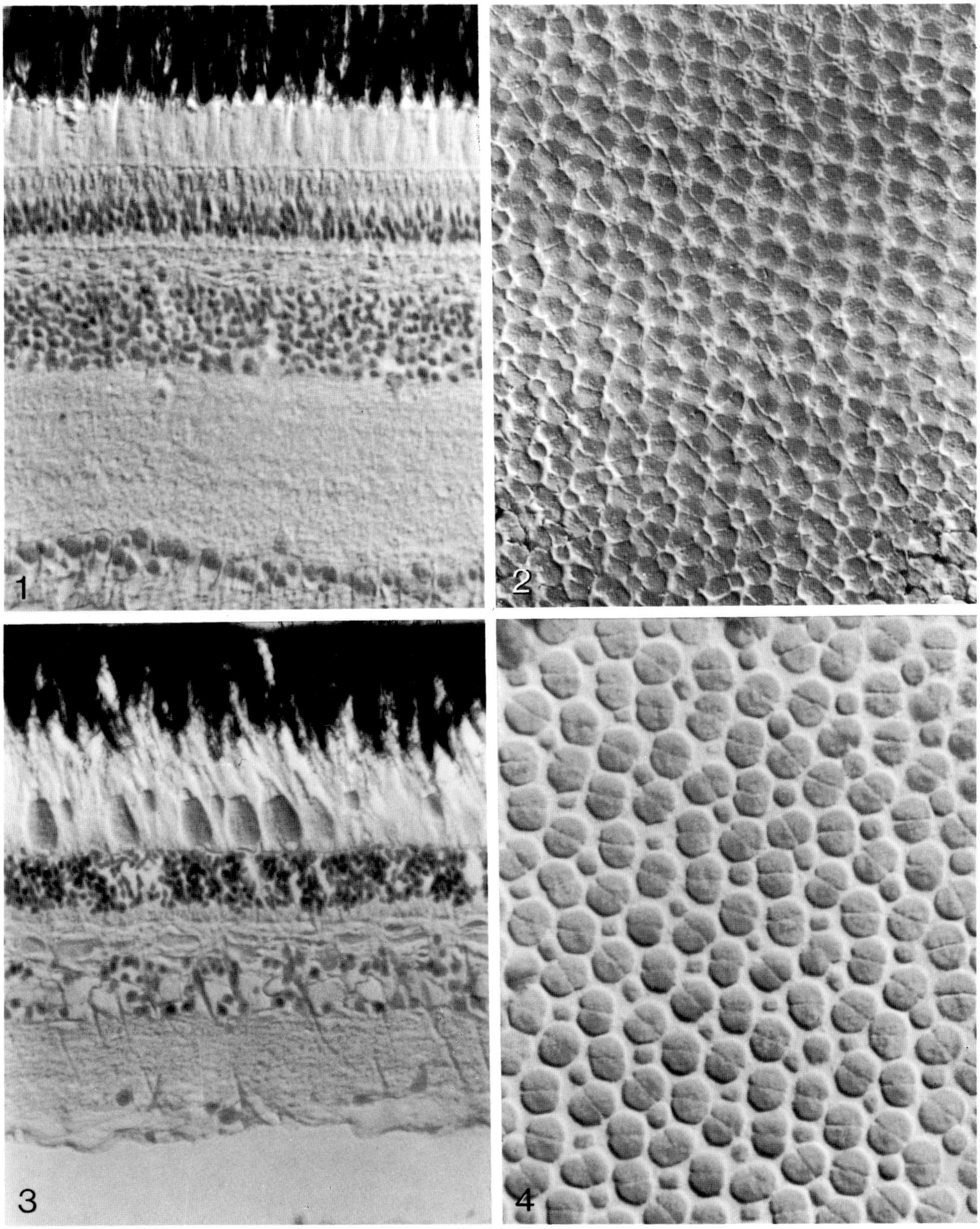

Plate 81.1-4. (Legend see opposite page)

Percidae (*Perches*)

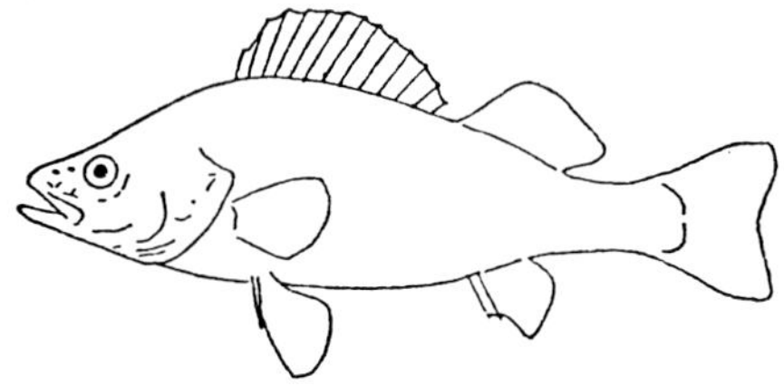

Acanthopterygii, Percomorpha, Perciformes, Percoidei

Yellowperch

Percids are elongate, spiny-rayed predaceous teleosts with two distinct dorsal fins and ctenoid scales. They inhabit lakes and rivers of North America, Europe, and Asia. The adult yellowperch, *P. flavescens,* usually moves about in schools and its diet includes animal plankton, aquatic insects, and fishes.
The yellowperch (*P. flavescens*) has a typical teleostean retina, similar to that found in diurnal fishes. The epithelial melanin pigment is abundant and distributed in processes (Plate 82.1). Rods and cones are present in large numbers. The cones are bulky with short outer segments, large ellipsoids, and short myoids (Plate 82.1). In tangential sections they are arranged in regular square mosaics in the ventro-temporal area retinae (Plate 82.3), or in modified quadrangular mosaics in other retinal regions (Plate 82.4). Triple cones are present in the fundus-retina (Plate 82.2, *arrow*). There are at least two layers of horizontal cells and the bipolar and amacrine cells are more or less numerous depending on retinal regions.

References: Eigenmann and Shafer (1900); Wunder (1925); Arey (1928); Lyall (1957); Engström (1963b); Ahlbert (1969, 1973); Wagner (1972)

Walleye

The walleye (*Stizostedion vitreum*) is a North American freshwater fish developing negative phototropism with age. As soon as it reaches its second year of life, the walleye is found in sheltered areas of clear waters during daytime, or in relatively turbid waters. This is a crepuscular fish, leaving its shelter at dusk in search of food and other activities. The retina of the walleye appears to be well adapted to dim-light conditions. The epithelial cells have very long processes filled with a somewhat granular material, and they constitute a well-developed retinal tapetum lucidum (Plate 83.1, 2). Some melanin pigment is present at the sclerad base of these cells and also at the vitread tip of the processes (Plate 83.1). This reflecting system is well-developed only in the ventral region

---

Plate 82.1-4. Transverse (1) and tangential (2)–(4) sections of the retina of the yellowperch, *P. flavescens.* (1) and (2) ×560; (3) and (4) ×1,120. See also Plate 5 in Introduction ▷

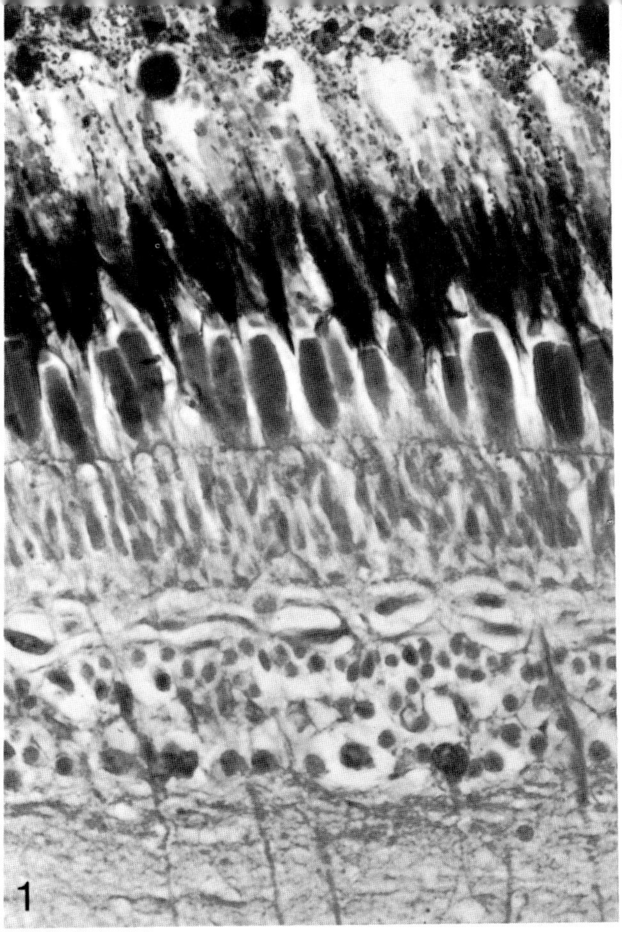
1

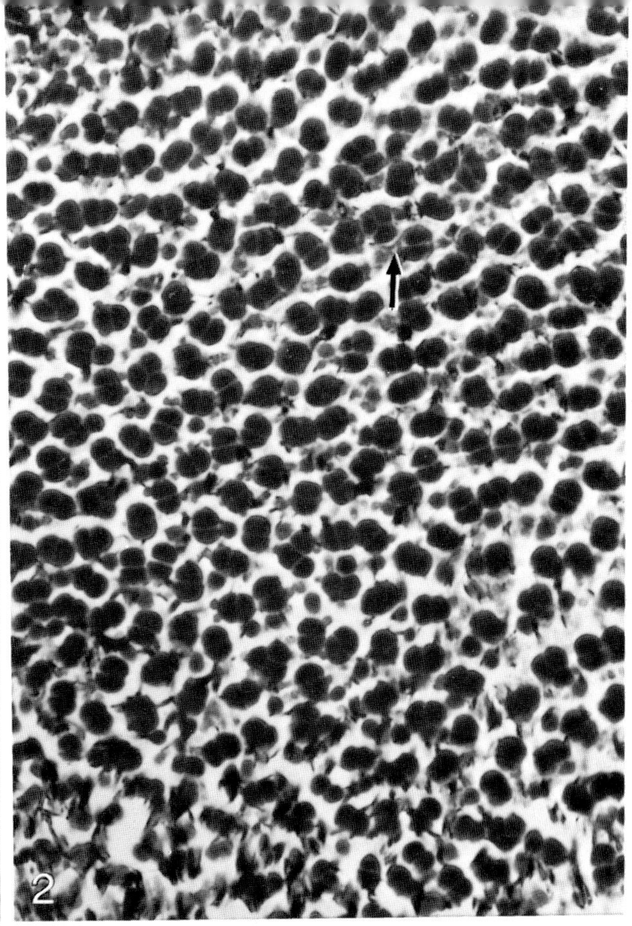
2

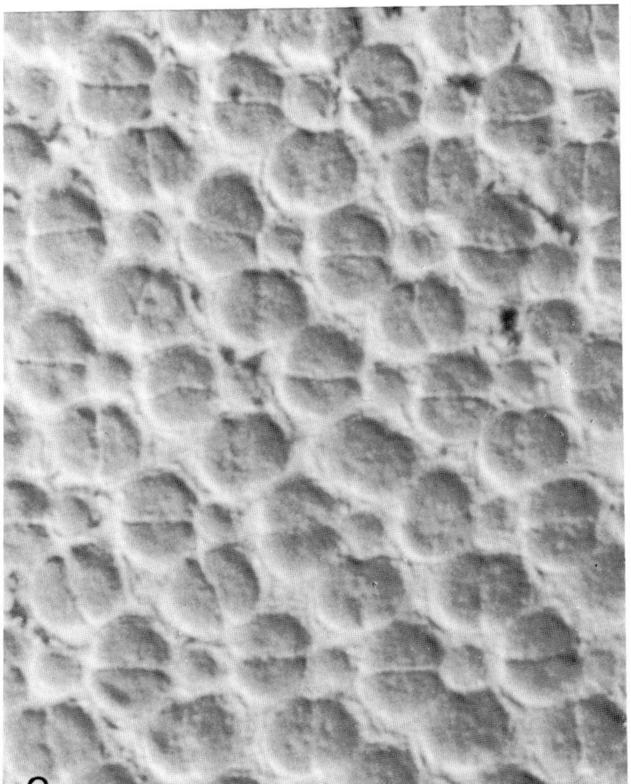
3

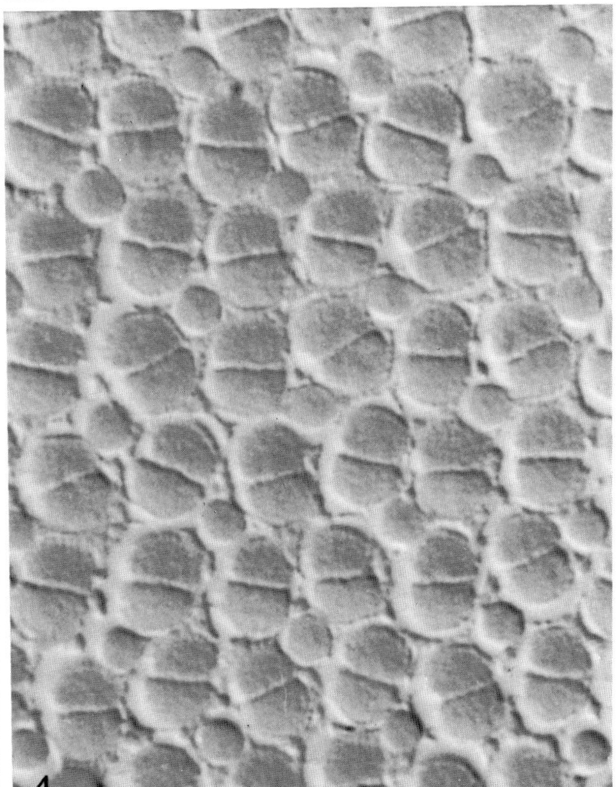
4

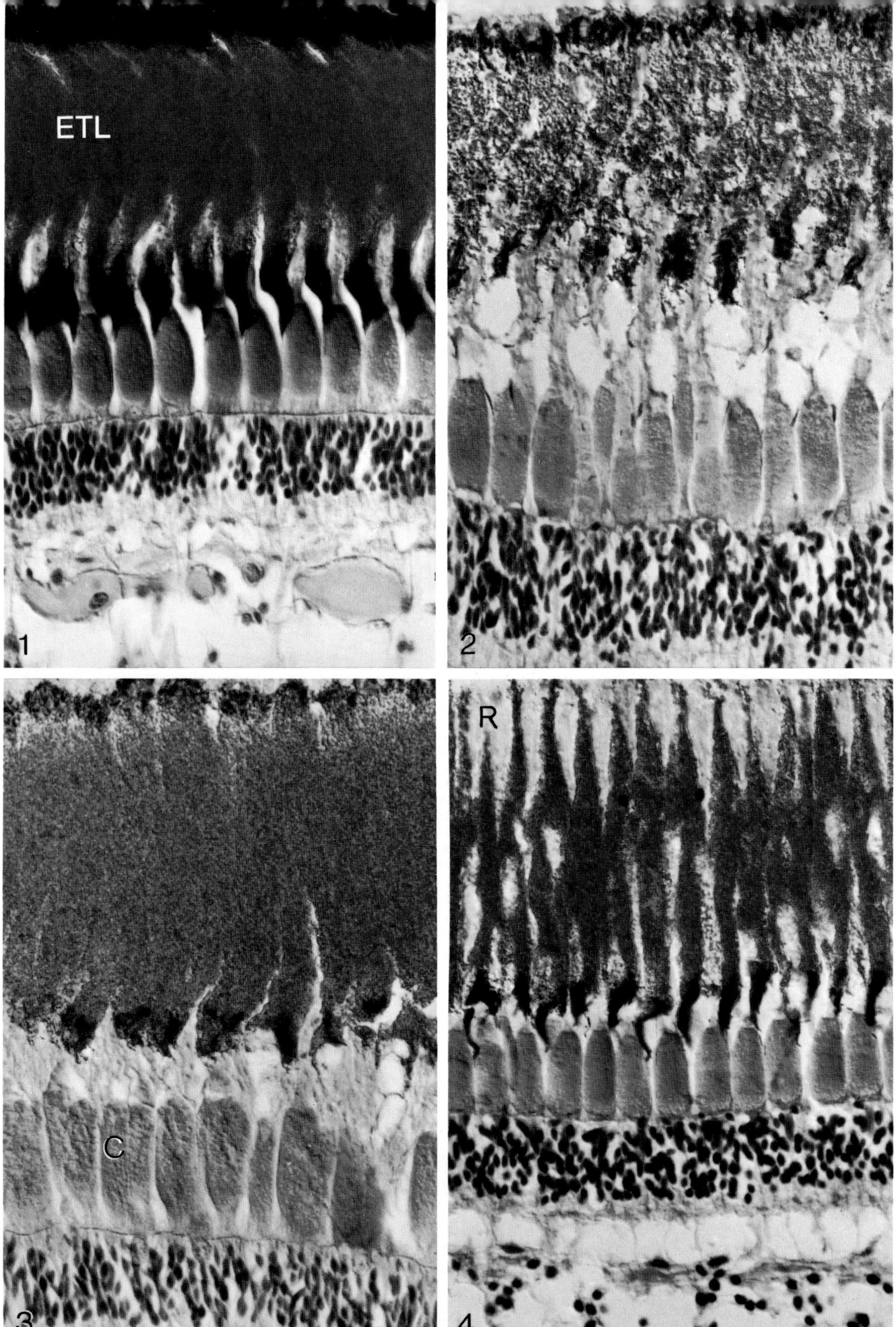
ETL
1
2
R
C
3
4

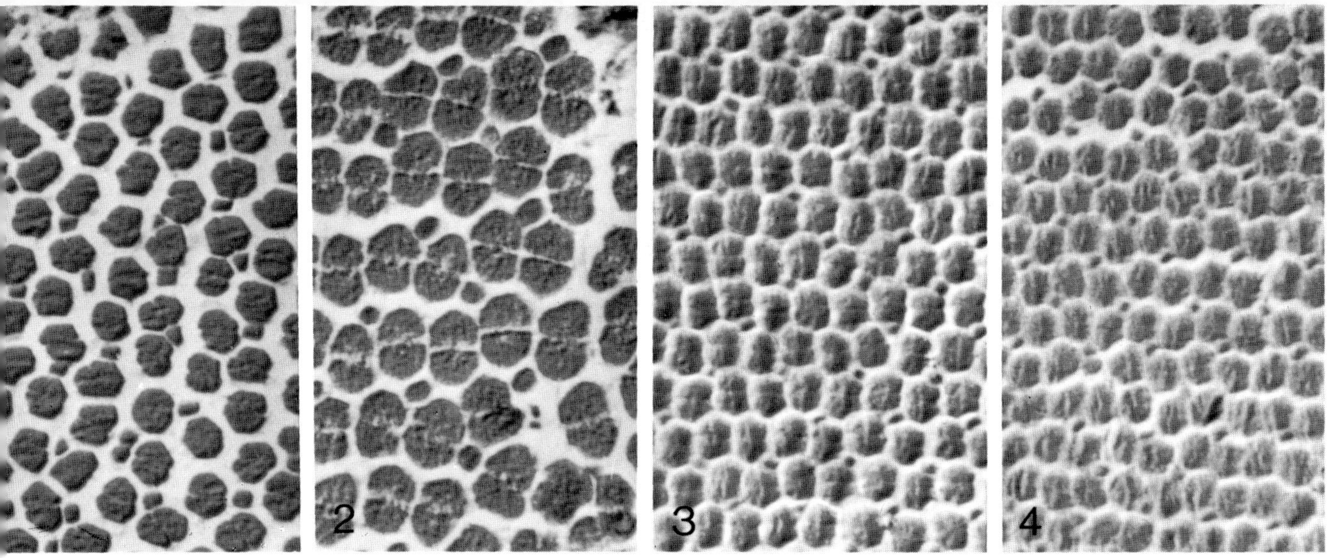

△ Plate 84.1-4. Tangential sections through the cone ellipsoids of the retina of the walleye (1) and (2) and sauger (3) and (4), ×1,120

of the retina (Plate 83.1), while in other regions (Plate 83.2, nasal) the tapetal granules are sparsely distributed and fewer. The rods are slender and very numerous compared to the large, stout cones. The single and double cones, whose segments are very short (Plate 83.1), are arranged in parallel rows in the tangential section (Plate 84.1, 2) and the area retinae is temporal (Plate 84.1). The horizontal cells are very large (Plate 83.1).

References: Ali and Wagner (1975a); Zyznar and Ali (1975)

Sauger

The sauger (*S. canadense*) like the walleye, develops negative phototropism with age, however it differs from the walleye in its habits. The sauger prefers extremely turbid waters and actively feeds throughout the day. Observations suggest that the sauger does not take advantage of physical shelter from light to the extent that the walleye does. Moreover when walleyes and saugers occur sympatrically an increase in turbidity may change the relative proportion of these two species in favour of the sauger. This is reflected in the differences in retinal structure between the two species, differences that favour the sauger in a dimmer (more turbid) environment. The sauger's ventral retina has a well-developed tapetum comparable to that of the walleye (Plate 83.3), but the other retinal regions, contrary to those of the walleye, have well organised tapetal structures (Plate 83.4, nasal). Note also that melanin pigment is slightly less abundant in the epithelial cells (Plate 83.3) compared to that of the walleye (Plate 83.1). There are two areae retinae in the sauger's retina: one is dorso-temporal (Plate 84.3), and the other is ventral (Plate 84.4).

References: Wunder (1930); Moore (1944); Baburina (1961b); Engström (1963b); Ali and Anctil (1968); Zyznar and Ali (1975); Ali and Wagner (1975a)

◁ Plate 83.1-4. Transverse sections of the dark (1) and (3) and light-adapted (2) and (4) retinas of the walleye, *S. vitreum,* and sauger, *S. canadense.* ×560. *C* cone, *ETL* epithelial tapetum lucidum, *R* rods

Theraponidae (*Shorefishes*)

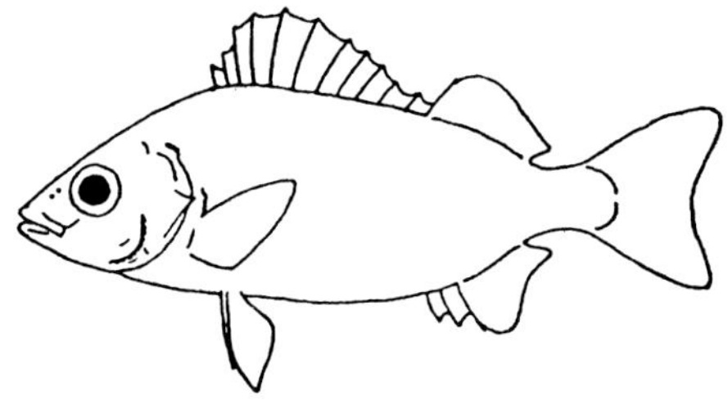

Acanthopterygii, Percomorpha, Perciformes, Percoidei

This is a small family of Indo-Pacific shorefishes of small to moderate sizes. A few species are limited to fresh water, while most are found in marine or brackish waters. They are closely related to the serranids.

The retina of *Therapon oxyrhinchus* possesses few cones compared to that of the serranids. Single and double cones are present. The area retinae is temporal.

Reference: TAMURA (1957)

Visual Pigments: A_1 max 497 (MUNZ and MCFARLAND, 1973)

Priacanthidae (*Bigeyes*)

Acanthopterygii, Percomorpha, Perciformes, Percoidei

These carnivorous shorefishes are characterised by very large eyes, small rough scales, and a bright red colour. They are nocturnal and are usually found in deep shore waters.

The retinal structure of the bigeye, *Priacanthus arenatus,* resembles that of the cardinalfishes. Sclerad to the choriocapillaris and the flat, poorly developed retinal epithelium, there is a well-developed, unpigmented reflecting tapetum lucidum (Plate 85.1). The rods are numerous and densely packed, whereas the cones, which are mostly of the double type, are bulky, few, and possess a very short outer segment (Plate 85.1, 2). There are two or three layers of moderately large horizontal cells (Plate 85.2) and few, sparsely distributed bipolar and amacrine cells. The area retinae is ventral.

References: TAMURA (1957); NICOL (1963); NICOL and ZYZNAR (1973); ALI and WAGNER (1975a)

Visual Pigments: A_1 max 496 (MUNZ and MCFARLAND, 1973)

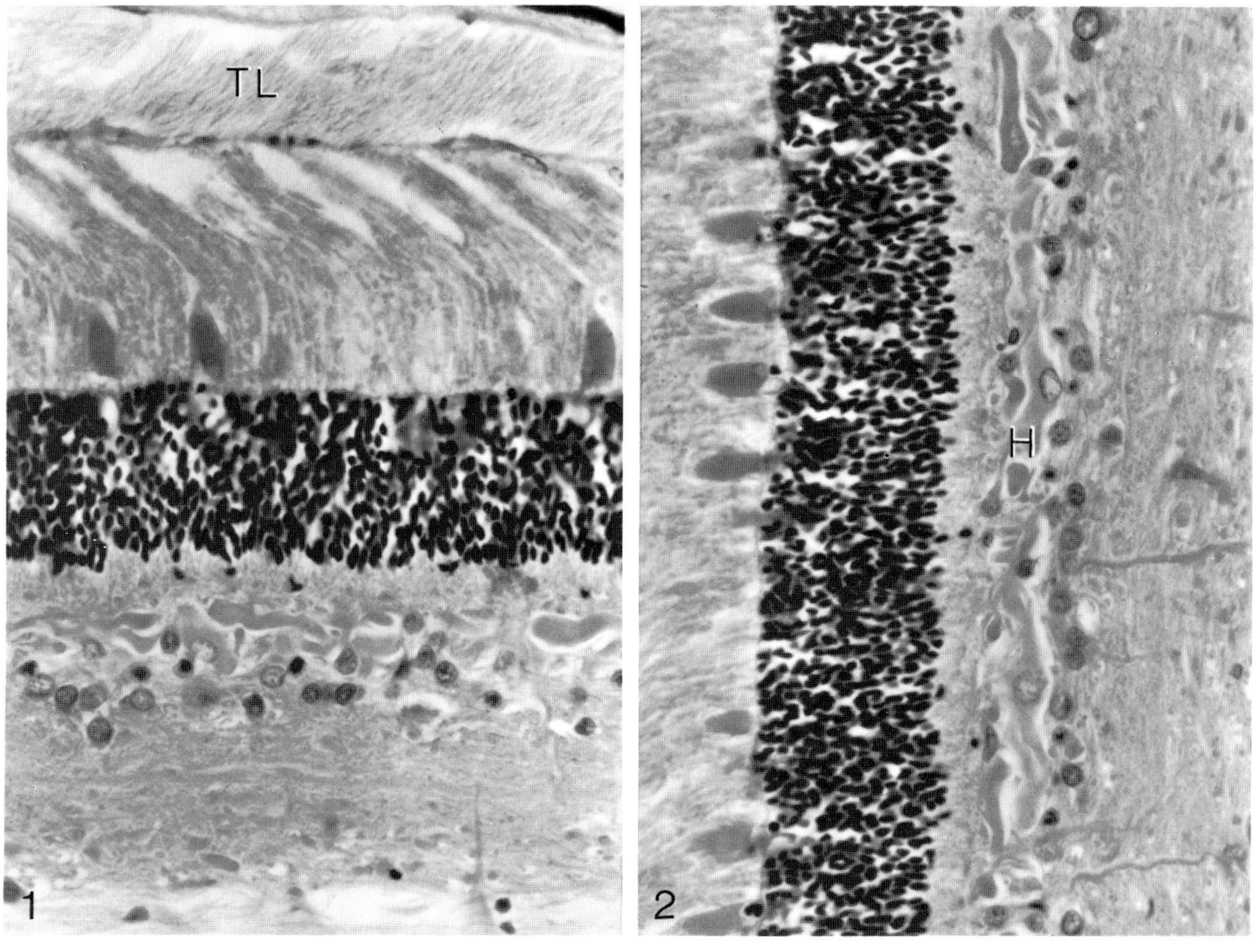

Plate 85.1, 2. Transverse sections of the retina of the bigeye, *P. arenatus*. × 560. *H* horizontal cell layers, *TL* tapetum lucidum

Apogonidae (*Cardinalfishes*)

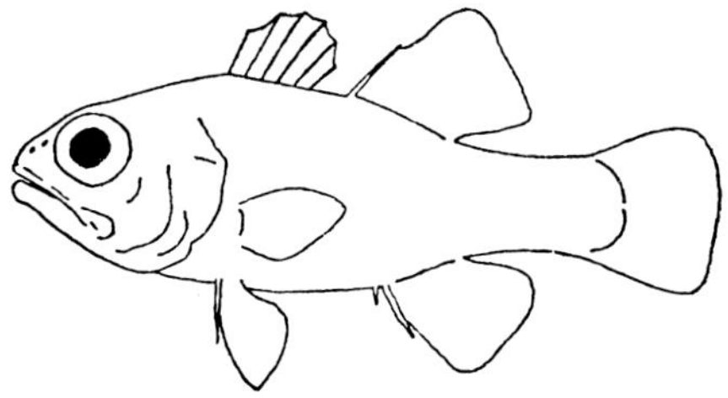

Acanthopterygii, Percomorpha, Perciformes, Percoidei

Cardinalfishes are small tropical shorefishes with bright colours, large eyes, and two separate dorsal fins. They frequent coral reefs and lagoons, and some types inhabit deeper shore waters. Cardinalfishes are notoriously nocturnal in their habits.

The retina of the flamefish is poorly developed. There are no processes in the heavily pigmented epithelium, the rods are short and numerous, the cones are bulky and few, the horizontal cells are indistinct; the bipolar, amacrine and ganglion cells are sparsely distributed (Plate 86.1). Both single and double cones are present (Plate 86.2), the latter being of the equal and unequal types.

References: TAMURA (1957, *Apogon lineatus*); KATO (1962)

Visual Pigments: A_1 max 482–495; $A_2 < 10\%$ (MUNZ and MCFARLAND, 1973)

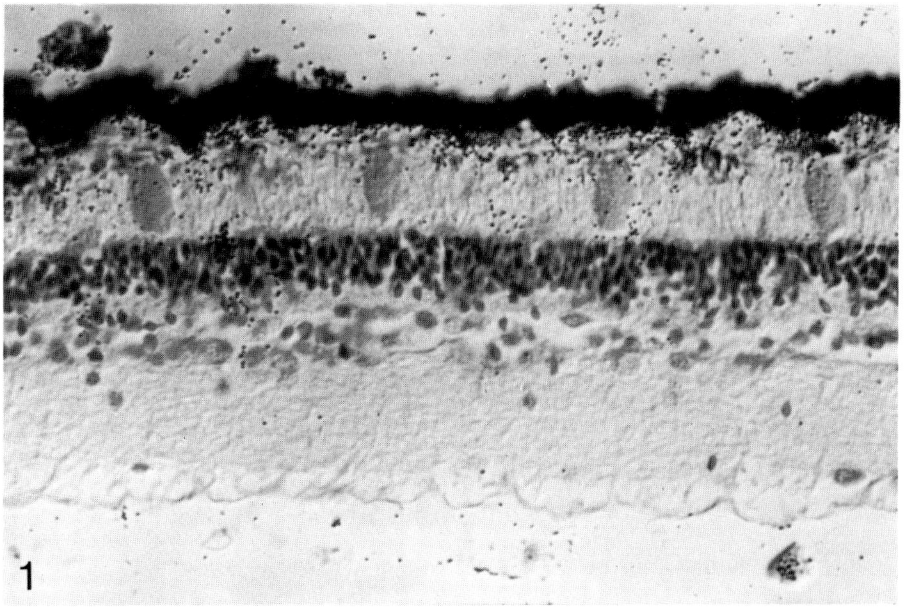

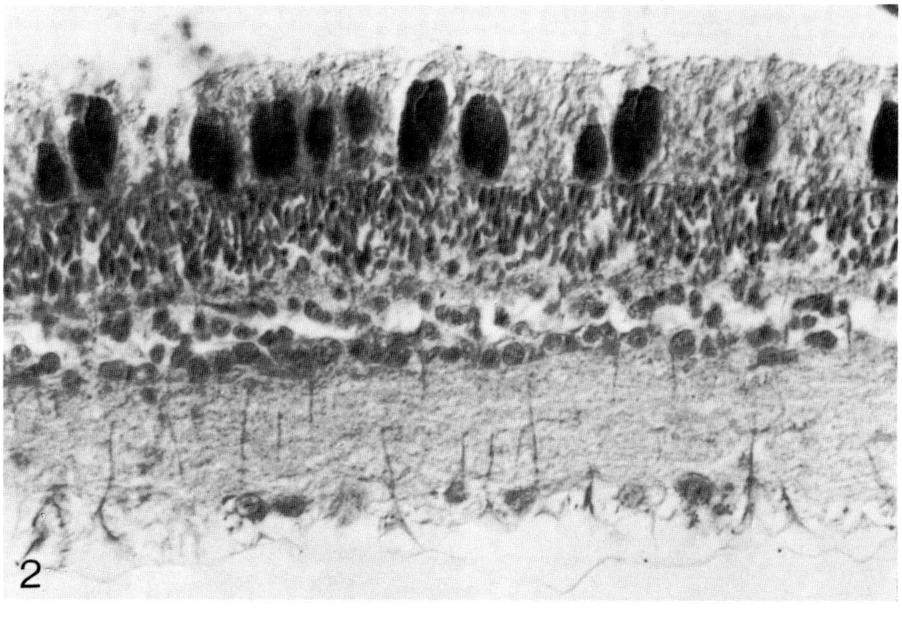

Plate 86.1,2. Transverse sections of the retina of the flamefish, *A. maculatus*. ×560

Pomatomidae (*Bluefishes*)

Acanthopterygii, Percomorpha, Perciformes, Percoidei

The bluefish is a fast-moving, schooling fish found in coastal marine waters. This fish is particularly voracious.

The retina of *Pomatomus saltatrix* possesses an abundant epithelial pigment, numerous rods and cones, all undergoing extensive retinomotor movements. Single as well as double cones are present. The horizontal cells are large and arranged in at least three layers.

Reference: OLLA and MARCHIONI (1968)

Coryphaenidae (*Dolphins*)

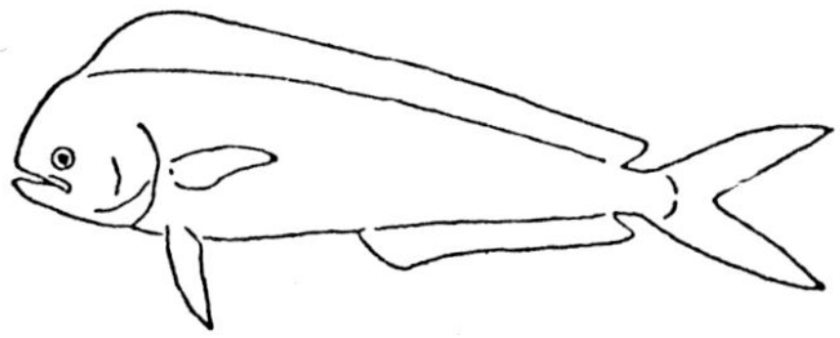

Acanthopterygii, Percomorpha, Perciformes, Percoidei

Dolphins are very fast-moving epipelagic percoid fishes, characterised by a long dorsal fin, and a forked tail like that of the carangids; the male has a large square head. They may be found singly or in schools, and they feed on a variety of fishes, including flyingfishes, and invertebrates.

There are about as many single cones as double cones in the retina of *Coryphaena hippurus,* thereby indicating that two types of single cones are present. The cone density is almost uniformly distributed throughout the retinal regions, with a peak in the temporal region (area retinae).

Reference: TAMURA and WISBY (1963)

Visual Pigments: $A_1:A_2$ 2:1 (ALI and HEUMANN, 1970); A_1 max 500.6 (MUNZ and MCFARLAND, 1973)

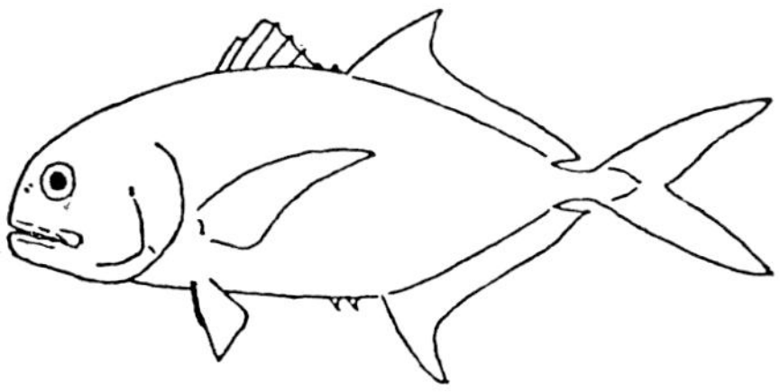

Carangidae (*Jacks and Pompanos*)

Acanthopterygii, Percomorpha, Perciformes, Percoidei

Carangids (jacks, pompanos) are fast-moving, coastal, pelagic teleosts. Some possess along each side of the caudal peduncle a sharp ridge made of a series of bony scutes or plates; many are torpedo-shaped.

Plate 87.1 illustrates a dark-adapted retina of the leatherjacket (*Oligoplites saurus*). The epithelial pigment is packed sclerally in the cell bodies, the long, slender, and densely packed rods occupy almost all the space of the visual cell layer, and the cones are moved sclerally. Single and equal double cones are present. There are three layers of moderately large horizontal cells Plate 87.1, the two vitread-most layers being shown in a tangential section in Plate 87.2. The sections of the retina of a tide pool fish from Brazil (*Trachinotus carolinus*) are shown in Plate 88.1, 2, 3. Note the four tiers of large horizontal cells (Plate 88.1, 3).

References: Verrier (1928a); Tamura (1957); O'Connell (1963); Tamura and Wisby (1963); Ali et al. (1973); Ali and Wagner (1975a)

Visual Pigments: A_1 max 495–500; $A_2 < 10\%$ (Ali and Heumann, 1970, 1972; Munz and McFarland, 1973; Ali and Wagner, 1975b)

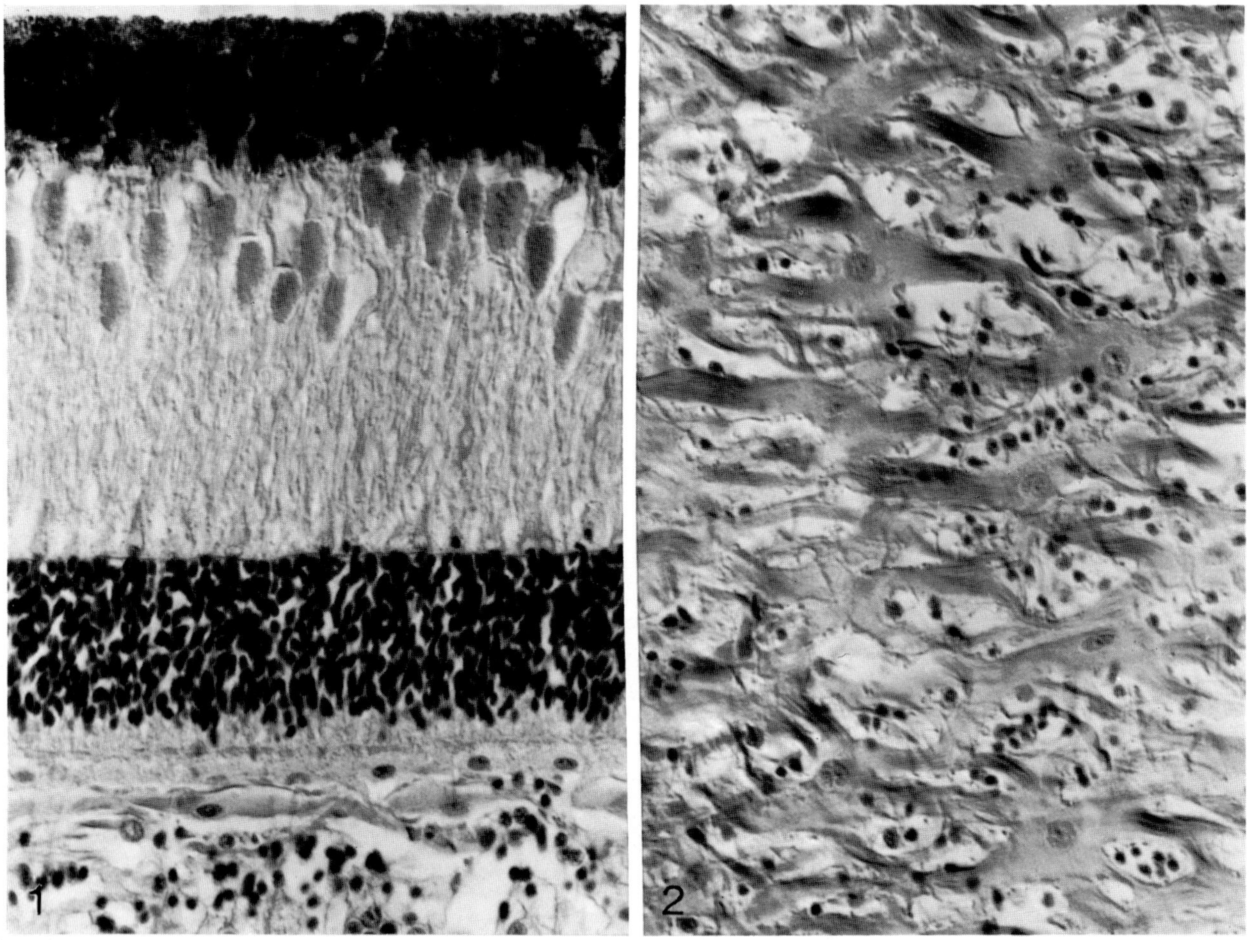

Plate 87.1, 2. Transverse (1) and tangential (2) sections of the retina of the leatherjacket *O. saurus.* × 560

Plate 88.1-3. Retina of *T. carolinus.* (1), transverse section of dorsal retina (× 700). (2), tangential section (× 1,380) of the dorsal retina. (3), transverse section of the ventral retina (× 550). Note the large horizontal cells arranged in four tiers in Plate 87.1 and Plate 88.1 ▷

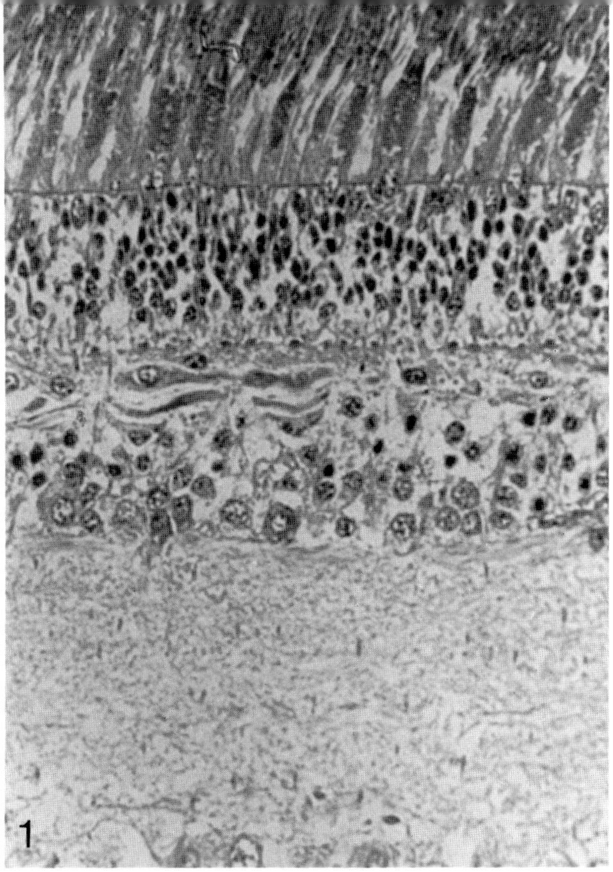

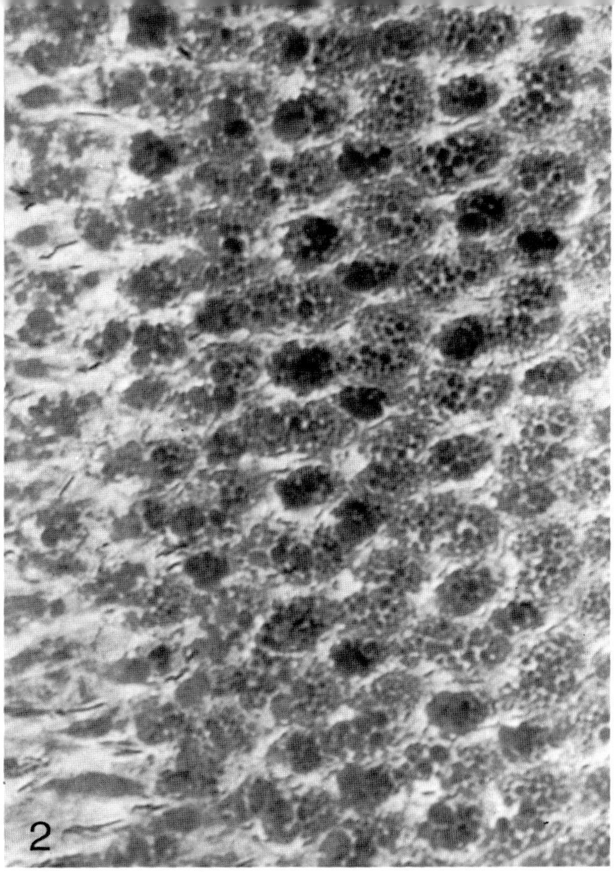

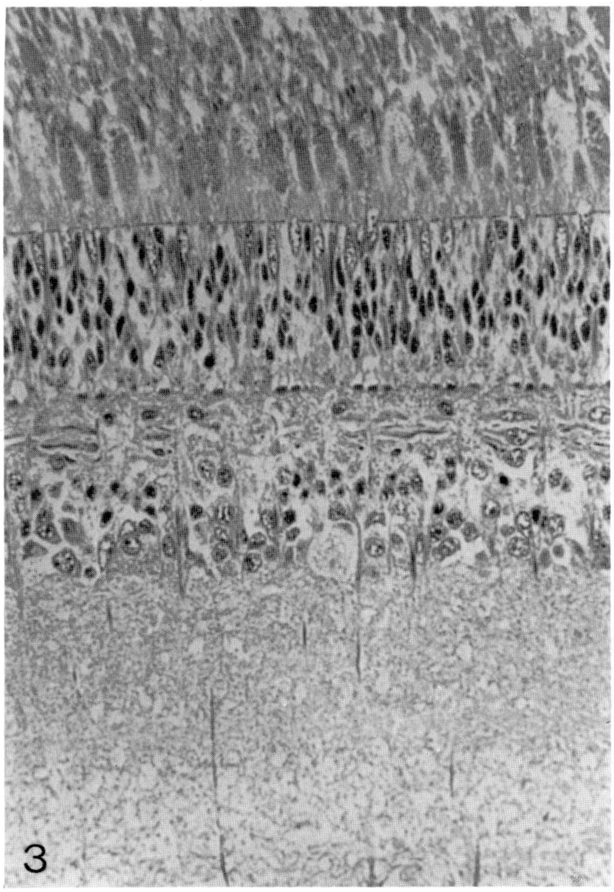

Plate 88.1-3. (Legend see opposite page)

Lutjanidae (*Snappers*)

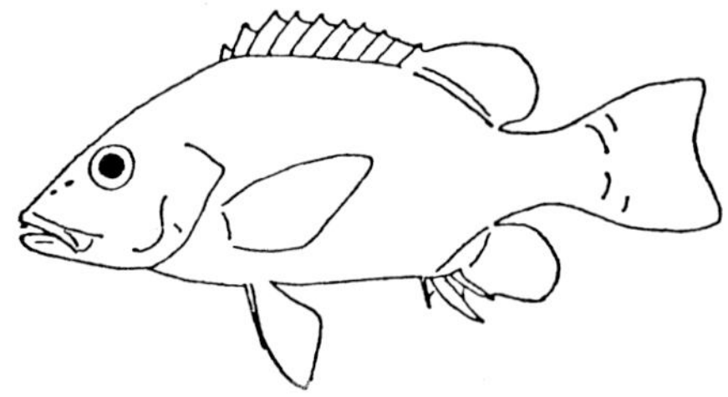

Acanthopterygii, Percomorpha, Perciformes, Percoidei

Snappers are marine, shallow-water, inshore fishes with distinctly enlarged jaw teeth and a snout with a flattened tip. They are carnivorous, occurring in large schools, and are essentially nocturnal. The lane snapper, *Lutjarus synagris,* is known to inhabit deep shore water and to feed on fishes and large invertebrates.

The retina is characterised by the retracted, dark-adapted epithelial pigment, by extremely numerous, long and slender rods (Plate 89.1, 2), and by single and equal double cones (Plate 89.2, 3), the centre of abundance being in the temporal region. There are three layers of small horizontal cells (Plate 89.1, 2). As seen in Plate 89(1), the bipolar, amacrines, and ganglion cells are well represented.

References: SELVIN DE TESTA (1966); O'DALY (1967); PARTHE (1967, 1972); ALI et al. (1973)

Visual Pigments: A_1 max 494–498 (ALI and HEUMANN, 1972; MUNZ and MCFARLAND, 1973)

Leiognathidae (*Slipmouths*)

Acanthopterygii, Percomorpha, Perciformes, Percoidei

These are small Indo-Pacific fishes, chiefly marine in habitat, characterised by deep, laterally compressed bodies, by extremely protrusible mouths forming tubes when extended, and by the presence of luminous organs around the base of the oesophagus. Slipmouths live inshore and are nocturnal.

Single cones and double cones are present in the retina of *Leiognathus argenteus*. The area retinae is found in the dorso-temporal and temporal regions.

Reference: TAMURA (1957)

Plate 89.1-3. Transverse (1) and (2) and tangential (3) sections of the retina of the lane snapper, *L. synagris*. (1) × 225; (2) and (3) × 560. *H* horizontal cell layer ▷

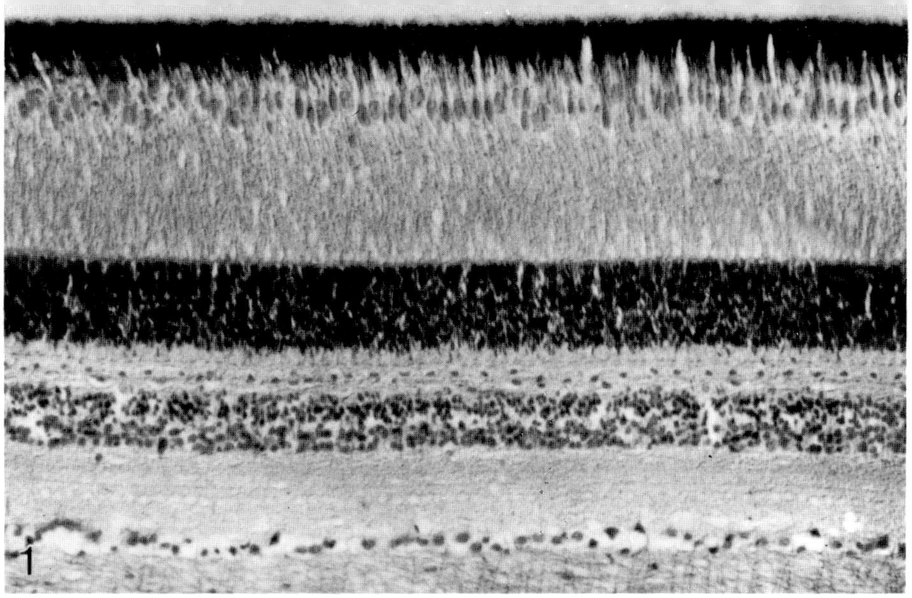
1

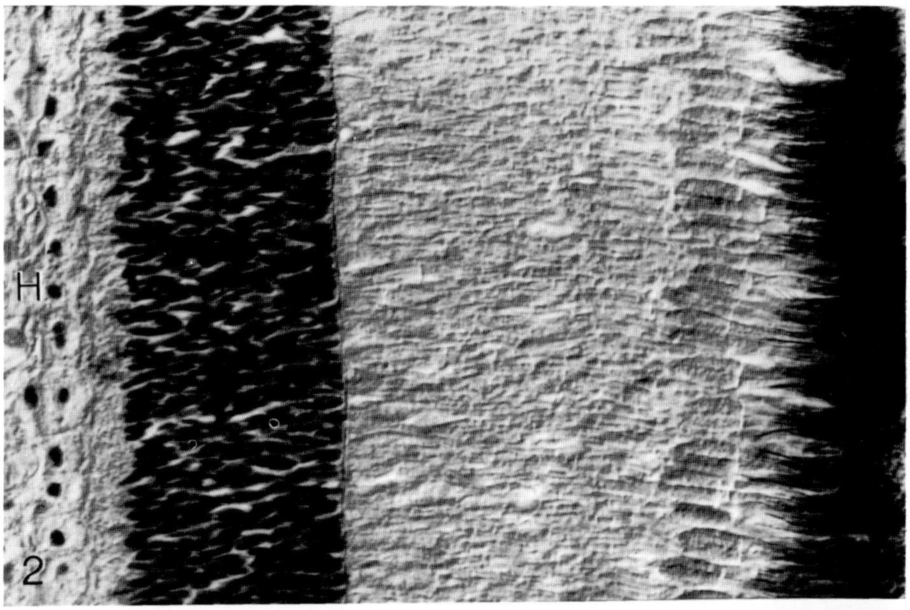
H
2

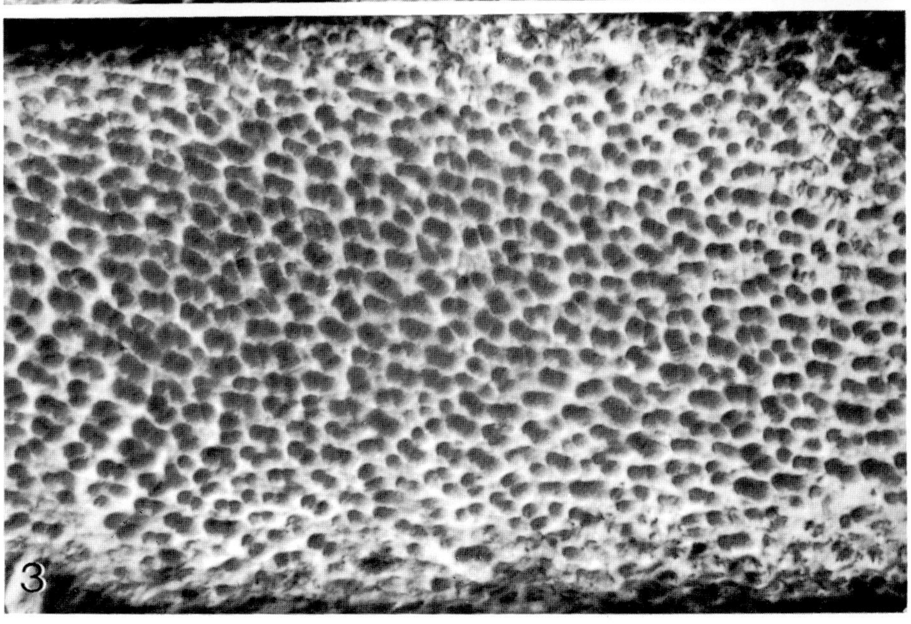
3

Toxotidae (*Archerfishes*)

Acanthopterygii, Percomorpha, Perciformes, Percoidei

The spitting archerfish is a freshwater fish, which adapts readily to brackish or salt water. It projects its water pellets by compressing the gill covers, which results in the fish's tongue pressing against a longitudinal groove on the palate. The archerfish shoots by sight with spectacular accuracy using as targets insects on which it feeds.

There is a sharply defined ventro-temporal area retinae (upper-fore visual axis) in the retina of *Toxotes jaculatrix*. Single and equal double cones are present.

Reference: LÜLING (1958)

Gerridae (*Mojarras*)

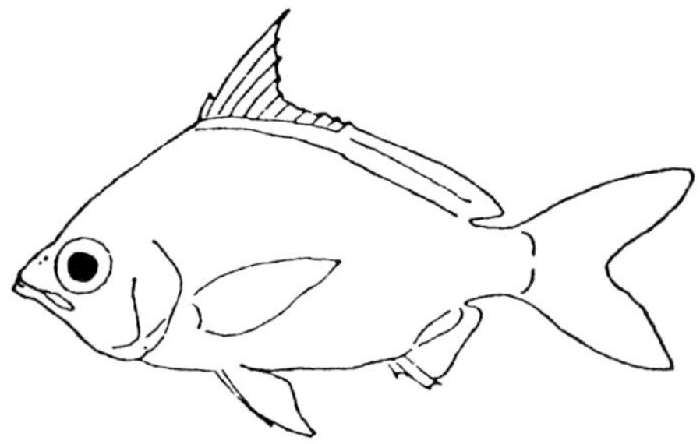

Acanthopterygii, Percomorpha, Perciformes, Percoidei

Mojarras are closely related to the leiognathids and are small, silvery shorefishes with protrusible mouths.

The retinal neurones of this species have been extensively studied by silver-impregnation techniques. Plate 90.1 shows a rod horizontal cell *(centre)* and large rod bipolars *(left)*. Plate 90.2 shows small *(centre right)* and large *(centre left)* cone bipolar cells. Plate 90.3 shows a small rod bipolar cell *(right arrow)*, a large cone bipolar cell *(centre arrow)*, and large rod bipolar cells *(left arrow)*. Plate 90.4 shows an oligopolar cell *(arrow)*. Higher magnifications are shown of rod and cone foot-pieces (Plate 91.1), rod pedicles contacting a rod horizontal cell (Plate 91.2), a rod pedicle making synapse with a small rod bipolar (Plate 92.1), and a large cone bipolar cell with dendritic ramifications on one side only (Plate 92.2, *arrow*).

References: SELVIN DE TESTA (1966); O'DALY (1967); PARTHE (1967, 1972); VANEGAS and EBBERSON (1973); LAUFER and VANEGAS (1974)

Visual Pigments: $A_1 : A_2$ 2:1 (ALI and HEUMANN, 1970, 1972)

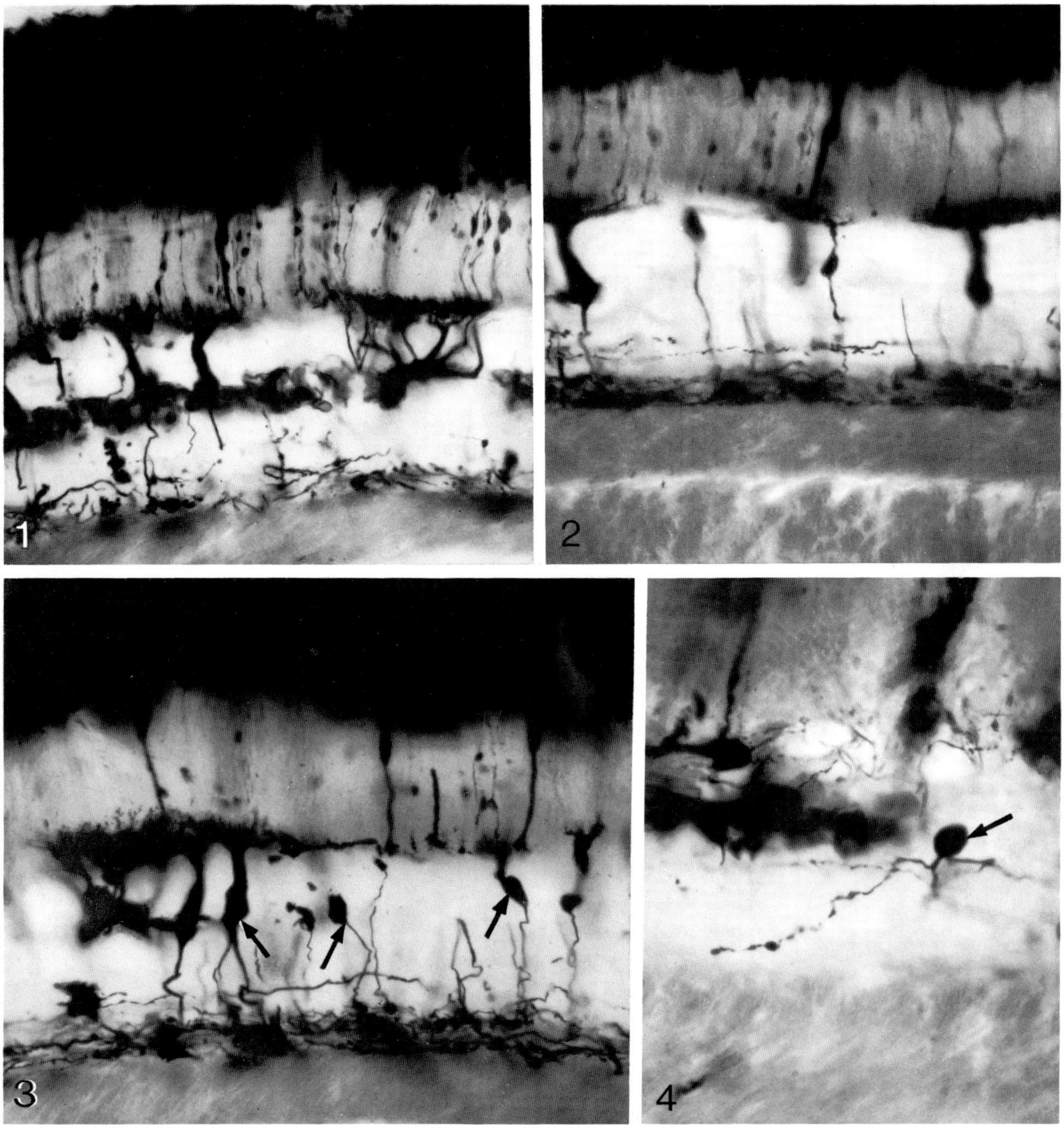

Plate 90.1-4. Plate 91.1, 2. Plate 92.1, 2. Thick, transverse sections of the Golgi impregnated retina of the mojarra, *Eugerres plumieri* (see text). Enlargements not available. *C* cone foot-piece, *R* rod foot-piece

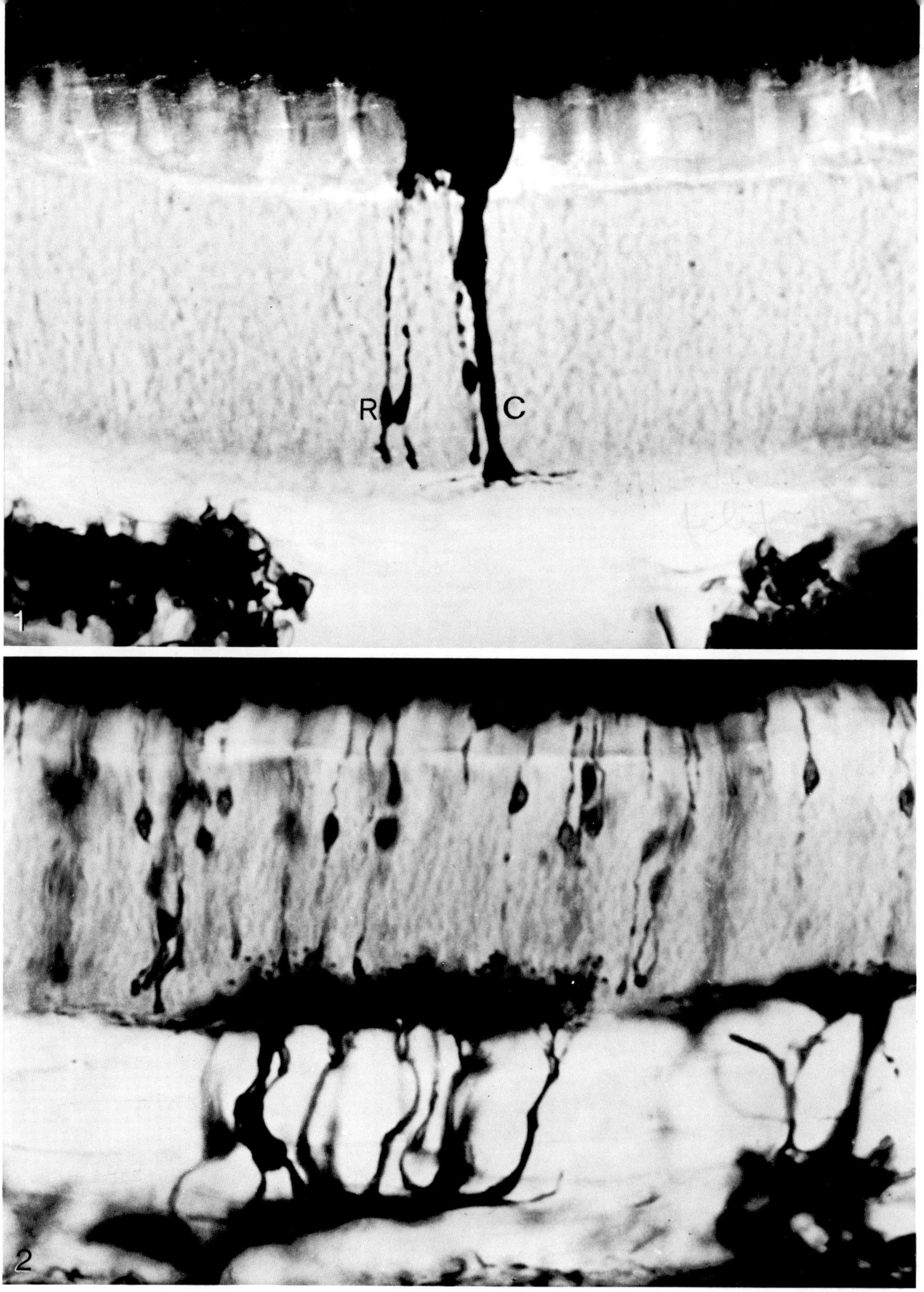
R
C
1
2

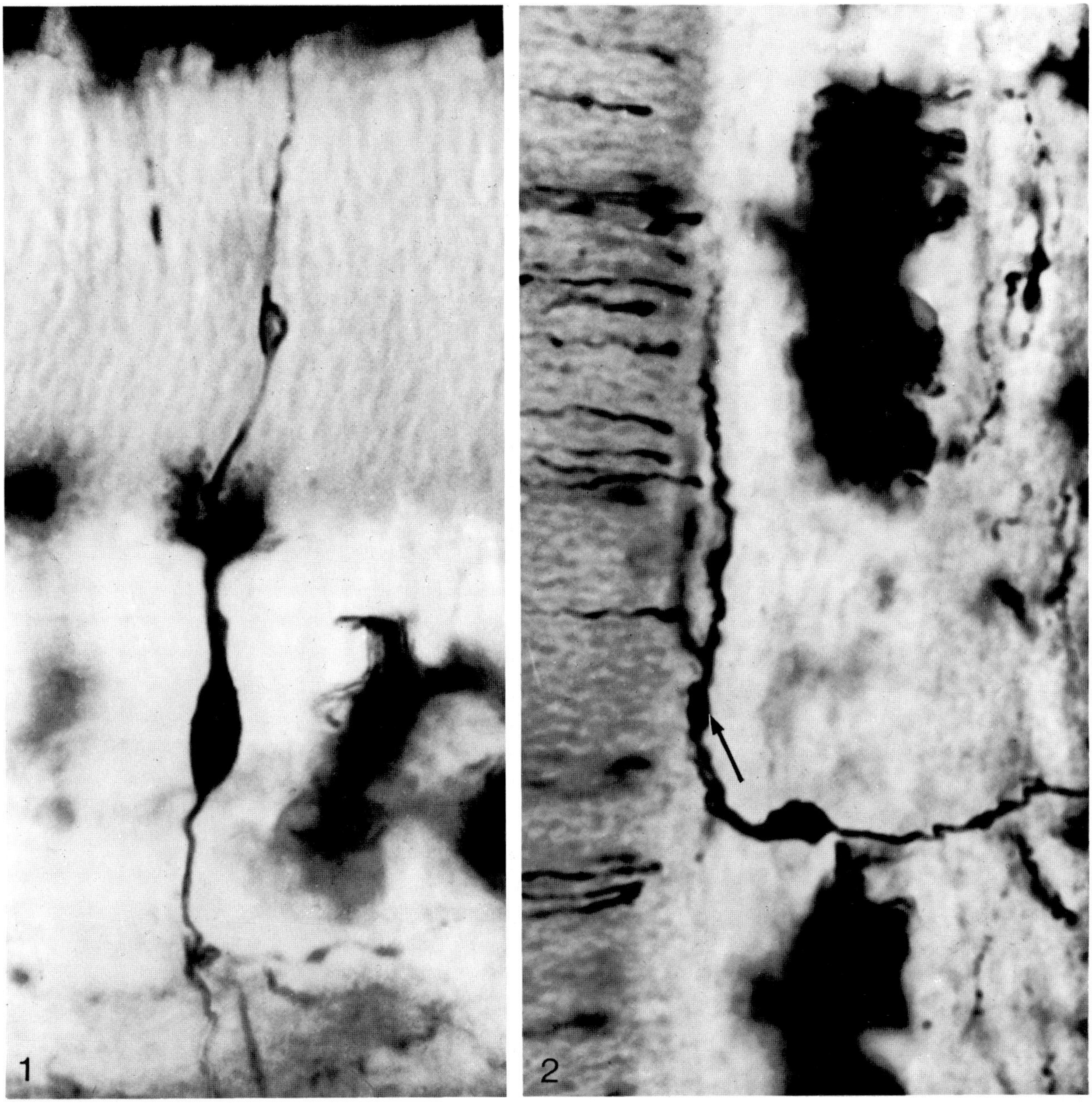

Plate 92.1,2. (Legend see page 187)

◁ Plate 91.1,2. (Legend see page 187)

Pomadadasyidae (*Grunts*)

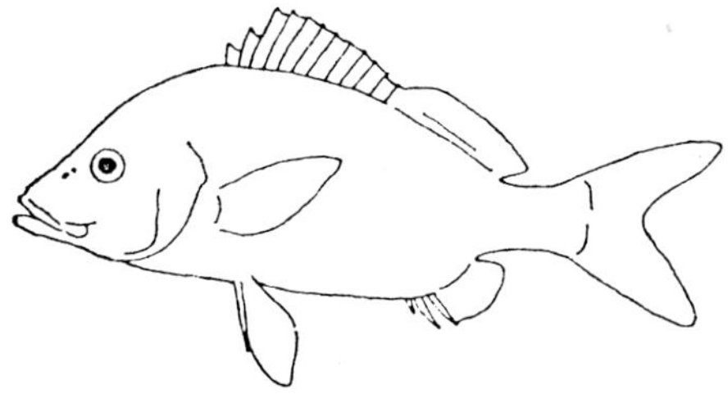

Acanthopterygii, Percomorpha, Perciformes, Percoidei

Grunts are shorefishes and their external appearance is much like the snappers (Lutjanidae). They produce sounds by grinding their pharyngeal teeth together, and the adjacent air bladder acts as an amplifier. They are nocturnal in habit, passing the day grouped in relatively inactive schools over rocky or sandy bottoms. At night, the schools disperse and each individual fish actively forages on the sand.

The retina of the grunt (*Pomadasys crocro*) is similar in structure to that of the snapper (Lutjanidae), mojarras (Gerridae), and snooks (Centropomidae). The rods are abundant, the single and double cones are also numerous, and three layers of horizontal cells are present.

References: SELVIN DE TESTA (1966); ZYZNAR and NICOL (1973)

Visual Pigments: A_1 max 500 (ALI and HEUMANN, 1970, 1972; BEATTY, 1973)

Sparidae (*Porgies and Sea Breams*)

Acanthopterygii, Percomorpha, Perciformes, Percoidei

Porgies and sea breams are deep-bodied percoid fishes, usually equipped with powerful incisor teeth in the jaw as well as strong grinding teeth. These marine fishes, which resemble the grunts in general appearance, occur in loose aggregations or in large schools along the sandy or rocky shores.

The retina of the sheepshead (sea bream) (*Archosargus rhomboidalis*) has the characteristic features of the retina of noctural teleostean shore fish: the epithelial pigment is distributed in short epithelial processes, the rods are extremely long and numerous, and the horizontal cells, which are set in four layers, are fairly large (Plate 93.1). Note in Plate 93.1, the fully dark-adapted state of the pigment and cones. The single and equal double cones are very small and arranged in square patterns (Plate 93.2). The area retinae is temporal.

References: TAMURA (1957); ALI et al. (1973); ALI and WAGNER (1975a)

Visual Pigments: A_1 max 493–518 (ALI and HEUMANN, 1972; BEATTY, 1973; ALI and WAGNER, 1975b)

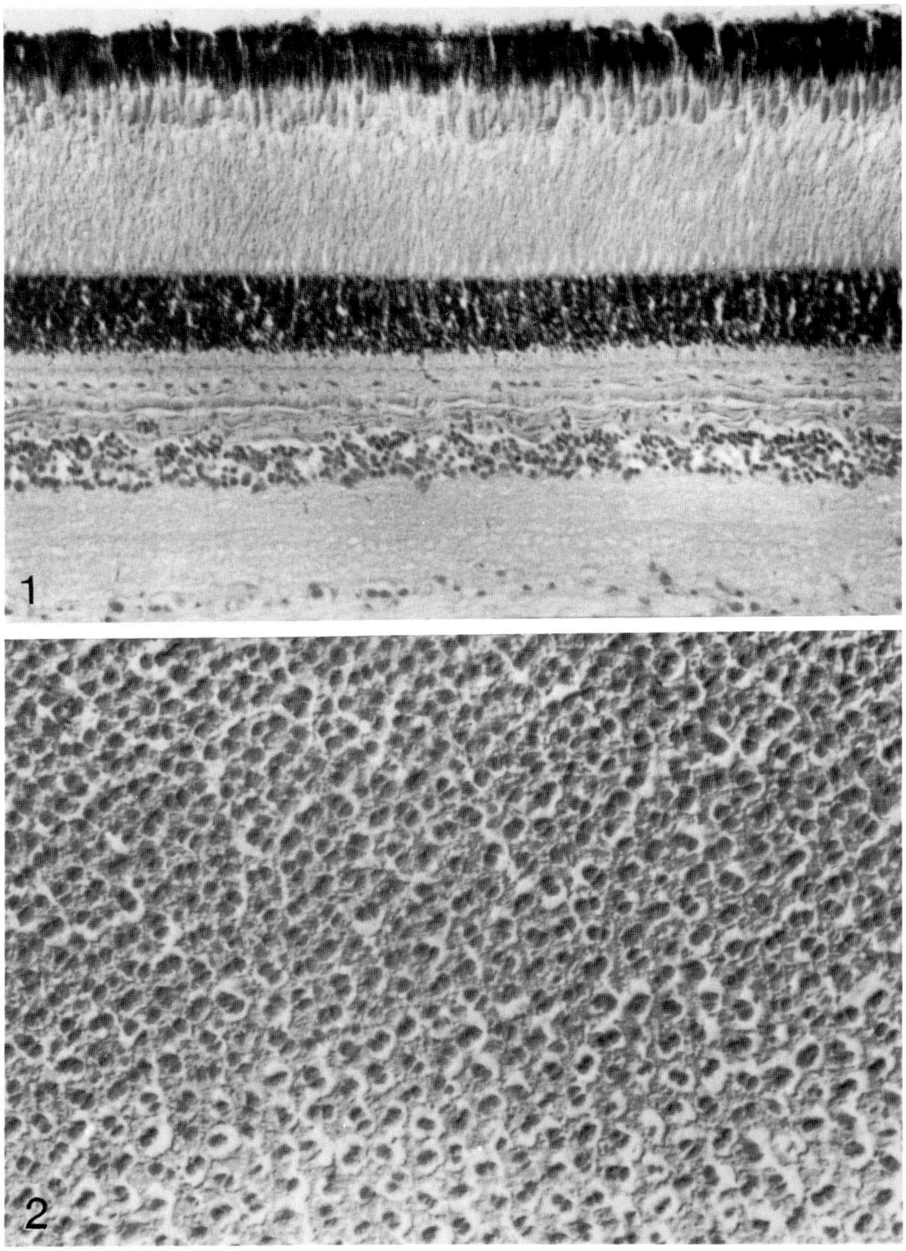

Plate 93.1, 2. Transverse (1) $\times 225$ and tangential (2) $\times 560$ sections of the retina of the sea bream, *Archosargus rhomboidalis*

Sciaenidae (*Croakers*)

Acanthopterygii, Percomorpha, Perciformes, Percoidei

Croakers are shallow-water fishes and mainly marine, being usually carnivorous and related to the basses. They are gregarious and associated with the benthos; they are noted for their ability to produce sound. They have nocturnal habits.

The retina is degenerate in the Red drum (blind croaker) *Sciaenops ocellata*. The retina of the seatrouts *Cynoscion arenarius* and *C. nebulosus* is characterised by the presence of a reflecting tapetal system in the pigment epithelium, made of lipid spherules. The visual cell layer of the seatrout retina is composed of single and double cones; and of extremely long and slender rods, which are all capable of retinomotor changes.

References: WARD and GUNTER (1962); ARNOTT et al. (1970, 1972); ALI and WAGNER (1975a)

Visual Pigments: A_1 max 500–504 (BEATTY, 1973; ALI and WAGNER, 1975b)

Mullidae (*Goatfishes*)

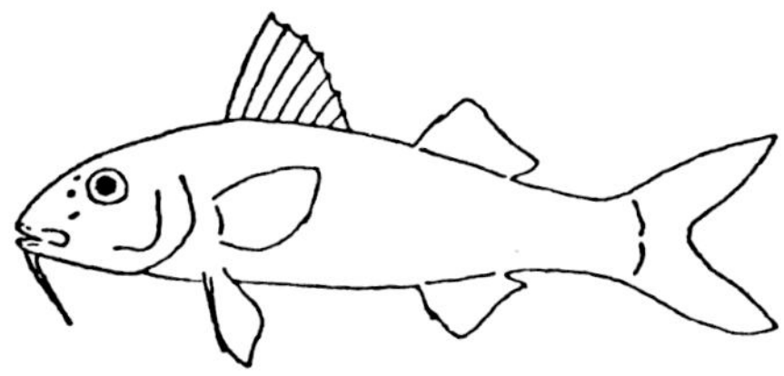

Acanthopterygii, Percomorpha, Perciformes, Percoidei

Goatfishes are benthic, living in shallow waters of the littoral zone. They are elongate and have two long tactile barbels under their chins which they drag over the bottom. They are carnivorous and feed principally on invertebrates.

Double cones have been reported in the retina of *Mullus barbatus*.

Reference: VERRIER (1928a)

Visual Pigments: A_1 max 484–495.6; $A_2 < 10\%$ (ALI and HEUMANN, 1972; MUNZ and MCFARLAND, 1973)

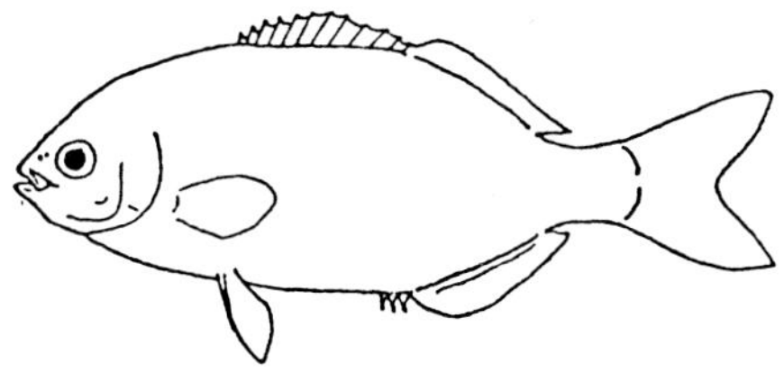

Kyphosidae (*Rudderfishes and Nibblers*)

Acanthopterygii, Percomorpha, Perciformes, Percoidei

Kyphosids are oval-shaped, schooling fishes with small mouths and fine teeth. They are abundant in shallow littoral waters around rocky areas, and in tide pools.

In *Microcanthus strigatus* and *Girella punctata,* the area retinae is located in the temporal or ventro-temporal regions of the retina. Single and double cones are arranged in regular square mosaics.

References: YAMANOUCHI (1956); TAMURA (1957)

Visual Pigments: A_1 max 493.9–497.8; $A_2 < 10\%$ (ALI and HEUMANN, 1972; MUNZ and MCFARLAND, 1973)

Ephippidae (*Spadefishes*)

Acanthopterygii, Percomorpha, Perciformes, Percoidei

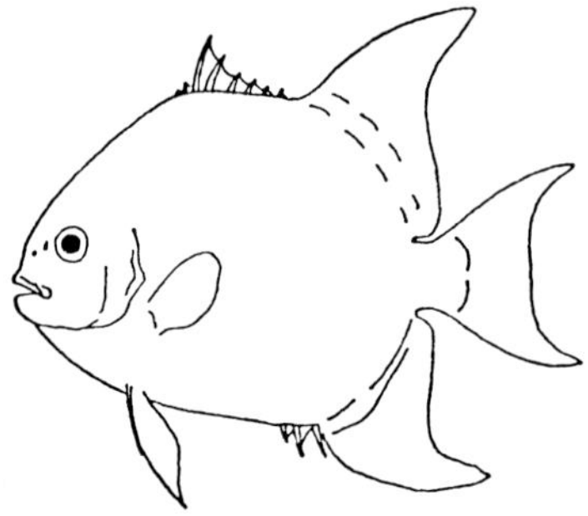

Spadefishes are deep-bodied shorefishes, which are somewhat similar in general appearance to butterflyfishes, but they differ from the latter in having two dorsal fins, one spiny and the other soft-rayed, and by the fact that most spadefishes are schooling fishes. The Atlantic spadefish, *Chaetodipterus faber,* hides during the day among the wrecks and pilings of the reefs or in deeper littoral waters; it is noted for its craving for shellfish.

The principal characteristics of their retinas are the lipid tapetal material in the epithelial cell processes (Plate 94), the long and densely packed rods, and the large cones that are capable of retinomotor changes (see the dark-adapted state in Plate 94). The single and equal double cones are arranged in square mosaics and are almost equally abundant in all the retinal regions. There are three layers of horizontal cells (Plate 94) and the internal nuclear layer is moderately developed.

References: Selvin de Testa (1966); Arnott et al. (1970, 1972); Ali et al. (1973); Ali and Wagner (1975a)

Visual Pigments: A_1 max 496.5–497; $A_2 < 10\%$ (Ali and Heumann, 1970; Munz and McFarland, 1973)

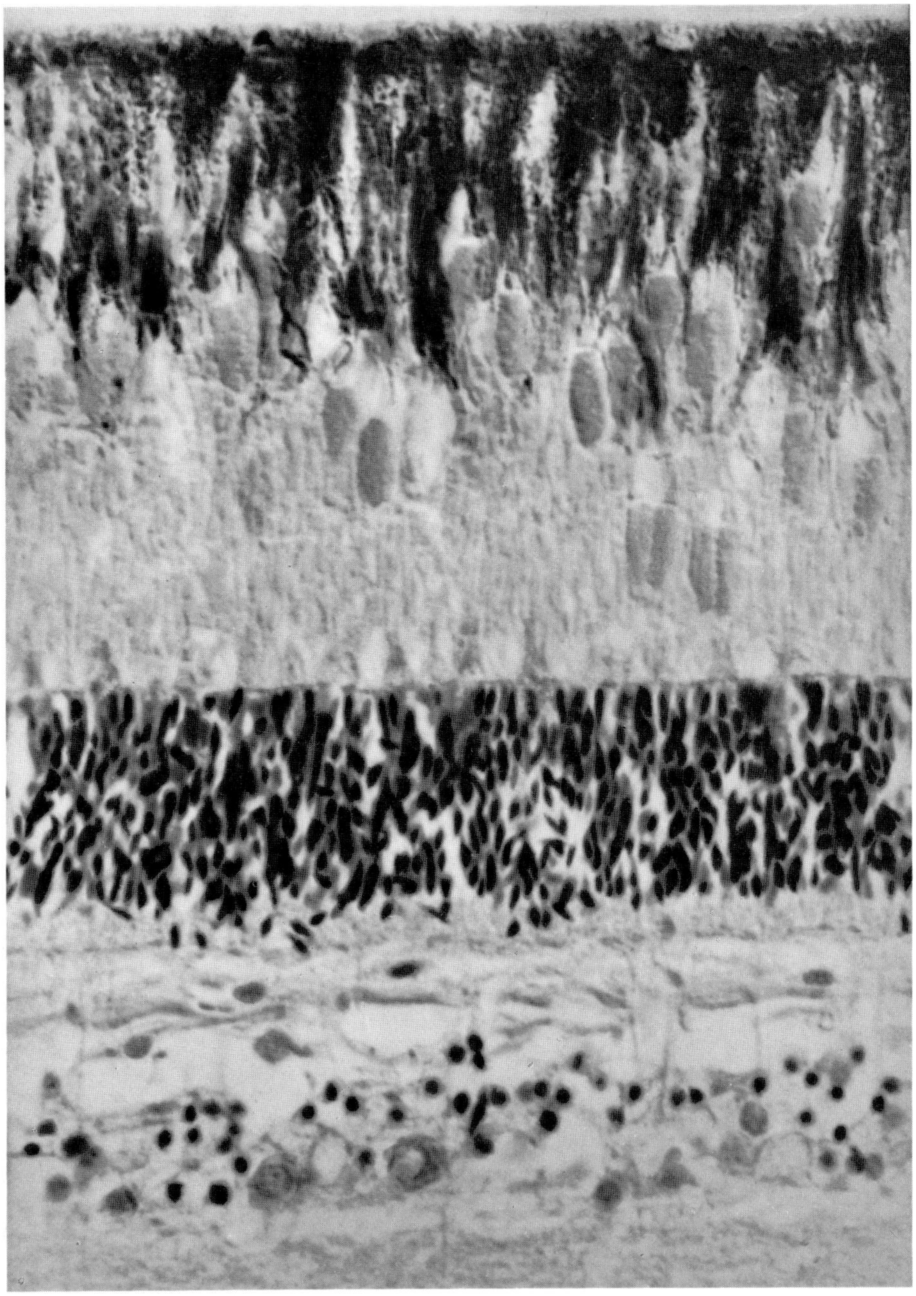

Plate 94. Transverse section of the retina of the Atlantic spadefish, *C. faber*. ×825

Chaetodontidae (*Butterflyfishes and Angelfishes*)

Acanthopterygii, Percomorpha, Perciformes, Percoidei

Butterflyfishes and angelfishes are small, oval-shaped, compressed fishes chiefly inhabiting shallow waters of the seashore. They possess small, extended snouts that are well adapted to picking up small invertebrates from cracks and crevices of the reefs. They are solitary and diurnal in habits.

The retina of this butterflyfish is well developed and its cells are small and densely packed (Plate 95.1). The epithelium is fairly well pigmented, the rods are small and relatively few, and the cones are tiny and arranged in more or less regular square mosaics of double and single cones (Plate 95.3, 4). The internal nuclear layer and ganglion cell layer are particularly well developed (Plate 95.1, 2).

Visual Pigments: A_1 max 491–495.6; $A_2 > 10\%$ (ALI and HEUMANN, 1970, 1972; MUNZ and MCFARLAND, 1973)

Nandidae (*Leaffishes*)

Acanthopterygii, Percomorpha, Perciformes, Percoidei

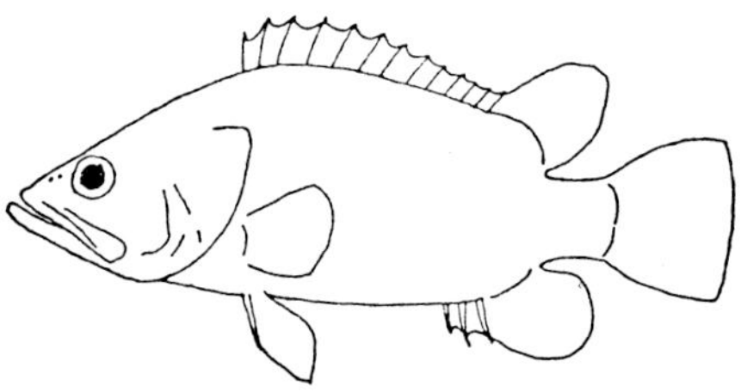

This is a small group of small, robust, perchlike teleosts. They are restricted to freshwaters of South America, West Africa, and Southern Asia. These voracious predators prefer habitats of low-level illumination. The leaffish, *Monocirrhus polyacanthus,* is a bad swimmer resting obliquely, head-down among plants in the water.

Regular square mosaics composed of central single cones and double cones have been reported in the retina of the leaffish. There are as many rods as cones, and two layers of horizontal cells.

Reference: WAGNER (1972)

Plate 95.1-4. Transverse sections (× 580) of the dorsal and ventral retinas of *Chaetodon striatus* are shown in (1) and (2) respectively. Tangential sections (× 1,380) of dorsal and ventral retinas in (3) and (4) respectively. Note the well-developed Müllerian fibres ▷

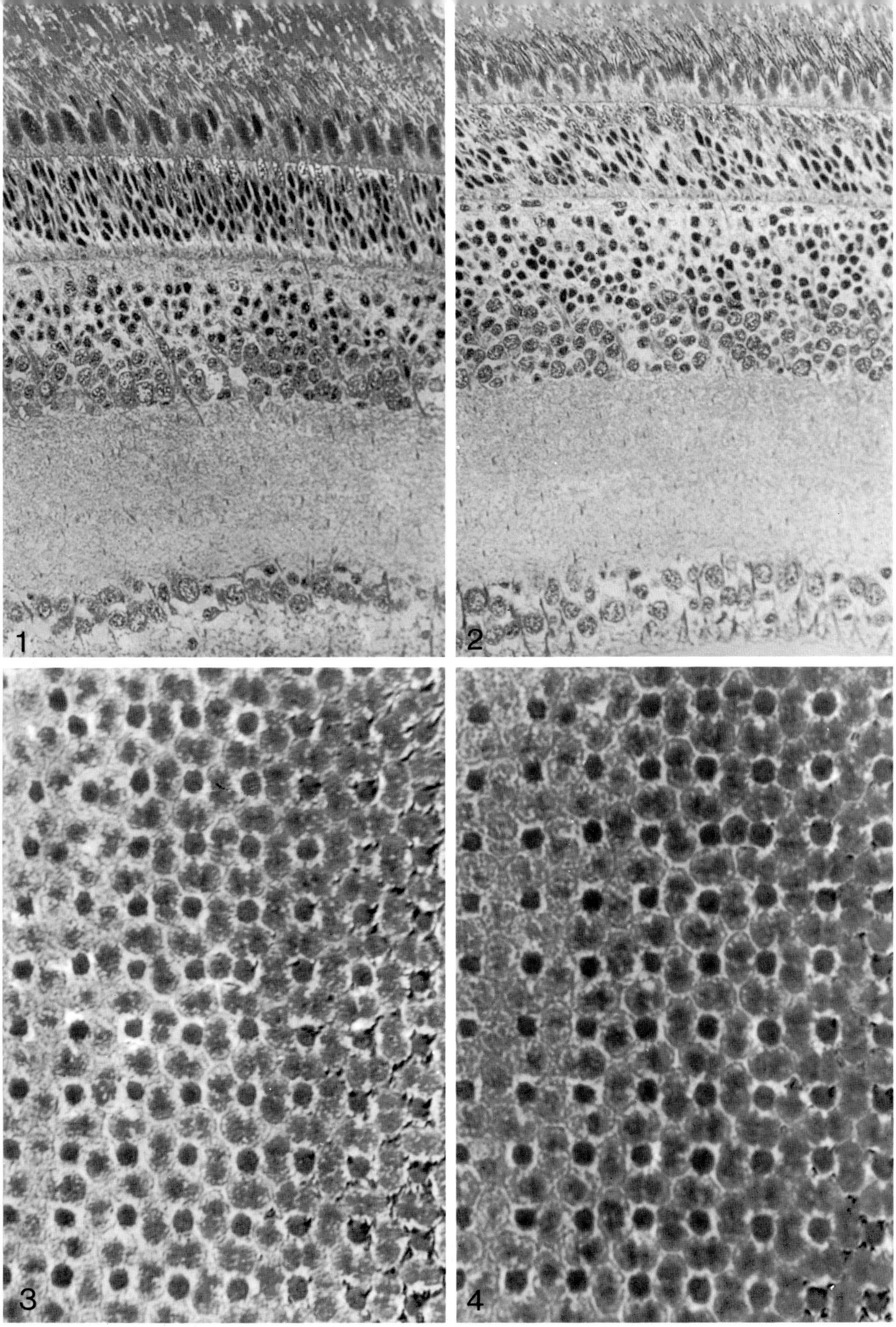
1
2
3
4

Cichlidae (*Cichlids*)

Acanthopterygii, Percomorpha, Perciformes, Percoidei

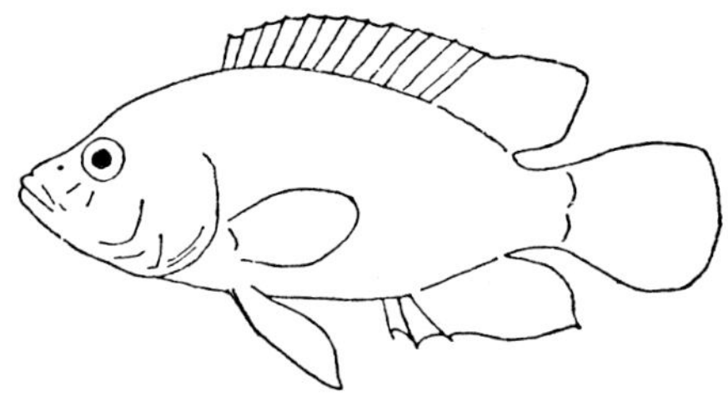

Cichlids are tropical perchlike fishes, and are distinguished from true perches (Percidae, Centrarchidae, etc.) by such characteristics as the presence of only one nostril on each side of the head. The cichlids frequent still or sluggish waters that are provided with a cover for hiding. Almost all cichlids (except *Geophagus* and *Tilapia*) are predatory fishes and feed predominantly upon smaller fishes, insect larvae, water-beetles, worms, etc.

Red Devil

The retina of the red devil (*C. labiatum*) is fairly well developed as seen in Plate 96.1. The epithelial cells send long processes vitreally, which contain pigment granules in the light-adapted state. Rods are more are more or less abundant while the cones are numerous throughout the various retinal regions and are of two types: the single, and equal double cones (Plate 96.2). The bipolars and amacrines are very numerous. The horizontal cells viewed tangentially possess regular mosaics like those of the cones (Plate 96.2,3). See also Introduction, Plate 2.

References: Braemer (1957); Engström (1963b); Kuenzer and Wanger (1969); Hibbard (1971); Wagner (1972, 1973, 1974)

Plate 96.1-3. Transverse (1) and tangential (2) and (3) sections of the retina of the red devil, *C. labiatum*. × 560

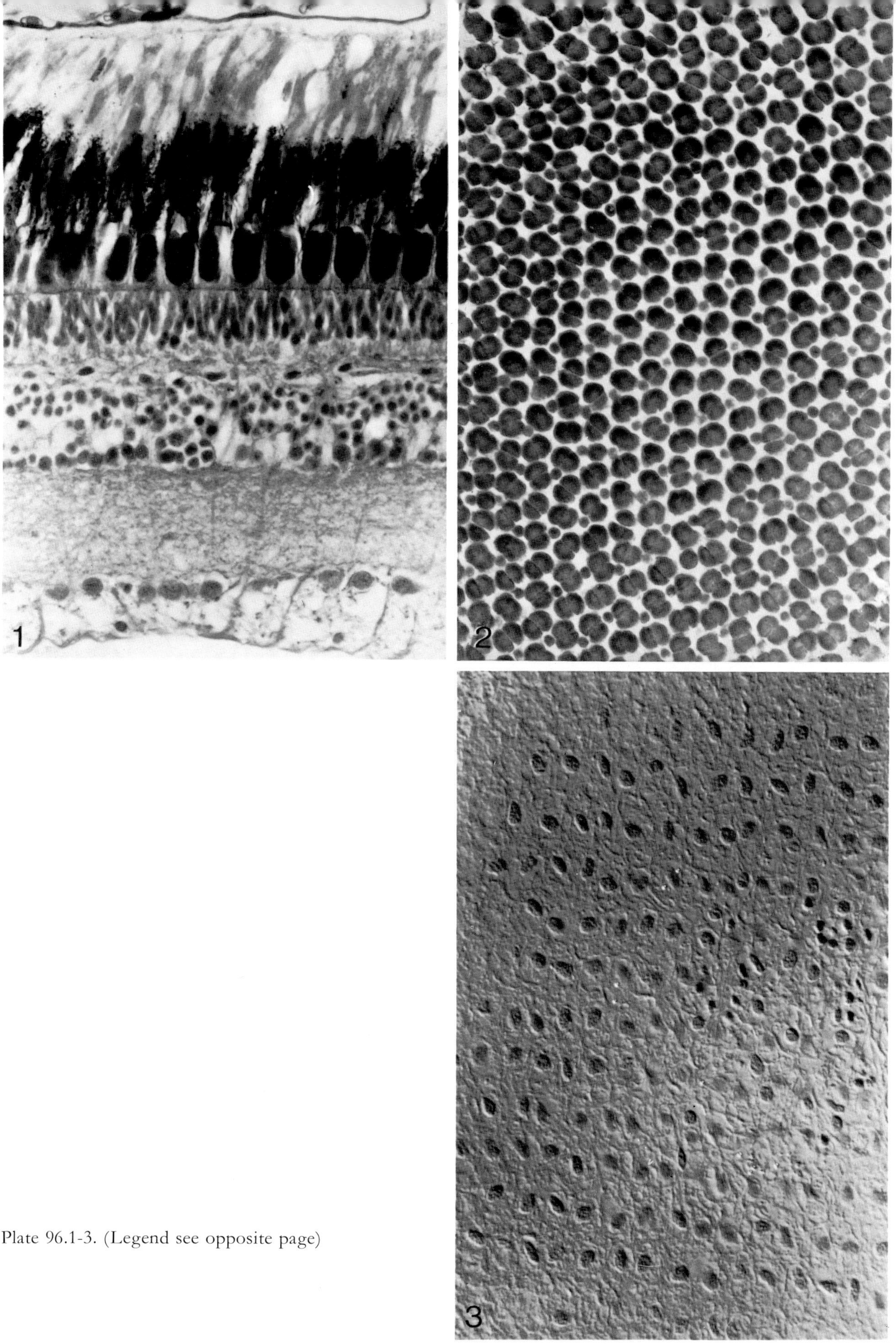

Plate 96.1-3. (Legend see opposite page)

Tilapia

The retina of *T. rhendali* is essentially similar in structure to that of *C. labiatum*. In the dark-adapted ventral retina (Plate 97.1), the pigment of the epithelial cells is more or less retracted sclerally, the rods form a distinct row close to the external limiting membrane, and the cones, whose myoids have elongated, have a distal (sclerad) position, near the epithelium. The dorsal, dark-adapted retina differs in the positions of the single and double cones (Plate 97.2); the single cones, which are shorter, form a distinct, vitreally placed layer in relation to the double cones. Note also (Plate 97.2) the three layers of small horizontal cells.

Visual Pigments: A_1 max 480–503; A_2 max 516–527 (Munz and McFarland, 1973; Ali and Wagner, 1975b)

Plate 97.1, 2. Transverse sections of the ventral (4), and dorsal (2) dark-adapted retina of *T. rhendali*. × 560

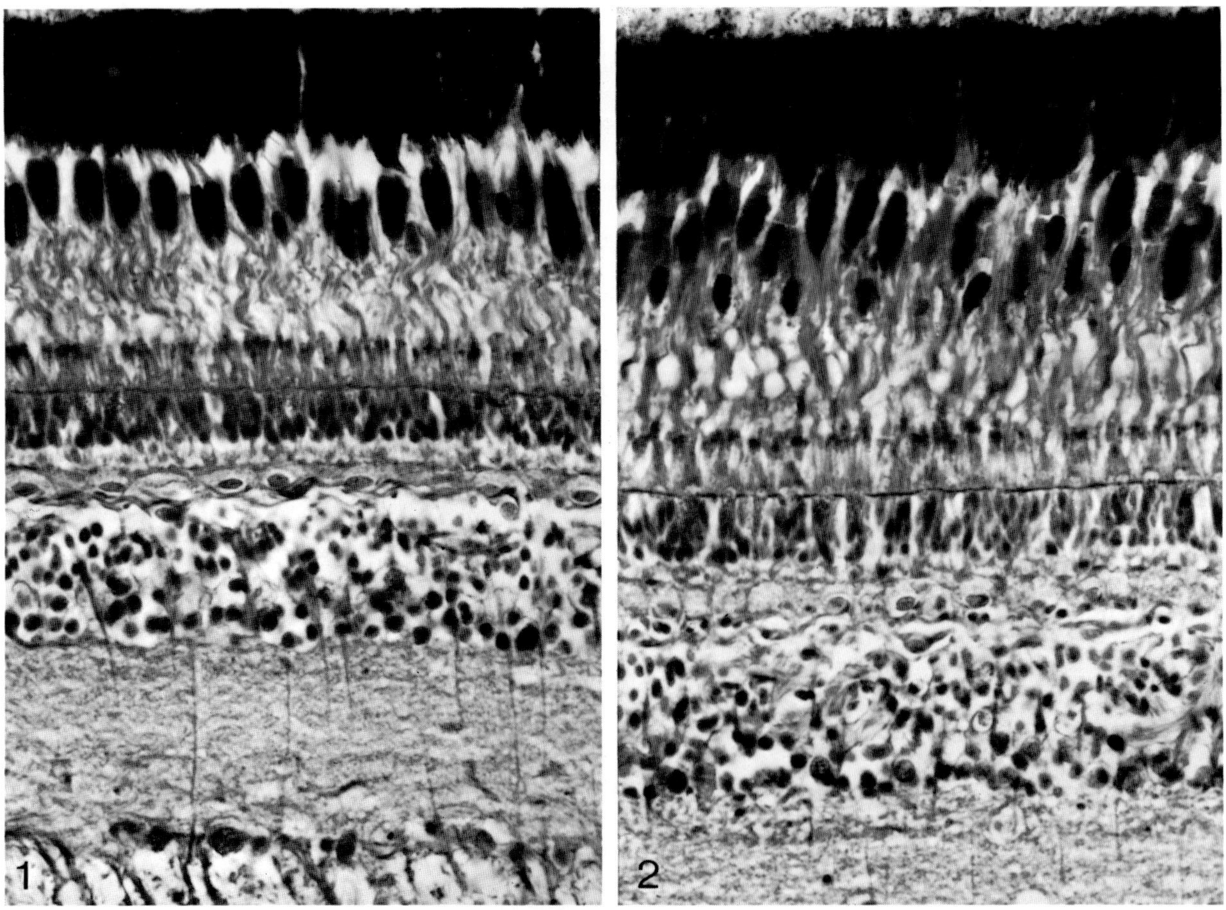

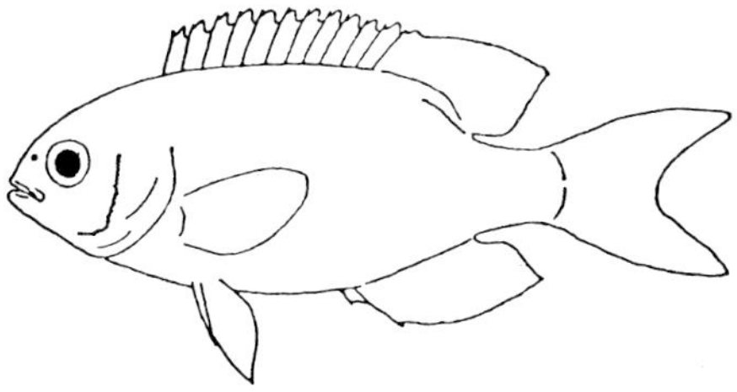

Pomacentridae (*Damselfishes*)

Acanthopterygii, Percomorpha, Perciformes, Percoidei

Damselfishes are typical inhabitants of sea shores and coral reefs. The round to oval fishes have compressed bodies, however, like the cichlids but unlike the chaetodontids and ephippids, they have only a single nostril on each side of the snout instead of two. The damselfishes display a great variety of shapes and colours, having a highly developed social behaviour. They are diurnal in habits and feed primarily on plant material and invertebrates.

The retina of the blue devil is characterised by the abundance of its epithelial pigment, by the large number of slender cones in comparison to rods, and by the very thick internal nuclear layers (Plate 98.1). Both single cones and equal double cones are present and arranged in more or less regular square mosaics; the area retinae is found in the ventro-temporal region (Plate 98.2).

Reference: LYTHGOE and LYTHGOE (1971)

Visual Pigments: A_1 max 491–497; A_2<10% (ALI and HEUMANN, 1970; MUNZ and MCFARLAND, 1973; ALI and WAGNER, 1975b)

Plate 98.1, 2. Transverse (1) and tangential (2) sections of the retina of the blue devil, *Eupomacentrus coeruleus*. (1) ×560; (2) ×1,400

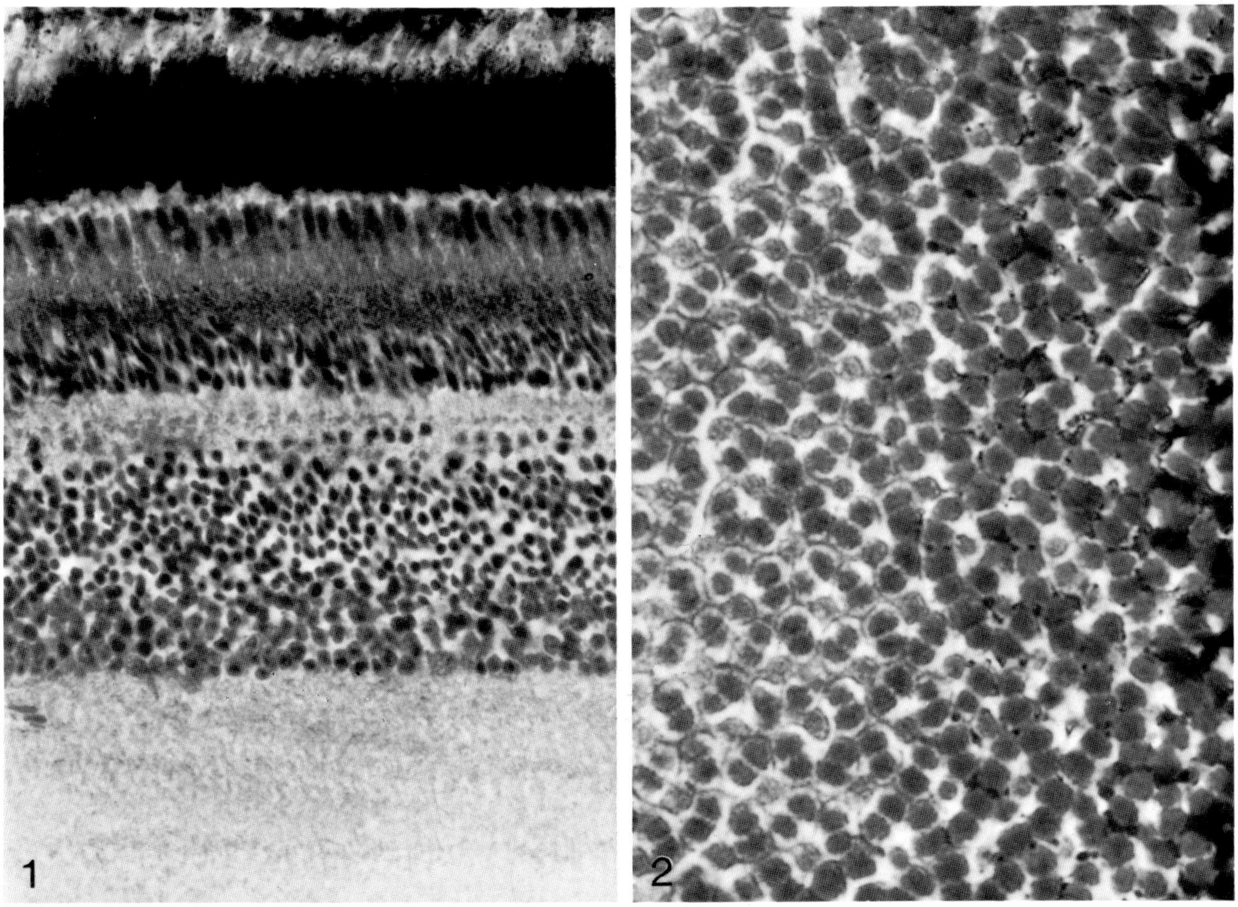

Labridae (*Wrasses*)

Acanthopterygii, Percomorpha, Perciformes, Labroidei

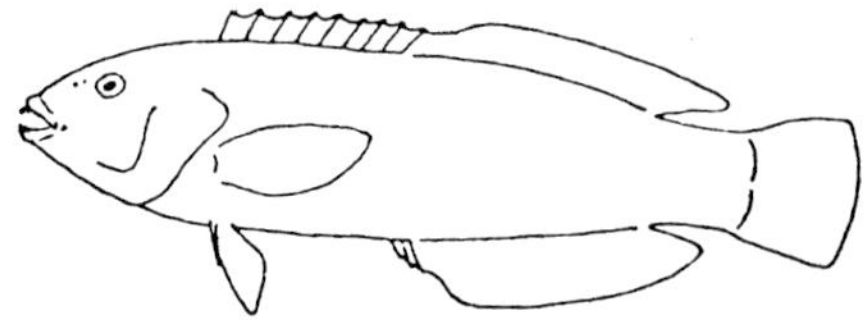

Wrasses are shorefishes with a single, long dorsal fin that is anteriorly spiny. This is a very successful and common group of tropical fishes with some temperate representatives. The cunner, *Tautogolabrus adspersus,* is an omnivorous fish which browses among seaweeds, stones, and dock piling in shallow waters. It spends most of its time resting quietly close to the substrate, or swimming slowly.

The retina of the cunner is characterised by the presence of abundant epithelial pigment (Plate 99.1), the large numbers of cones in all the retinal regions, and the relatively low rod-densities. The equal double cones and the long and short single cones are arranged in regular square mosaics with central and accessory single cones (Plate 99.2). The internal nuclear layer is well developed (Plate 99.1).

References: VERRIER (1928a); ENGSTRÖM (1963a, 1963b); ANCTIL (1969); KAWAMURA and TAMURA (1973)

Visual Pigments: A_1 max 484.6–527; A_2 max 510–513 (ALI and HEUMANN, 1972; MUNZ and MCFARLAND, 1973; ALI and WAGNER, 1975b)

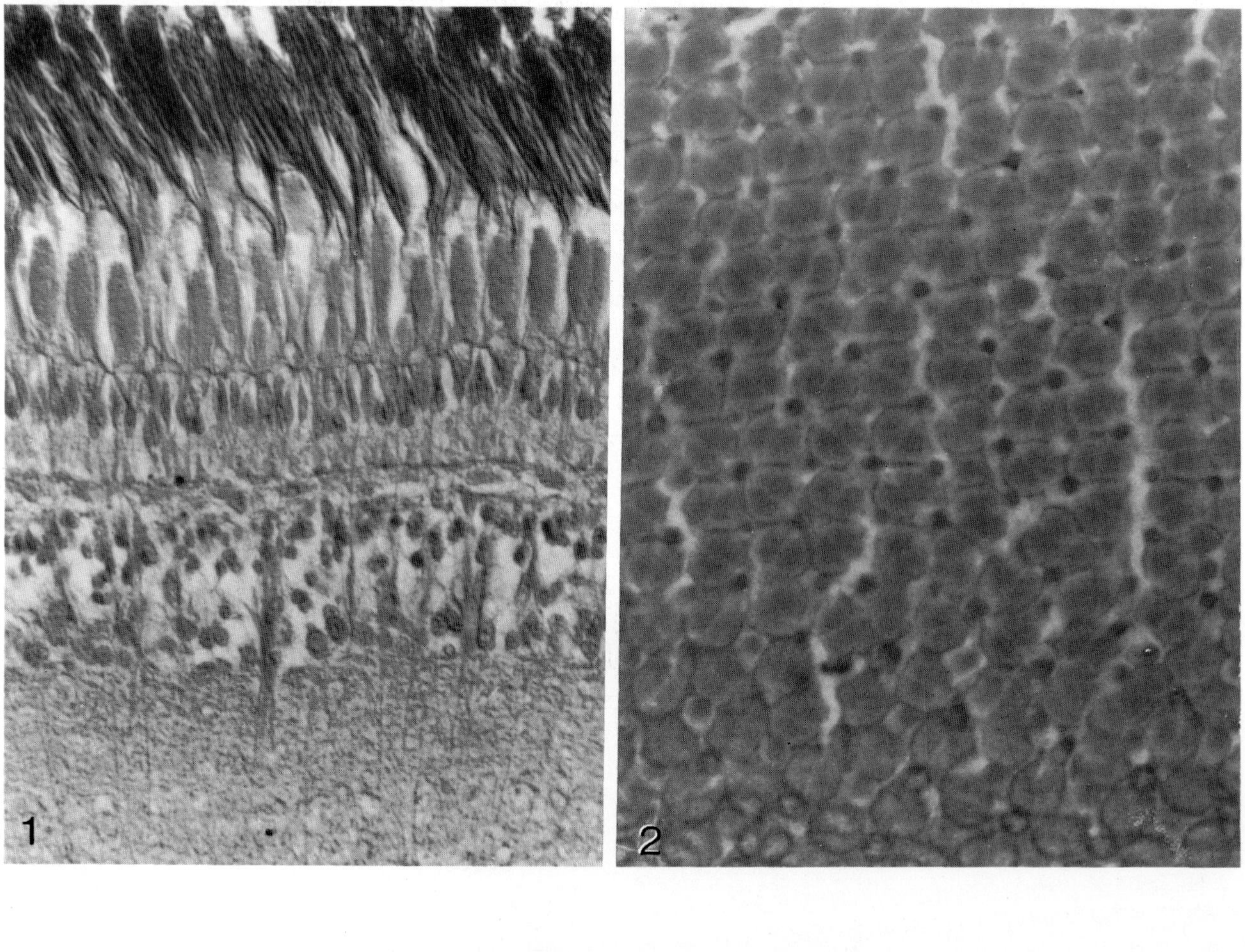

Plate 99.1, 2. Transverse (1) and tangential (2) sections of the retina of the cunner, *T. adspersus.* (1) × 560; (2) × 700

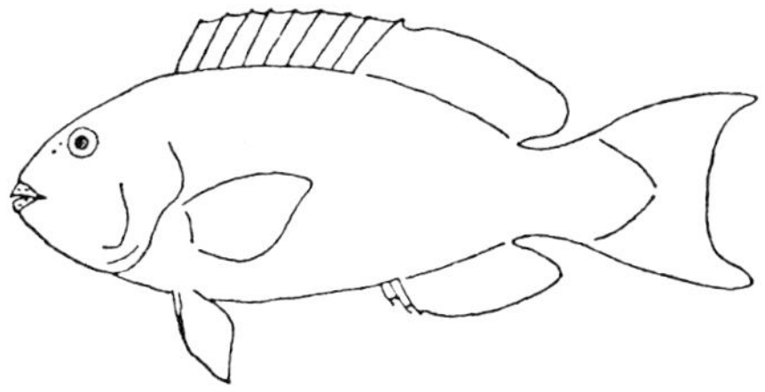

Scaridae (*Parrotfishes*)

Acanthopterygii, Percomorpha, Perciformes

Parrotfishes are reef fishes whose beak-like, chisel-sharp jaw teeth in front appear as an adaptive feature for coral feeding. Some scarids show a strong homing instinct and use the sun as a compass for orientation. The bluelip parrotfish, *Cryptotomus roseus,* is a small wrasse-like fish dwelling in shallow grassy waters and has diurnal feeding habits. At night it buries itself in the sand like a wrasse, and there secretes a mucous envelope.

The presence of a temporal foveal depression (Plate 100.1), the exceptional development of the internal nuclear and ganglion cell layers, together with the large amount of retinal epithelial pigment are the main features of the retina. Note the extremely slender cones in the visual cell layer (Plate 100.2).

References: Ali et al. (1973); Ali and Wagner (1975a)

Visual Pigments: A_1 max 483–484.6; A_2 max 520 (Ali and Heumann, 1972; Munz and McFarland, 1973; Ali and Wagner, 1975b)

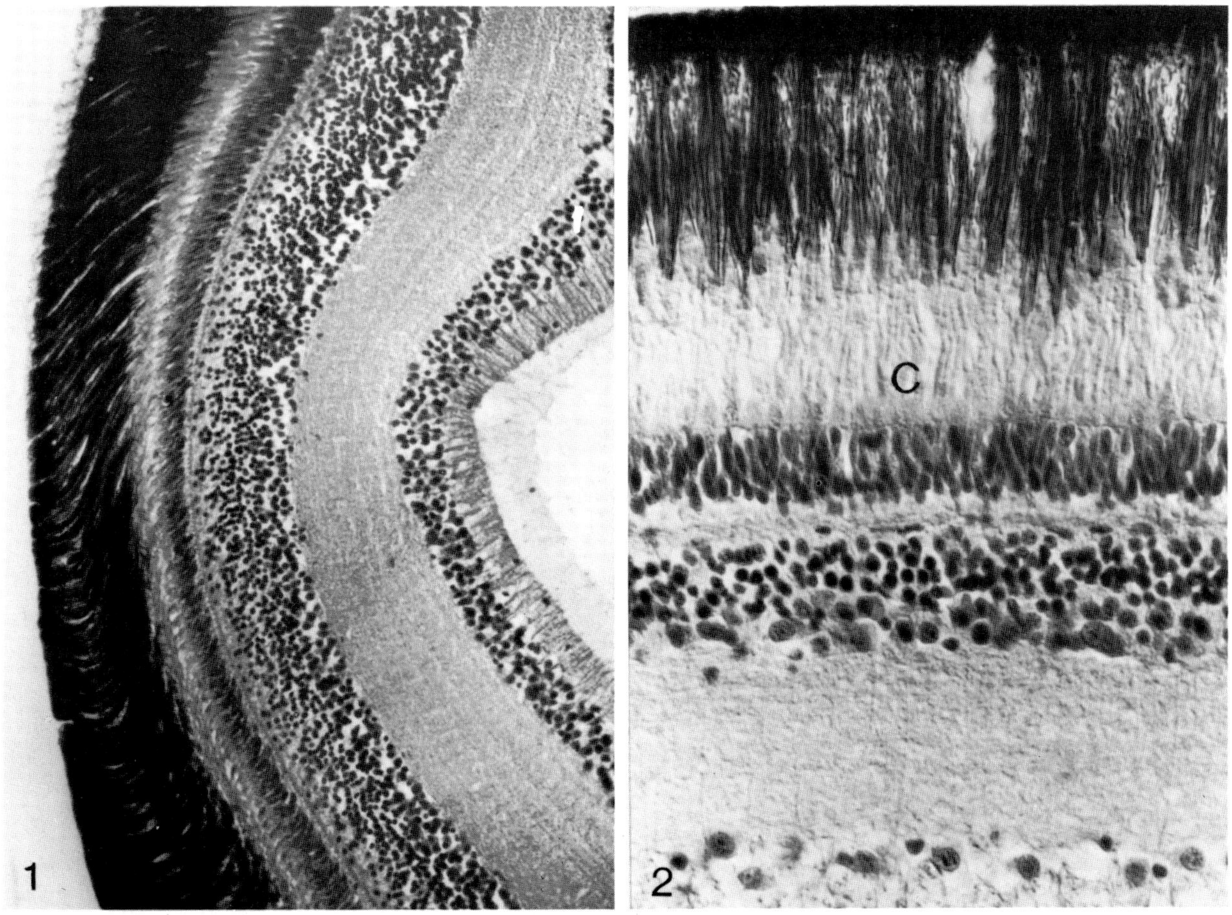

Plate 100.1,2. Transverse sections of the tangential (1) and nasal (2) retinas of the bluelip parrotfish, *C. roseus.* (1) × 225; (2) × 560. *C* cone layer

Mugilidae (*Mullets*)

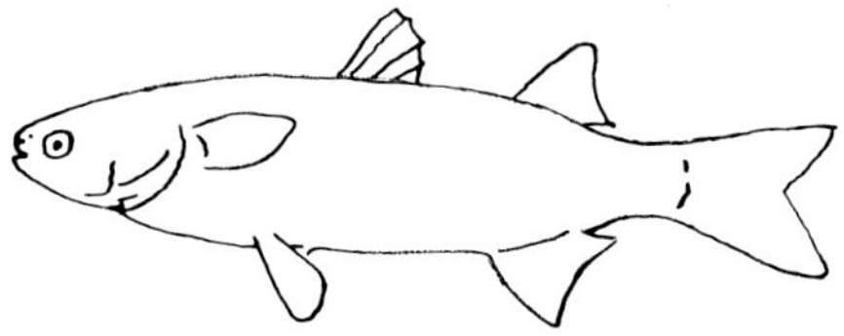

Acanthopterygii, Percomorpha, Perciformes, Mugiloidei

Mullets are torpedo-shaped, marine schooling fishes found in shallow water usually over sandy or muddy bottoms with detritus, where they normally grub for food. Their feeding consists of gulping mouthfuls of dirt and sand and spitting out the coarser particles. There is an adipose eyelid and a vertically elliptical pupil.

The retina of the white mullet, *Mugil curema,* is characterised by the poor development of the epithelial melanin pigment (Plate 101.1), by extremely abundant rods which are very long and slender (Plate 101.1), by relatively large single and equal double cones particularly numerous in the temporal area retinae (Plate 101.2), and by the large horizontal cells which are set in four layers (Plate 101.1), the vitread-most layer consisting of rod horizontal cells (Plate 101.3).

References: Friis (1879); Verrier (1928a); Selvin de Testa (1966); Parthe (1967, 1972); Wagner (1972); Ali et al. (1973); Ali and Wagner (1975a)

Visual Pigments: A_1 max 499; A_2 max 522 (Munz and McFarland, 1973; Ali and Wagner, 1975b)

Sphyraenidae (*Barracudas*)

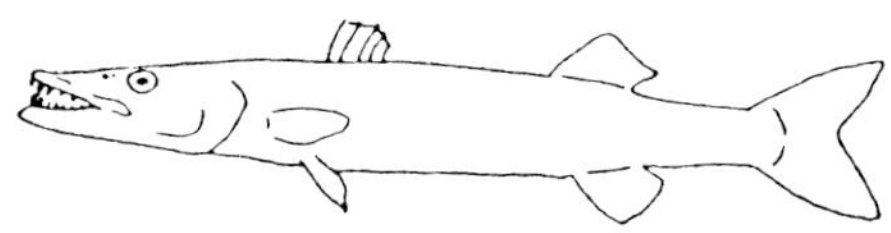

Acanthopterygii, Percomorpha, Perciformes, Sphyraenoidei

Barracudas are pelagic, torpedo-shaped fishes with a jutting lower jaw and fanglike teeth. These are active predators, and feed apparently by sight rather than by smell.

Although the cone density does not vary significantly among the retinal regions of *Sphyraena barracuda,* the nasal direction of lens accommodation indicates that the area retinae is roughly located in the temporal region. Single cones and double cones are present in the proportion 1:3.

Reference: Tamura and Wisby (1963)

Visual Pigments: A_1 max 497.6–498 (Munz and McFarland, 1973)

Plate 101.1-3. Transverse (1) and tangential (2) and (3) sections of the retina of the white mullet, *M. curema.* ×560

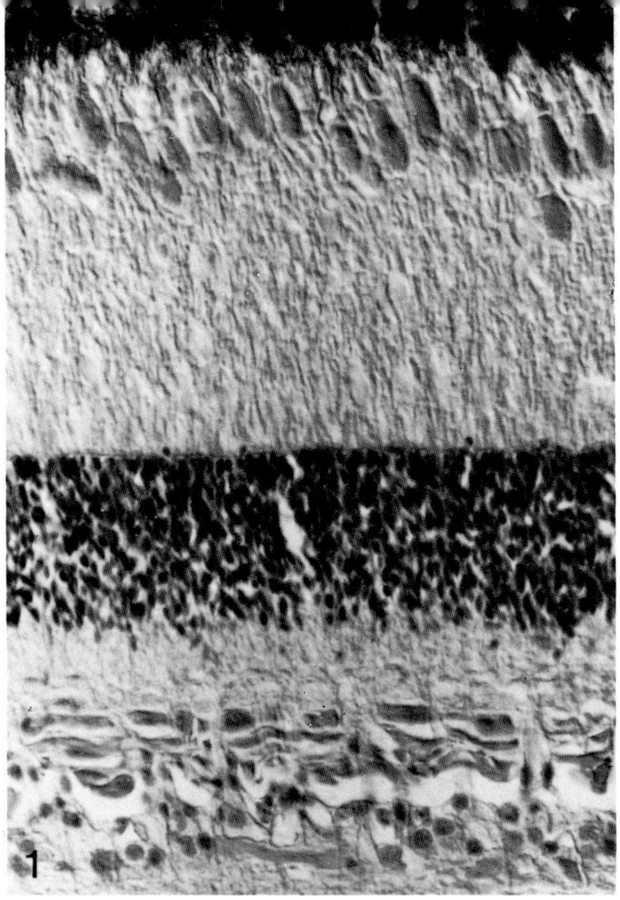

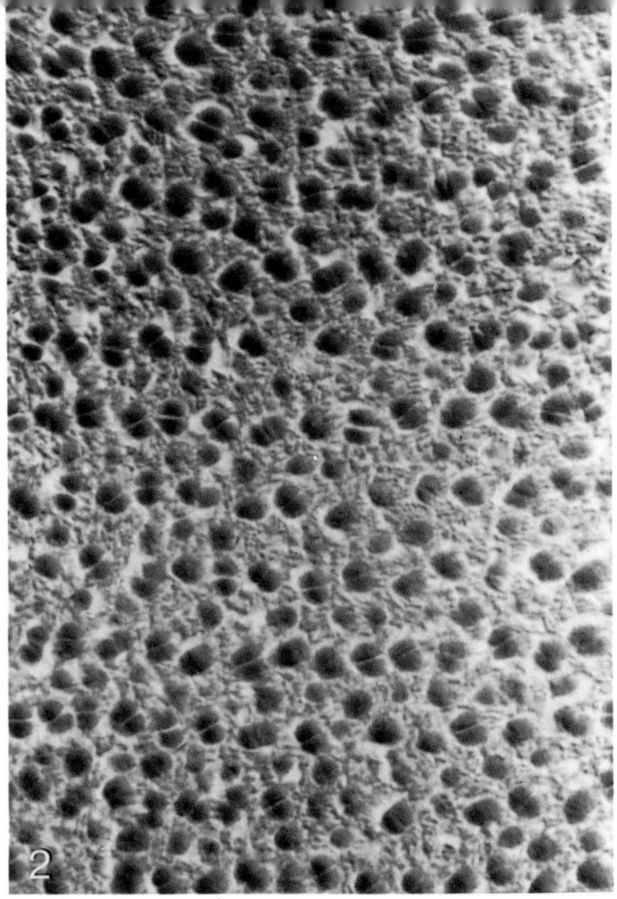

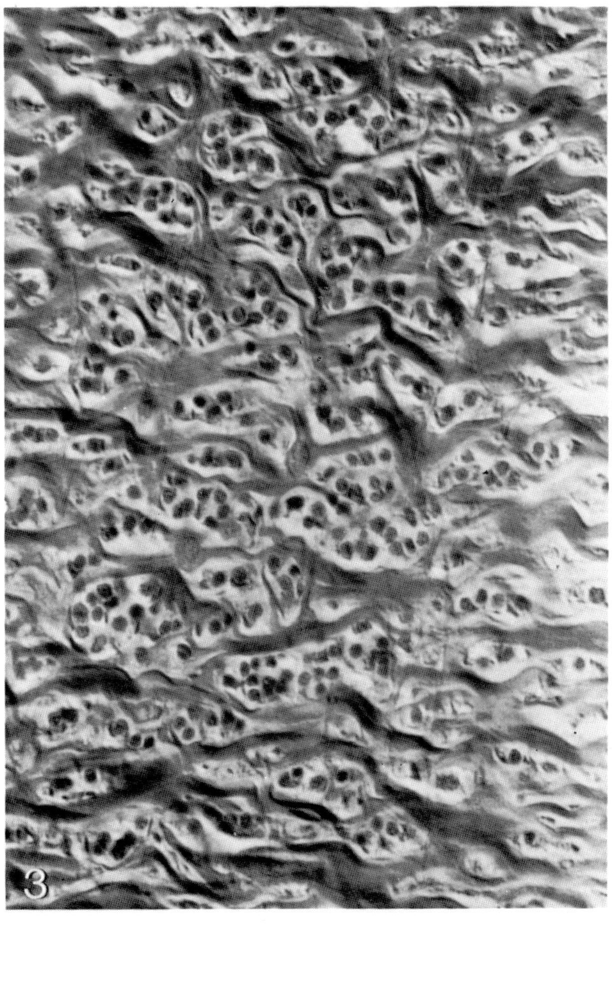

Plate 101.1-3. (legend see opposite page)

Mugiloididae (*Mugiloidids*)

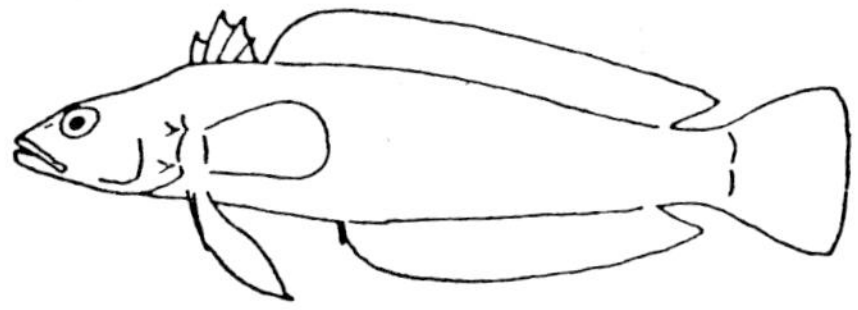

Acanthopterygii, Percomorpha, Perciformes, Trachinoidei

These are little-known marine coastal fishes.

Rods, single cones, and equal and unequal double cones have been reported in the retina of *Pinguipes* sp. The rods are more or less numerous. The unequal double cones are largely dominant in number over the two other cone types. All the cones are slender and densely packed, and their myoids are well developed.

Reference: SVERDLICK (1940, 1955)

Trachinidae (*Weeverfishes*)

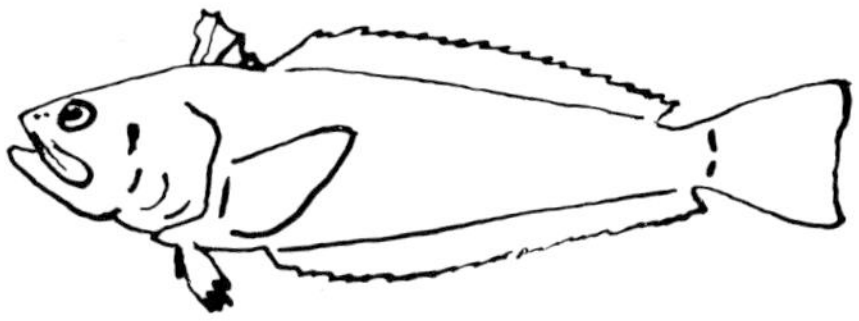

Acanthopterygii, Percomorpha, Perciformes, Trachinoidei

The weever family includes only four species of small, elongated marine fishes with long anal and second dorsal fins, which are noted for their poisonous spines. These fishes bury themselves in the sand; some, like *Trachinus draco,* live in deep shelf waters.

Double cones have been reported in the retina of *T. Draco.*

Reference: VERRIER (1928a)

Dactyloscopidae (*Sand Stargazers*)

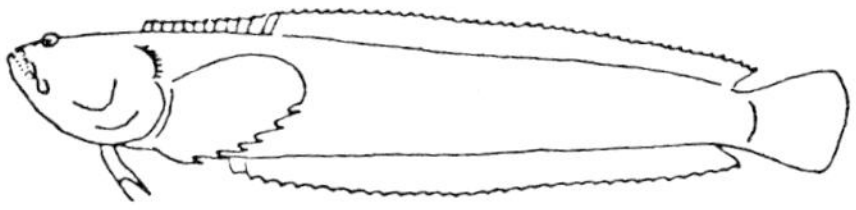

Acanthopterygii, Percomorpha, Perciformes, Trachinoidei

The sand stargazers are shorefishes dwelling near or buried in sandy bottoms. They feed on bottom-living animals, mostly small crustaceans and small fishes. Both their eyes and mouth are directed upwards (Plate 102.1).

The retina of *Dactyloscopus tridigitatus* is very well-developed. The epithelium possesses long, heavily pigmented processes (Plate 102.3). The visual cell layer contains both rods and cones. The rods are large and relatively few (Plate 102.3). The single and double cones are numerous and slender, with granular ellipsoids, short myoids, and short outer segments (Plate 102.3). The internal nuclear layer is thick and packed with bipolar and amacrine cells, and the ganglion cell layer is composed of two rows of cells (Plate 102.1,2). Note the grouping of rods (Plate 102.3, *arrow*).

Reference: WAGNER et al. (1976)

Plate 102.1-4. (1) ×100, cross section of the entire eye of a juvenile specimen of *D. tridigitatus.* (2) ×700, transverse section of the ventral retina. (3) ×1,380, transverse section of the dorsal retina. Oblique section (×1,770) of the central retina is shown in (4) ▷

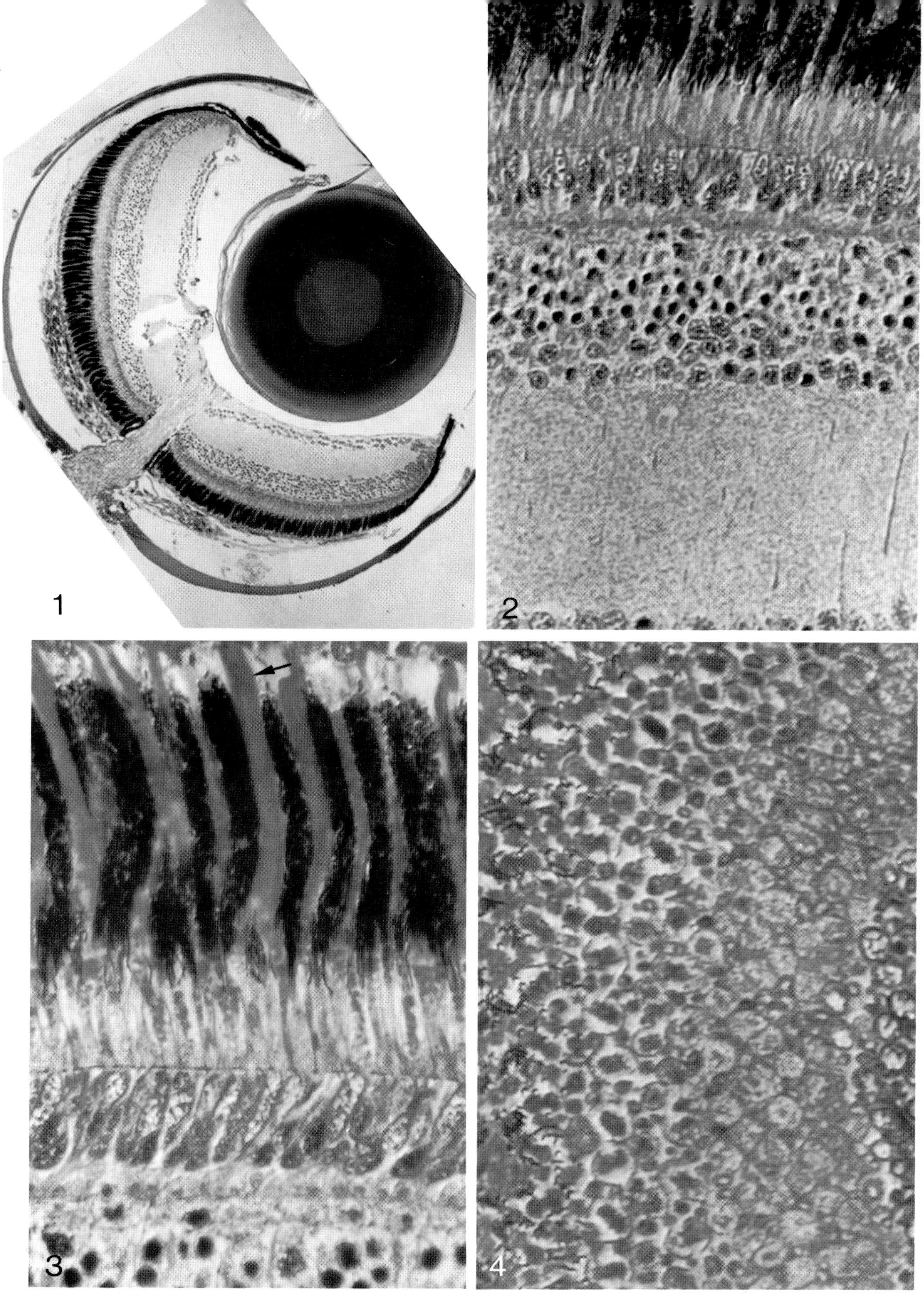
1
2
3
4

Uranoscopidae (*Electric Stargazers*)

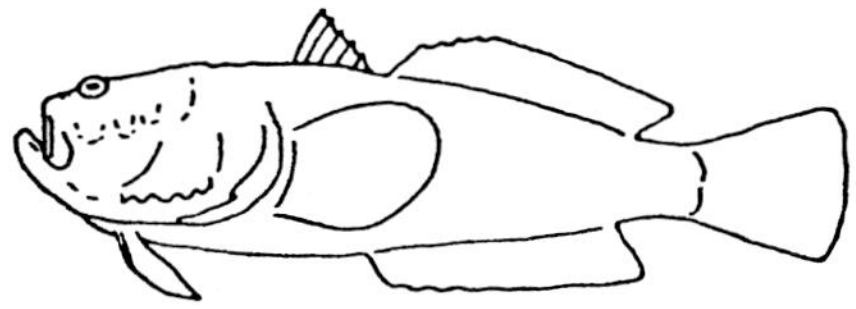

Acanthopterygii, Percomorpha, Perciformes, Trachinoidei

Electric stargazers are benthic fishes, being characterised by the presence of ocular muscles modified into electrogenic organs, by the small eyes on the top of the head, and by the mouth that is in a nearly vertical position. These stargazers may be found in shallow to very deep shelf waters and bury themselves in the sand. Double cones have been reported in the retina of *Uranoscopus scaber*.

Reference: VERRIER (1928a)

Clinidae (*Clinids or Klipfishes*)

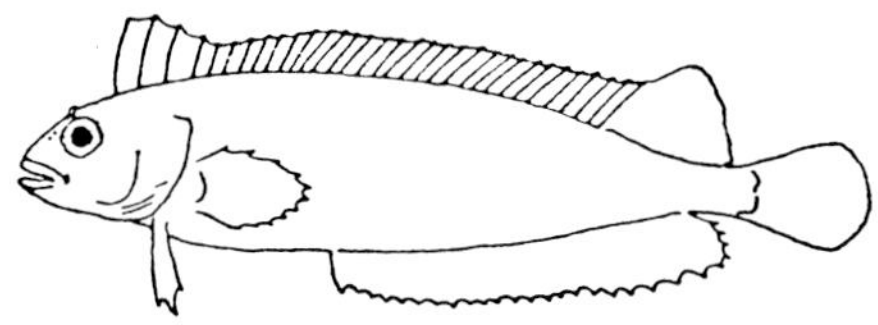

Acanthopterygii, Percomorpha, Perciformes, Blennioidei

Clinids constitute a large family of small inshore fishes which differ from the blenniids by the presence of scales on the body, a protractile mouth, and a dorsal fin in one, two, or three parts. *Dialommus fuscus* is a pot-hole dweller, living in deep depressions at some distance from all but the highest tides; this fish spends a considerable amount of time out of water.

D. fuscus is a "four-eyed" fish, with two lens muscles per eye to permit amphibious foveal vision and two flattened corneal windows per eye to permit emmetropic vision in air. *L. nuchipinnus* from a tide pool in São Paulo state, Brazil, shows an interesting combination of oil droplets in the cone ellipsoids (Plate 103.3), and grouped rods (Plate 103.1, 3), which end in a role as light guides for the cones ellipsoids (Plate 103.4).

The retina is well-developed. Rods as well as single and double cones are present and arranged in fairly regular square mosaics (Plate 103.5). The cones are particularly slender and densely packed in the area temporalis retinae. There is a shallow foveal depression in the dorso-temporal part of the retina.

References: MUNK (1969a); WAGNER et al. (1976)

Plate 103.1-5. *L. nuchipinnus* retina. Transverse sections (1) and (2) of dorsal and ventral retinas at ×700. (3) shows the photoreceptors at a higher (×1,300) magnification. Note the oil droplets *OD*, grouped rods *R* and the reflecting layer *ET*. Tangential sections (4) and (5) of the dorsal retina show two levels. One (4) across the outer segments, and the other (5) across the ellipsoids. Note the light guiding role of the ellipsoids in (4). (4) and (5) at ×1,380 ▷

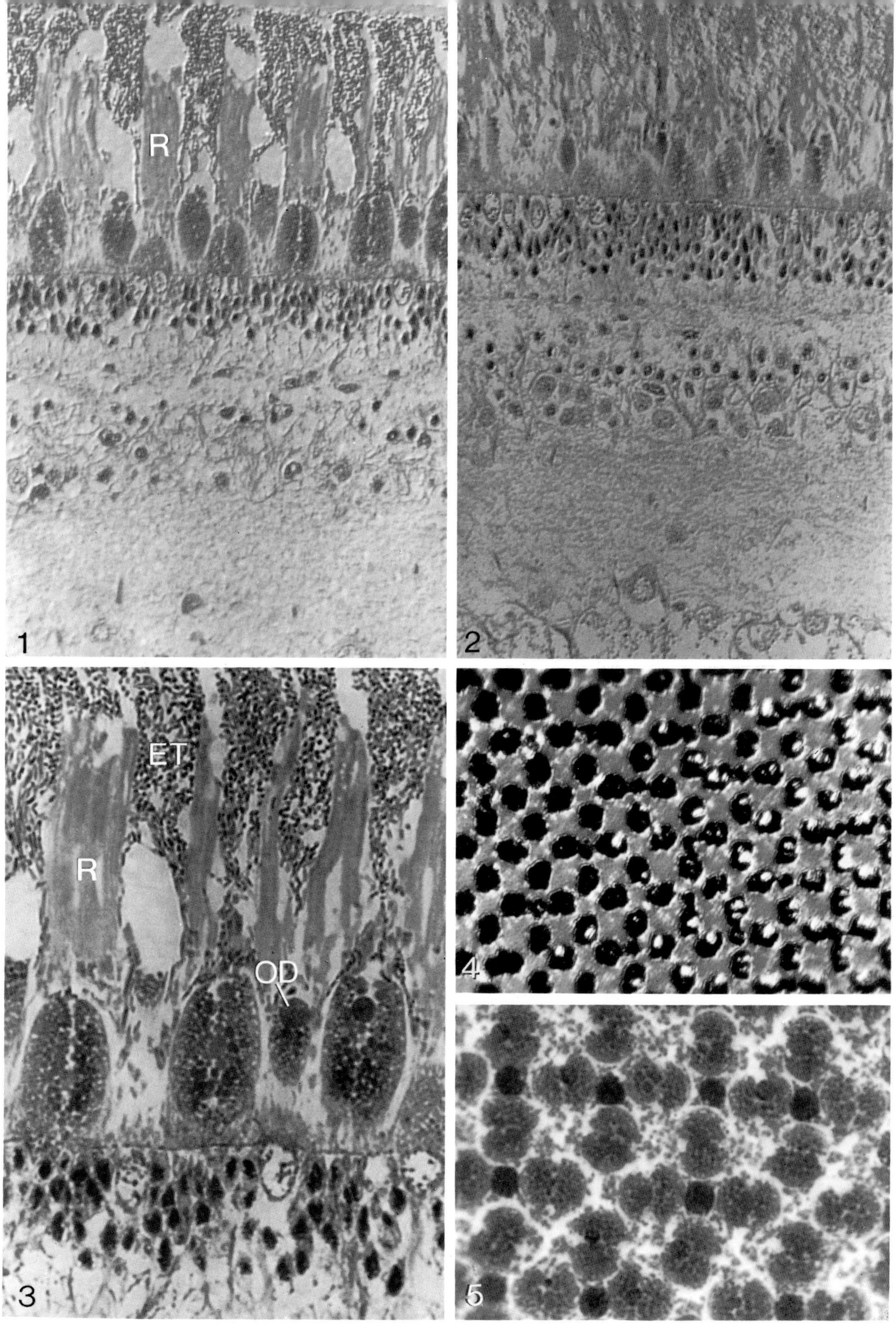
R
1
2
ET
R
OD
3
4
5

Blenniidae (*Combtooth Blennies*)

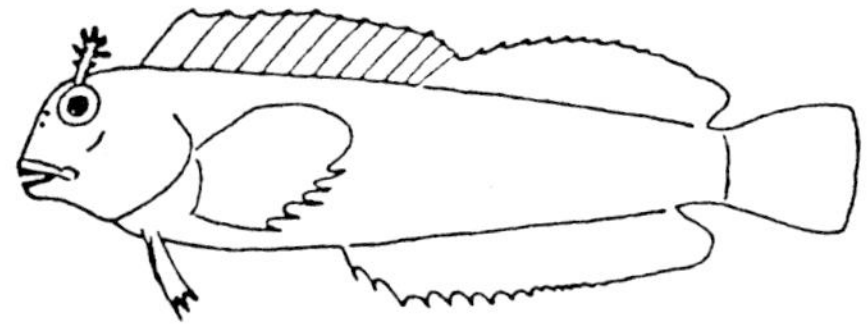

Acanthopterygii, Percomorpha, Perciformes, Blenniodei

The blenny family comprises about two hundred species of small shorefishes. They are found in rocky shore waters and tide pools. Their skin lacks scales and possesses many slime glands protecting them against abrupt changes in salinity and temperature. Blennies feed on small crustaceans.

Scaleless tropical blennies are small, rather elongated, littoral fishes with reduced, or entirely absent, pelvic fins. They are also characterised by the presence of crests, ridges, and fringes on their heads. This is a very successful group of carnivorous or omnivorous, bottom-dwelling fishes.

The retina of several *Blennius* spp. and *Salarias enosimae* have been studied. Single and double cones are present; they are arranged in square mosaics with central single cones in *B. gentilis*, and with accessory single cones only in *B. vulgaris* (Plate 104.1, 2). Two horizontal cell layers are present. There is a temporal fovea in the retina of *B. basilicus*.

The retina of *B. cristatus*, is well-developed with characteristically thick internal nuclear and ganglion cell layers (Plate 105.1, 2). The rods are fairly large and numerous cones are arranged in two types of square patterns (Plate 105.3,4). Horizontal, bipolar, amacrine, and ganglion cells are very abundant. Note the prominent area dorsalis (Plate 105.1). See also Plate 7.

References: Beer (1894); Karsten (1923); Rochon-Duvigneaud and Roule (1927); Verrier (1928a); Bathelt (1970); Wagner (1972); Wagner et al. (1976)

Visual Pigments: A_1 max 496 (Munz and McFarland, 1973)

Synbranchidae (*Swamp Eels*)

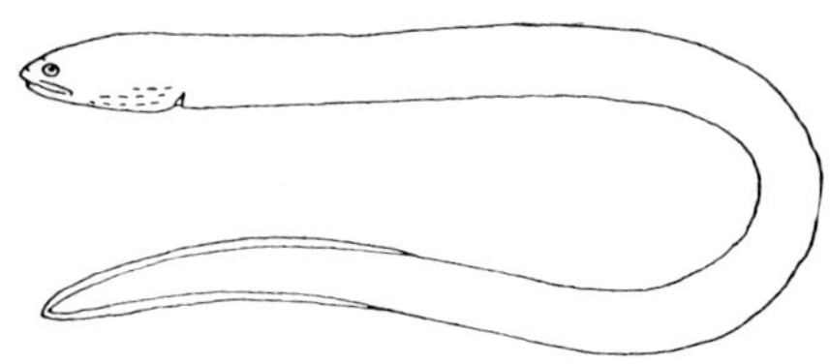

Acanthopterygii, Percomorpha, Synbranchiformes, Synbranchoidei

Synbranchids, which are not true eels, are air-breathing freshwater teleosts that lack the pectoral and pelvic fins, and whose dorsal and anal fins are reduced to a ridge. Only one species is marine.

The retina of *Synbranchus* sp. has been reported to possess primarily single cones which are relatively long and slender; only a few double cones have been observed. Rods are very long, slender, and numerous.

Reference: Sverdlick (1940, 1955)

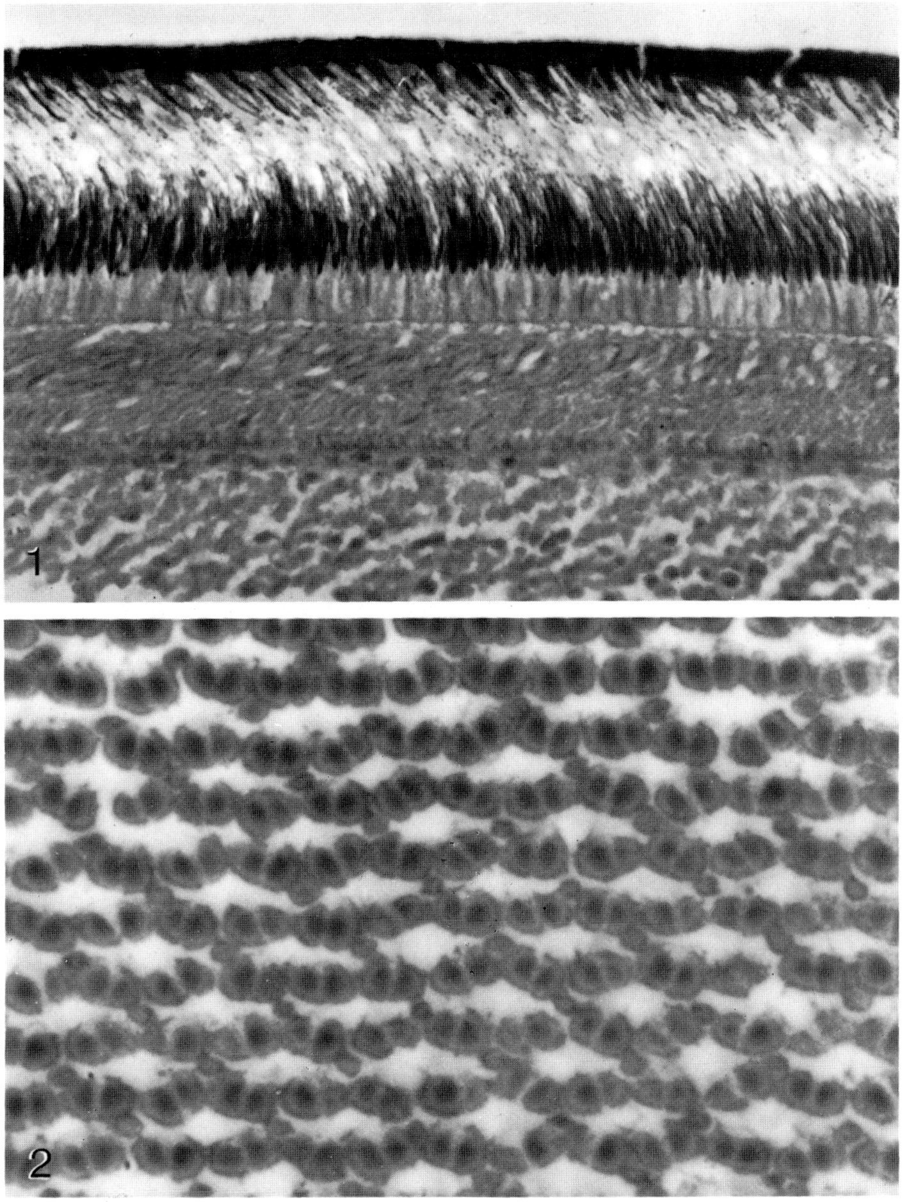

Plate 104.1,2. Transverse (1) and (2) and tangential (2) sections of the temporal retina of the blenny, *B. vulgaris*. (1) × 560; (2) × 1,400

Plate 105.1-4. Transverse sections of area dorsalis (× 310) and ventral retina (× 700) in *B. cristatus* (1) and (2). Tangential sections of dorsal (3) and ventral regions (4) are also shown. (3) and (4) are at × 1,380. Note the additional single cones in the square mosaics in the dorsal region and central single cones in the ventral region ▷

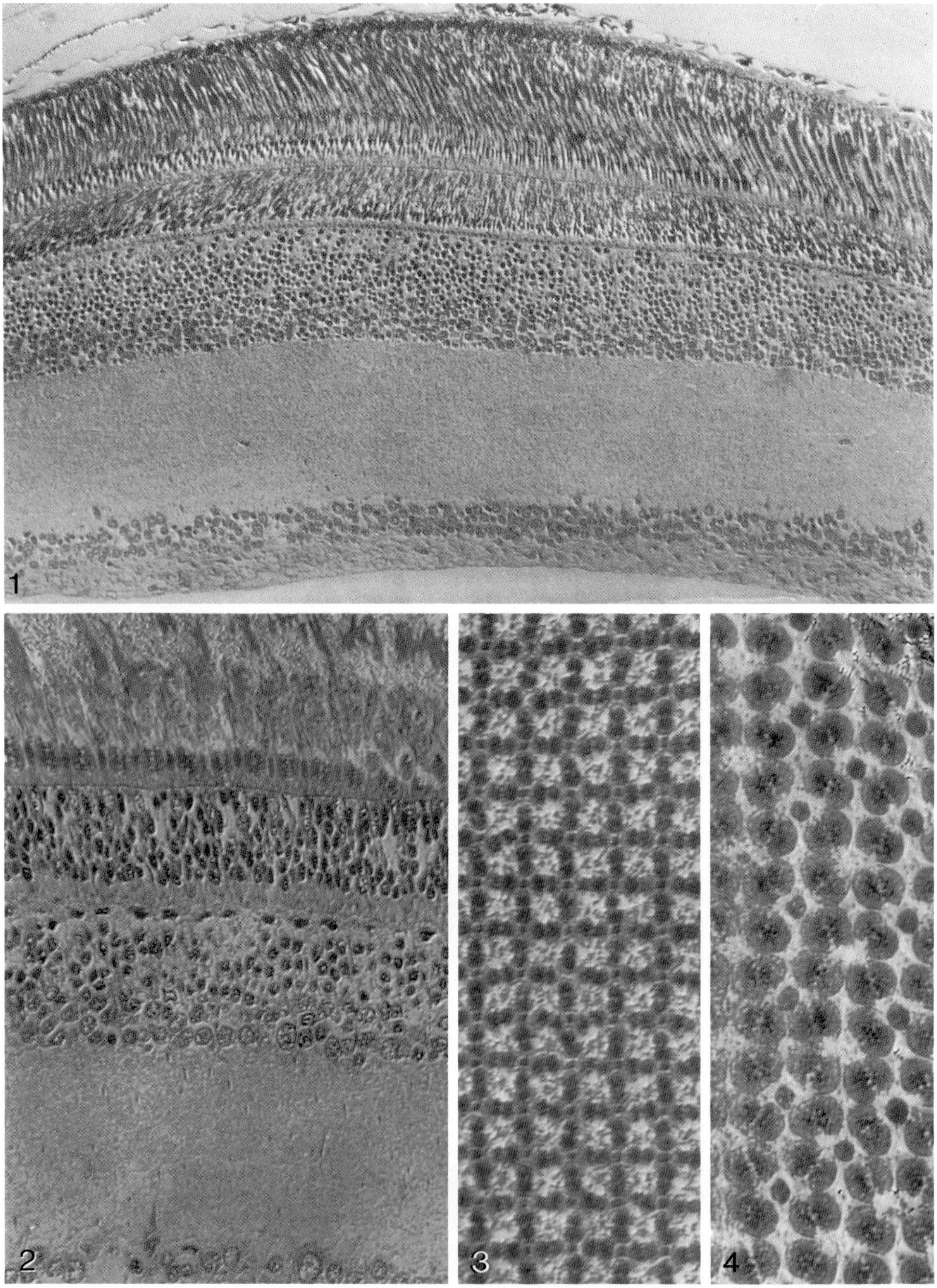

1
2
3
4

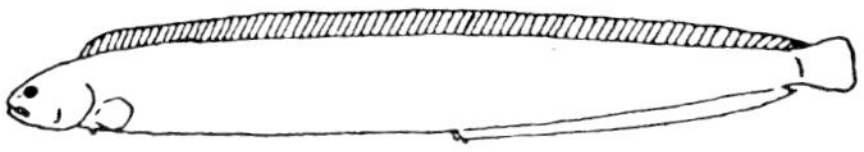

Pholidae (*Rock Eels or Gunnels*)

Acanthopterygii, Percomorpha, Perciformes, Blennioidei

Rock eels or gunnels are fish resembling the blenny with a single, long, spiny dorsal fin. These active, slippery fishes inhabit the littoral zones where they hide among stones and seaweed, feeding on various molluscs and crustaceans.

The retina of the rock eel is well-developed. Plate 106 shows the peripheral part of the retina. Note the abundance of epithelial pigment and processes, the well-defined layer of horizontal cells, and the outstanding development of the bipolar and amacrine layers. The single and double cones, which are arranged in quadrangular mosaics, are very numerous, especially in the fovea temporalis. The rods are comparatively few.

References: Verrier (1931); Engström (1963b); Ali (1968)

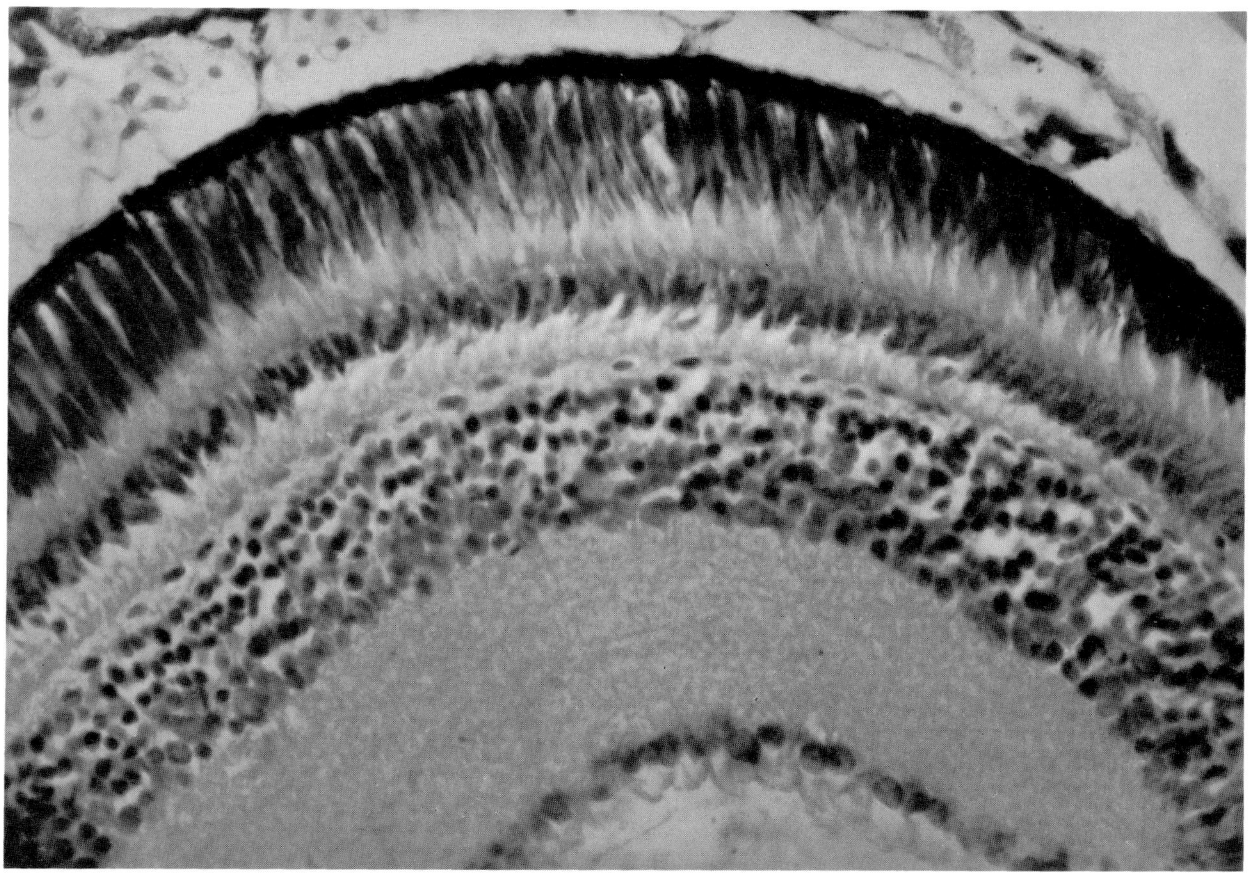

Plate 106. Transverse section of the peripheral retina of the rock eel, *Pholis gunnellus.* × 825

◁ Plate 105.1-4. (Legend see page 211)

Anarhichadidae (*Wolffishes*)

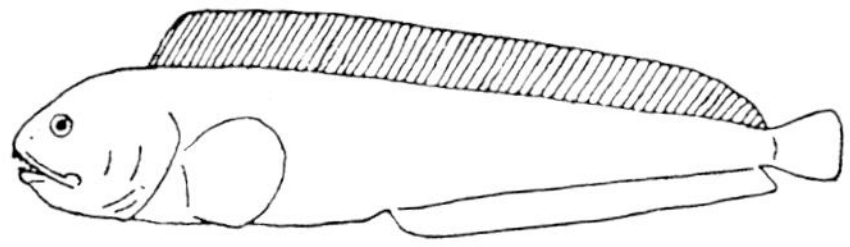

Acanthopterygii, Percomorpha, Perciformes, Blennioidei

Wolffish

Wolffishes are closely allied to the blennies, but differ by the presence of large molar teeth and canine tusks, the absence of pelvic fins, and being also larger size. The wolffish (*Anarhichas lupus*) is solitary, living in close association with hard bottoms where it hides among seaweeds and rocks or searches sluggishly for its food (hard-shelled molluscs, crustaceans, echinoderms). *A. lupus* is distributed from below the tidal zone down to moderate shelf depths (150–200 m).

The retina of this wolffish is moderately to poorly developed (Plate 107.1). The pigmented epithelium possesses moderately long processes; the rods are very short and few, while the cones are relatively large with short outer segments. The bipolar and amacrine cells are few and sparsely distributed. The single and equal double cones are arranged in square mosaics and the area retinae is found in the dorsal region.

References: ENGSTRÖM (1963b); ALI (1968); ALI and WAGNER (1975a)

Spotted Wolffish

The spotted wolffish lives deeper in the continental shelf than *A. lupus*: over bottoms as deep as 400–500 m.

The structural pattern of its retina resembles essentially that of *A. lupus*. The only significant difference lies in the larger number of rods, which are more elongate (Plate 107.2), and the lesser number of cones where the ellipsoid is fairly bulky.

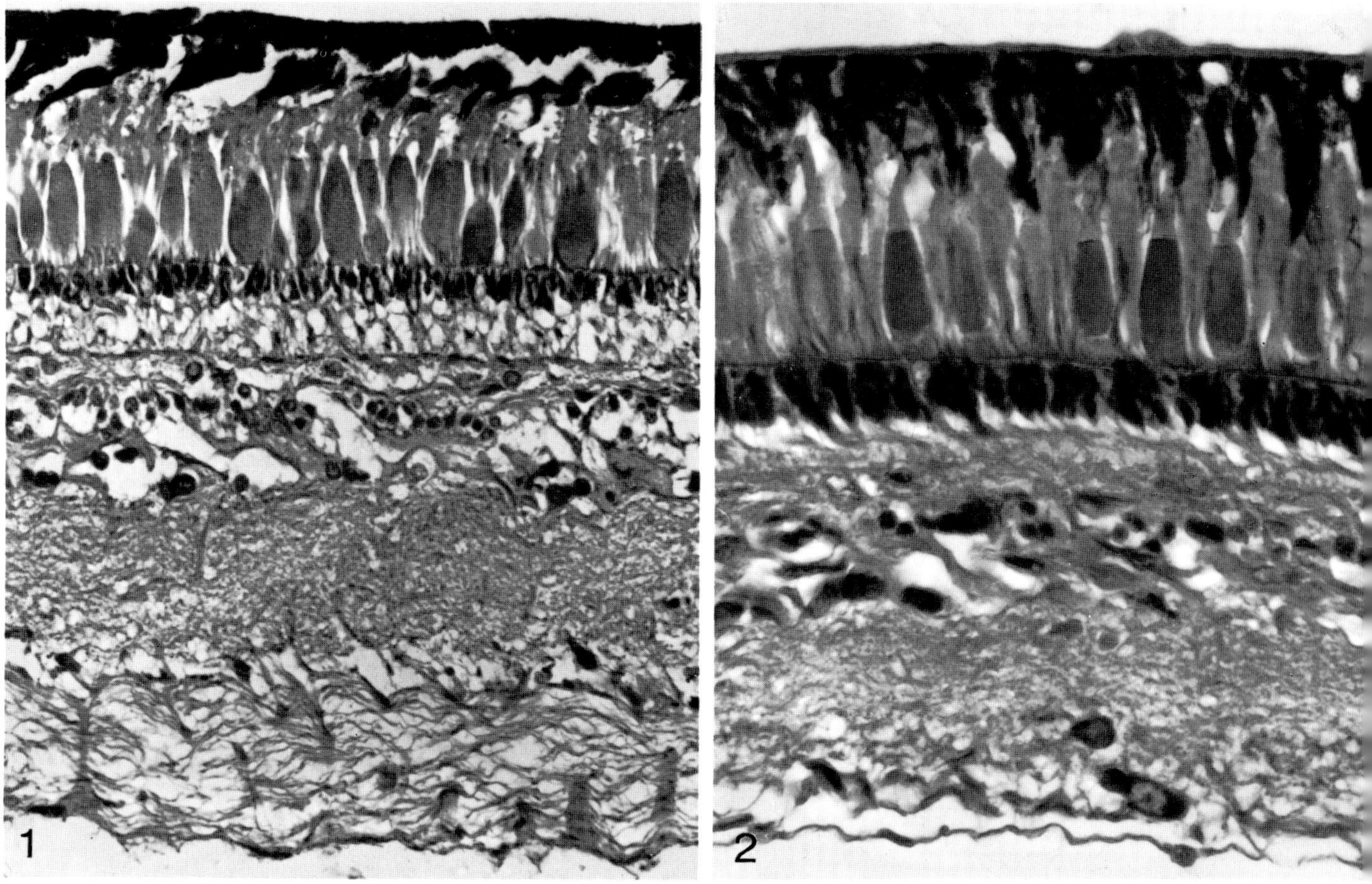

Plate 107.1, 2. Transverse sections of the retinas of two wolffishes, *A. lupus* (1) and *A. minor* (2). ×460

Stichaeidae (*Pricklebacks*)

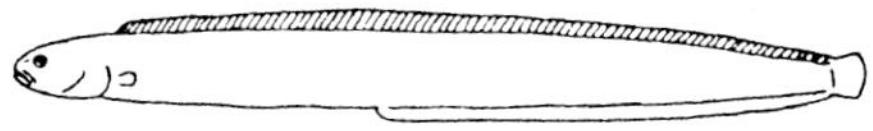

Acanthopterygii, Percomorpha, Perciformes, Blennioidei

Prickleback

Pricklebacks are generally elongate blennioid teleosts with a poorly developed lateral line. They are closely related to gunnels (Pholididae). Unlike gunnels, pricklebacks are not confined to the littoral or intertidal zone, but also occur in deep coastal marine waters. The fourline snakeblenny (*E. praecisus*) is found at depths of 20–100 m, and its lateral line system is exceptionally well developed.

The retina of the pricklebacks is characterised by the large numbers of cones, bipolar, amacrine, and ganglion cells represented in it (Plate 108.1). In spite of this, the heavily pigmented epithelial cells lack the processes found generally in well-developed retinas. The single and double cones are arranged in more or less regular quadrangular mosaics (Plate 108.2). There is a temporal area retinae. The rods are moderately numerous as seen transversely (Plate 108.1) and tangentially (Plate 108.2).

Arctic Shanny

The retina of the Arctic shanny, an active coastal fish, is better developed than that of *E. praecisus*. The epithelial cell processes are long and well pigmented in the light-adapted state (Plate 108.4). The cones are slender and densely packed. The bipolar and amacrine cells are very numerous and equally well represented.

Snakeblenny

The snakeblenny is the most slender of all the stichaeids, and the one with the largest eyes. This fish inhabits mud or clay bottoms from the lower tidemark to depths of 100 m. It tends to burrow in the mud or clay. Its diet includes small crustaceans.

The retina of *L. lumpretaeformis* resembles that of the fourline snakeblenny (*E. praecisus*), except that the epithelial cell processes are better developed here (Plate 108.3).

Radiated Shanny

The radiated shanny is a bottom fish, living among seaweed and stones from the low-tide mark down to 60 m or deeper.

The retina of this shanny follows the structural pattern of other stichaeids. The pigmented epithelium has very short processes as in *E. praecisus* (Plate 108.5). The rods are very long whereas the cones are very short, and the internal nuclear layer is less developed than in the other pricklebacks.

References: Engström (1963b); Ali et al. (1968); Anctil (1969)

Plate 108.1-5. Transverse (1), (3)–(5) ▷ and tangential (2) sections of the retinas of various stichaeid species (see text). (1) ×140; (2) ×340; (3) ×400; (4) and (5) ×420

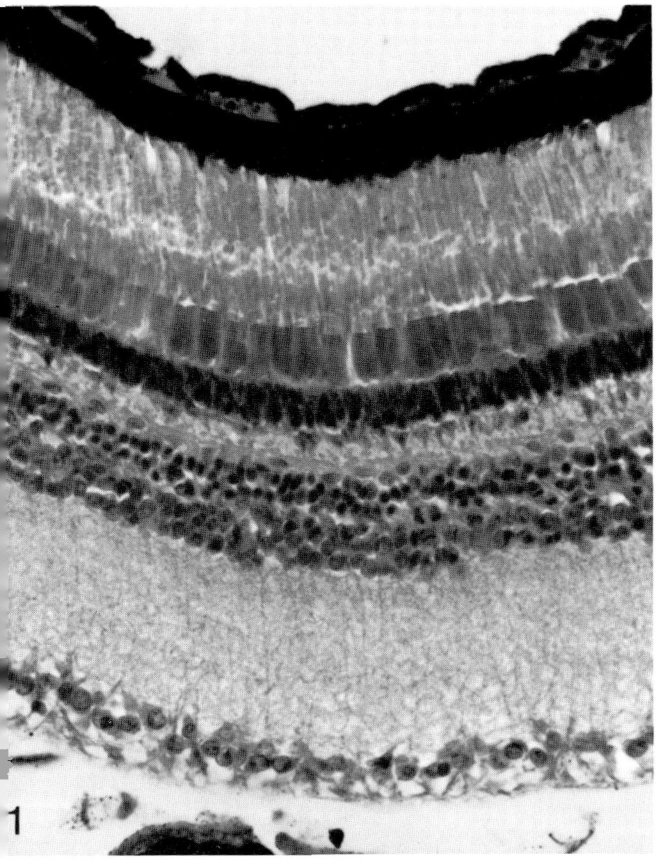
1

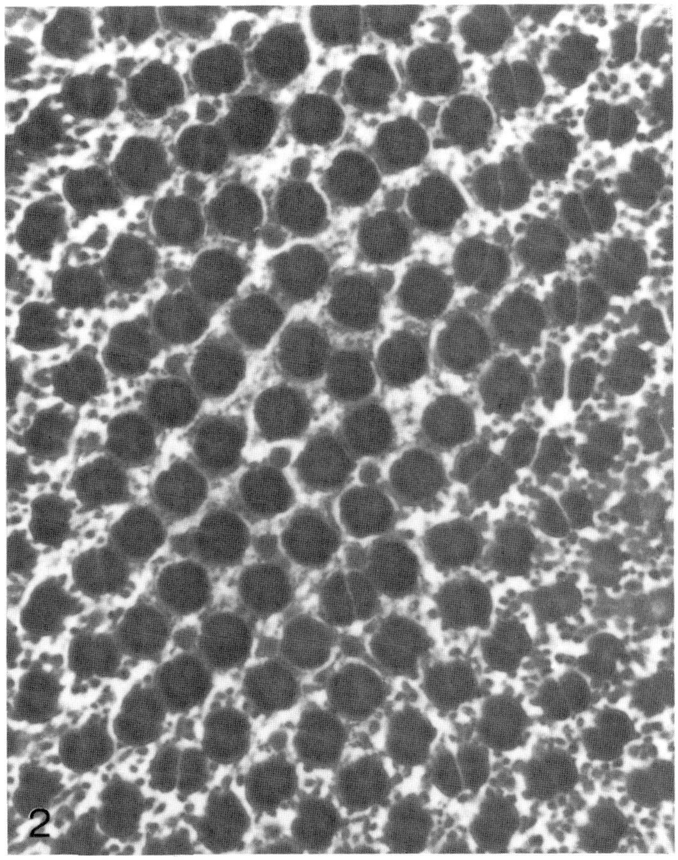
2

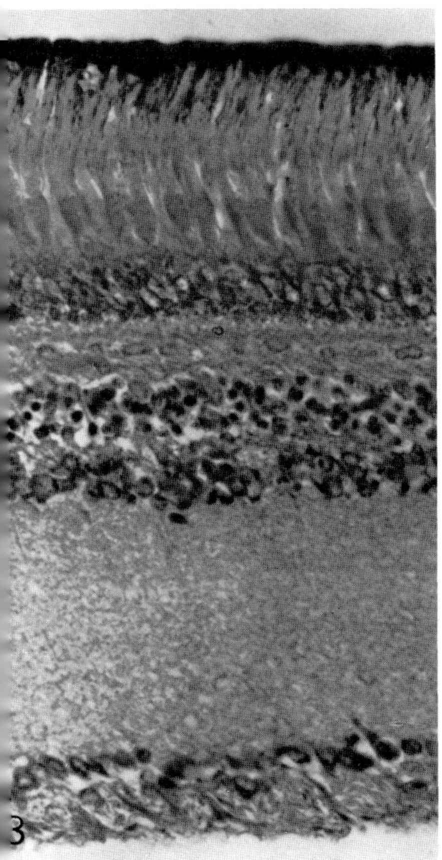
3

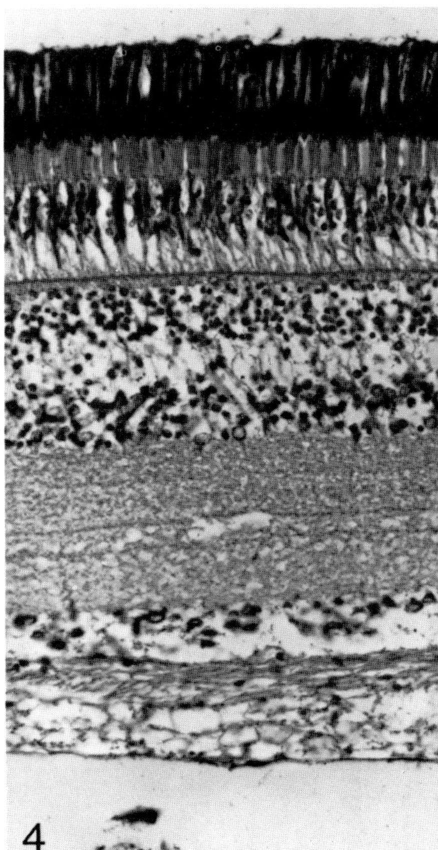
4

5

Ammodytidae (*Sand Lances*)

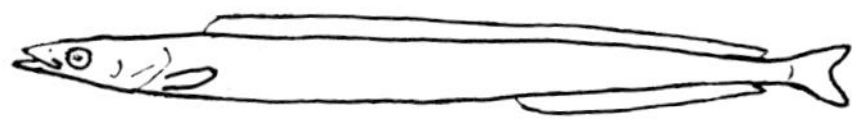

Acanthopterygii, Percomorpha, Perciformes, Ammodytoidei

Sand lances are small, elongate, carnivorous fish, somewhat eel-like in general appearance; they frequent inshore marine waters and often bury themselves in the bottom sand. They differ from many percoids in that the spinous dorsal fin is missing, and the soft-rayed dorsal is low and long. The northern sand lance, *Ammodytes dubius,* is found inshore and offshore on the continental shelf and occurs in large schools. The retina of the sand lance is well-developed (Plate 109.1). The epithelial pigment is abundant and distributed in long processes. The cones are numerous and the rods are large (Plate 109.2) and comparatively few. The equal double cones are dominant, but single and triple cones are also present; the double cones are arranged in parallel rows (Plate 109.3). There are two layers of small horizontal cells (Plate 109.1), and many bipolar, amacrine, and ganglion cells.

Reference: Friis (1879)

Plate 109.1-3. Transverse (1) and (2) and tangential (3) sections of the retina of the northern sand lance, *A. dubius.* (1) and (3) ×560; (2) ×720

▷

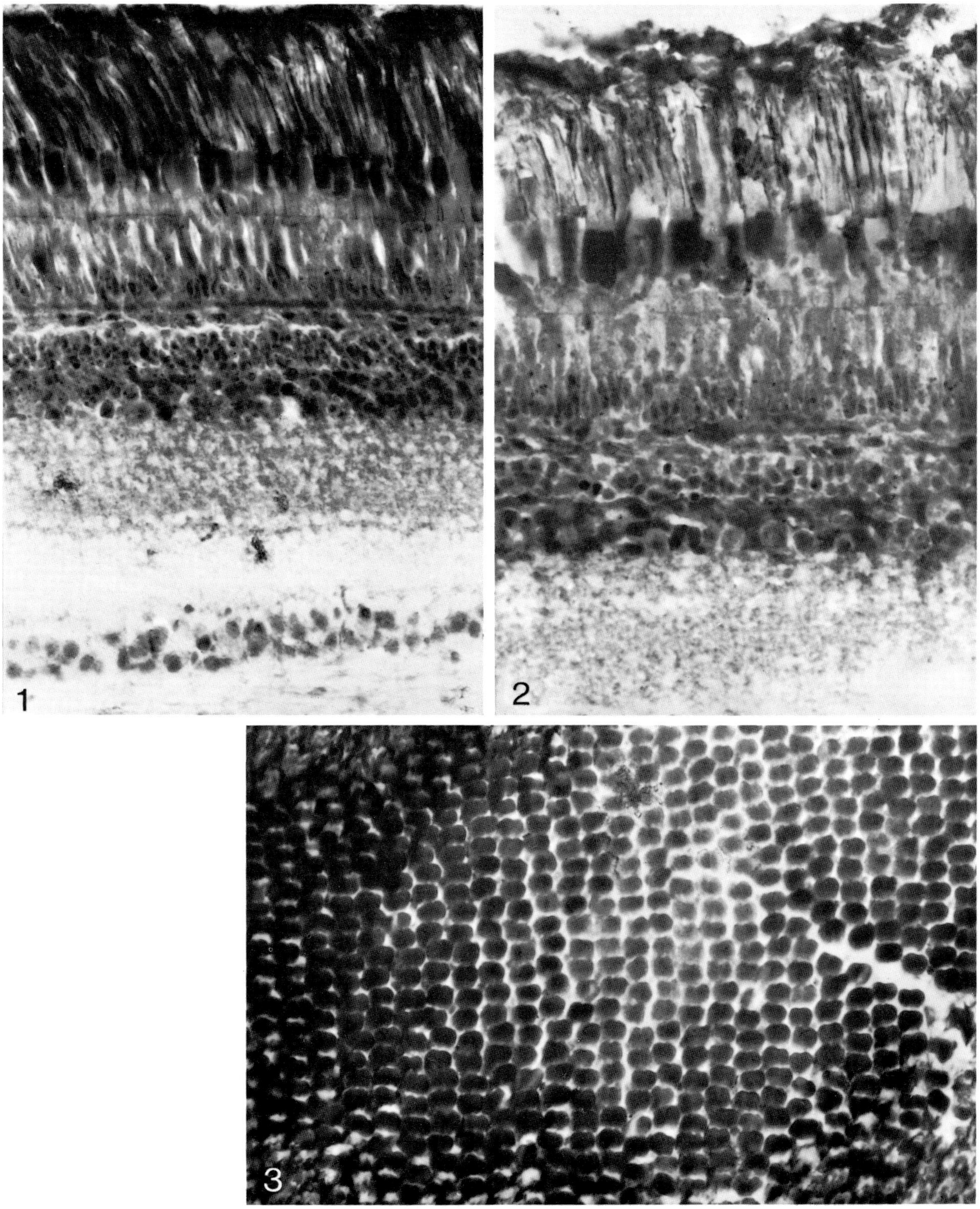

Plate 109.1-3. (Legend see opposite page)

Callionymidae (*Dragonets*)

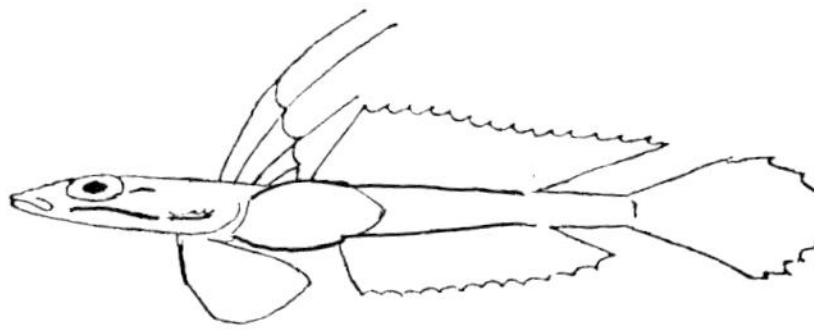

Acanthopterygii, Percomorpha, Perciformes, Callionymoidei

Dragonets comprise a group of small fishes with slender, anteriorly flattened bodies, a sharp preopercular spine usually in the form of a hook, and a very small gill opening on the upper part of the head. They live on the shelf floors of tropical and temperate seas.

The dorsal part of the retina of *Callionymus lyra* is pure-cone (Plate 110.1), while the ventral retina (Plate 110.2) has both rods and cones, with inner (vitread) layers much less developed. The single and equal double cones are arranged in regular square mosaics, (Plate 110.3) and there is a perfectly regular three-dimensional grid made of bipolar axons in the internal plexiform layer. The degree of convergence between the cone, bipolar, and ganglion cells (1:3:1) is remarkably low in the dorsal retina. Note the two giant amacrine cells (Plate 110.4) whose expansions lie in a single horizontal plane in the tangential section.

References: VILTER (1947a, 1947b); VRABEC (1966)

Visual Pigments: A_1 max 485–511 (ALI and WAGNER, 1975b)

Plate 110.1-4. Transverse (1) and (2) and tangential (3) and (4) sections of the retina of the dragonet, *C. lyra*. (4) Golgi-impregnated preparation. (1) ×260; (2) and (4) ×350; (3) ×1,000

▷

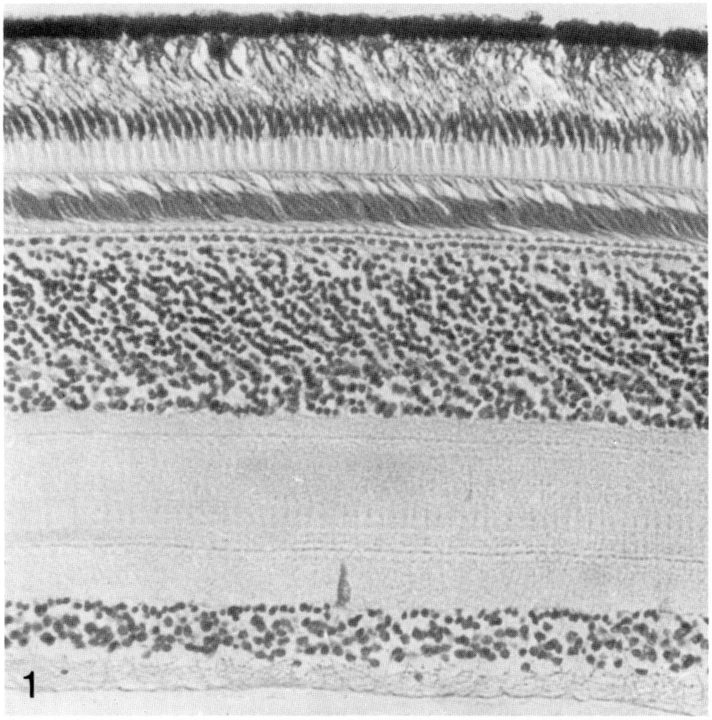

1

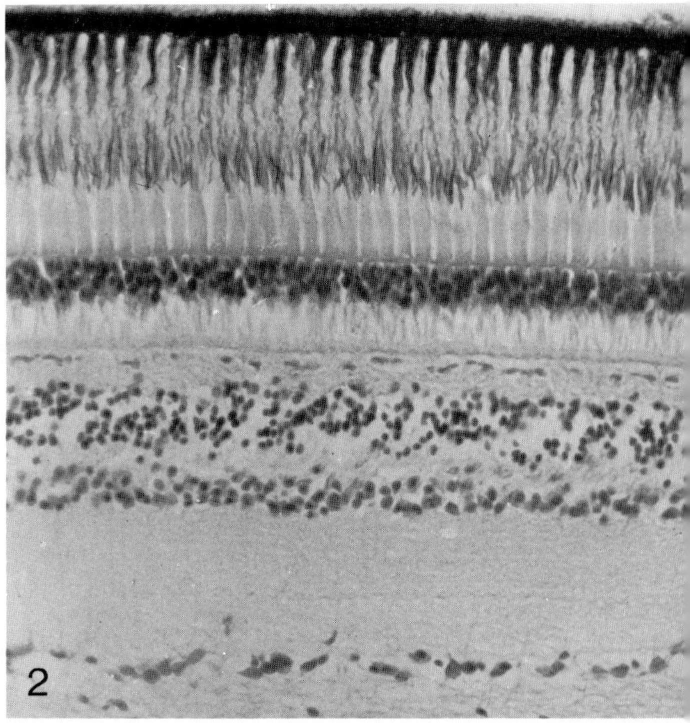

2

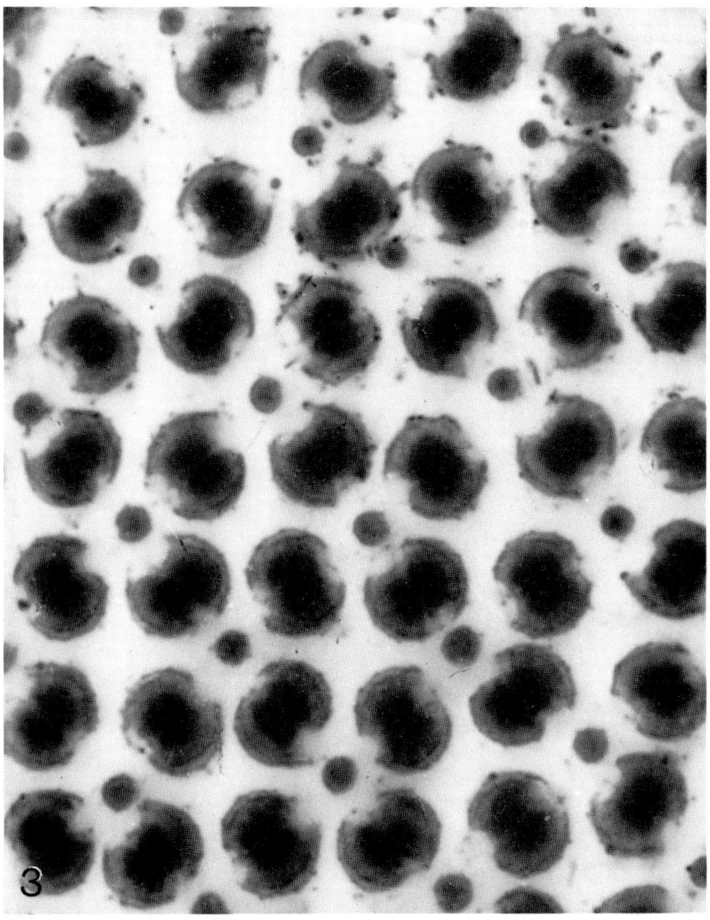

3

4

Gobiidae (*Gobies*)

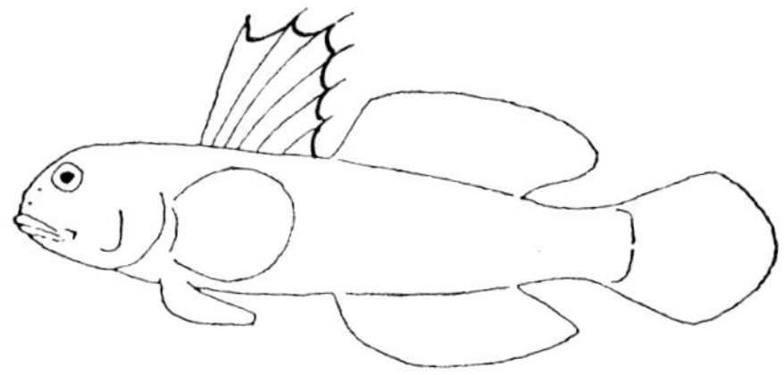

Acanthopterygii, Percomorpha, Perciformes, Gobioidei

Gobies are bottom fishes of world-wide distribution, whose main characteristics are the usually complete, sometimes partial, fusion of the pelvic fins, producing a basin-shaped sucker. Most gobies inhabit the shallow inshore waters of tropical seas. Some of them, the mudskippers, have adapted themselves to spend periods of time out of the water in moisture-saturated air.

The retinas of the mudskippers *Periophthalmus sobrinus* (Plate 111.1) and *Boleophthalmus sculptus* (Plate 111.2) are very well-developed and characterised by the presence of a horizontal band-shaped area retinae. This area extends across the retina slightly dorsal to the centre of the fundus, and is marked by a thickening of the retina (Plate 111.1,2). Both rods and cones are present in the visual cell layer; the cones are very slender in the area retinae (Plate 112.1,2). The bipolar and ganglion cells are extremely numerous in the area retinae (Plate 111.1,2). The retina of a Brazilian goby (*Barbulifer ceuthoecus*) from a tide pool is depicted in Plate 113.1,2,3. The change in the mosaics from dorsal to ventral regions is noteworthy.

References: Franz (1910); Karsten (1923); Verrier (1928a); Engström (1963b); Munk (1970); Bathelt (1970); Moskalkova (1971); Wagner (1972); Kunz and Kelly (1973)

Visual Pigments: A_1 max 500–512 (Ali and Heumann, 1972; Ali and Wagner, 1975b)

Trypauchenidae (*Eel Gobies*)

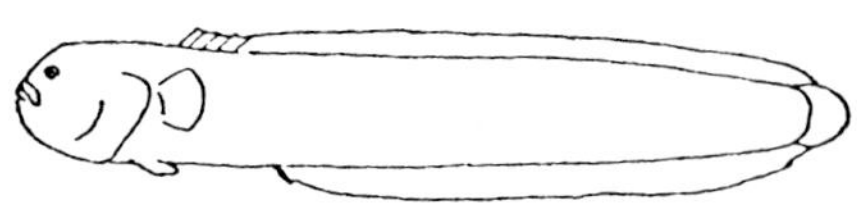

Acanthopterygii, Percomorpha, Perciformes, Gibioidei

Elongated eel gobies have representatives in fresh and salt water. Like the true gobies (Gobiidae), they usually have a sucking disc under the body, but they differ by the presence of a spiny-rayed dorsal fin that is continuous with the soft-rayed portion. The eyes of the eel gobies are small or vestigial.

Pure-cone retinas were reported in the cave-dwelling gobies *Trypauchen wakae* and *Trypauchenophrys anotus*. On the other hand, *T. vagina* is said to possess a pure-rod retina.

References: Franz (1910); Eggert (1931)

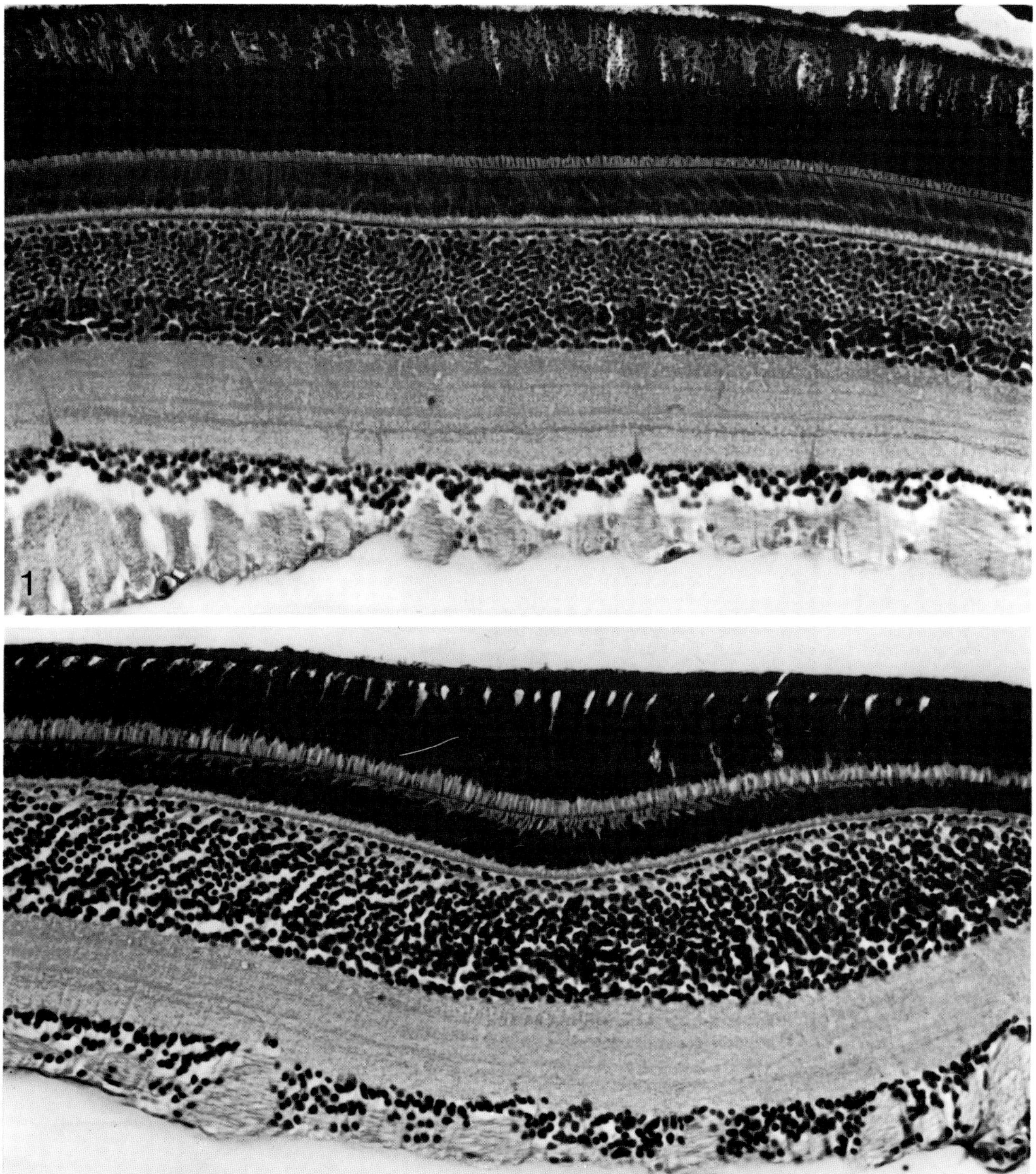

Plate 111.1, 2. Photomicrographs of the transversely cut retinas of *P. sobrinus* (1) and *B. scultus* (2) in the region of the band-shaped area retinae. ×350

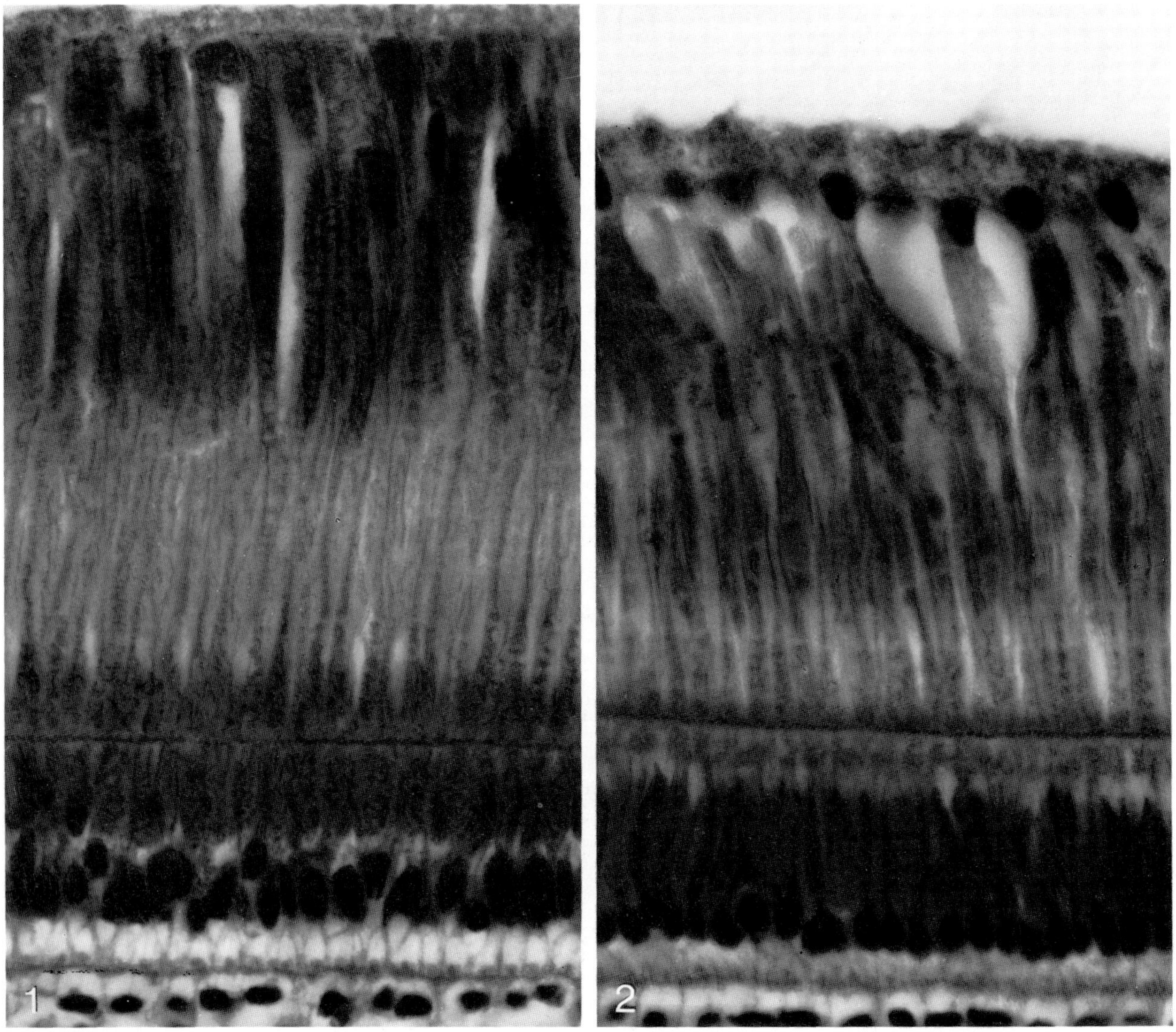

Plate 112.1, 2. Transverse sections (× 1,400) of the depigmented retinas of *P. sobrinus* (1) and *B. sculptus* (2)

Plate 113.1, 2. Transverse section (× 450) of the central region of the retina of *B. ceuthoecus* (1). (2) (× 525) and (3) (× 1,380) show tangential sections. Note the square mosaics and the change in them from dorsal (2) to ventral (3) regions. In the dorsal region the mosaics contain additional single cones while in the ventral there is a central single cone ▷

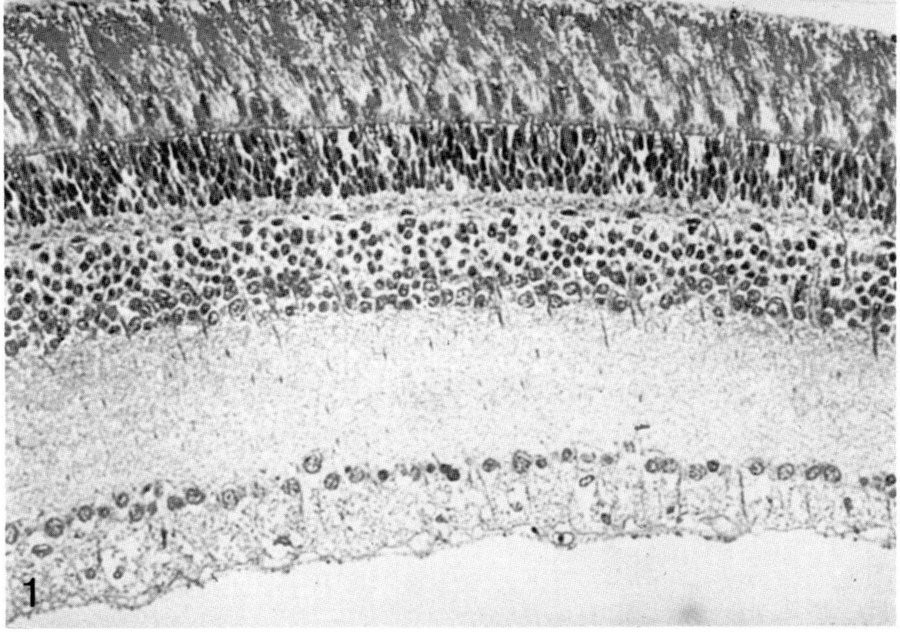

1

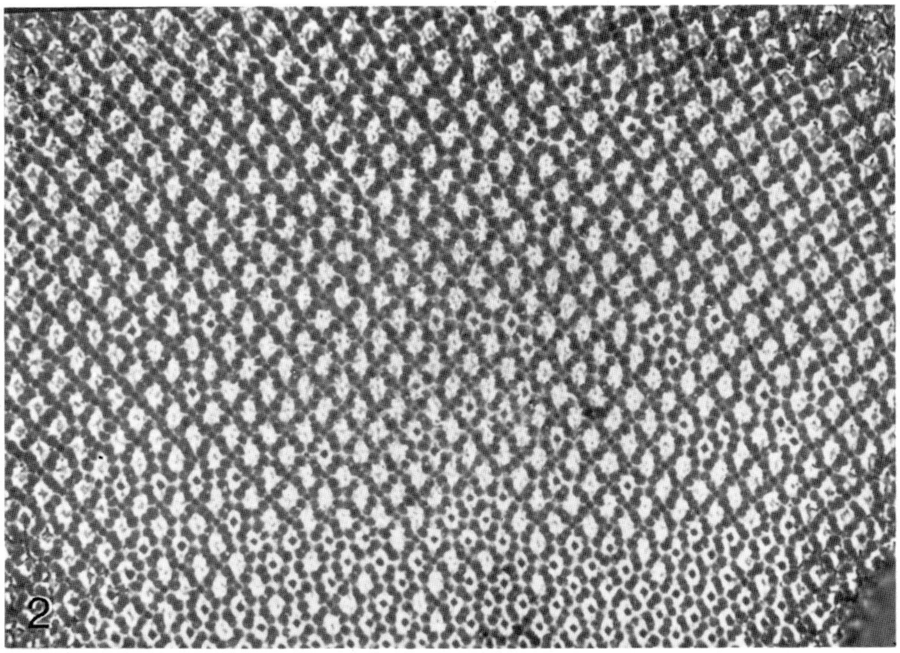

2

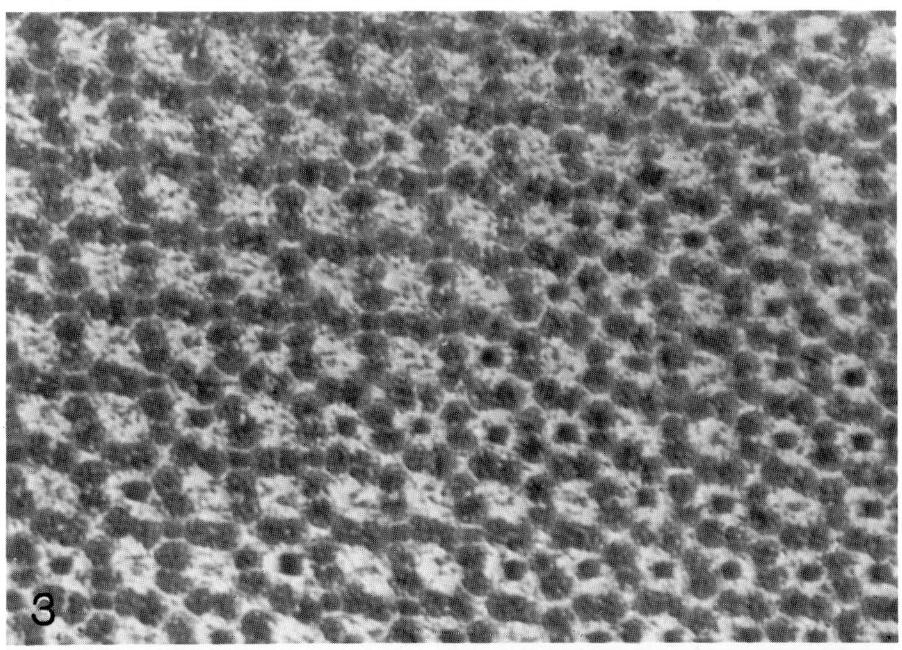

3

Acanthuridae (*Surgeonfishes*)

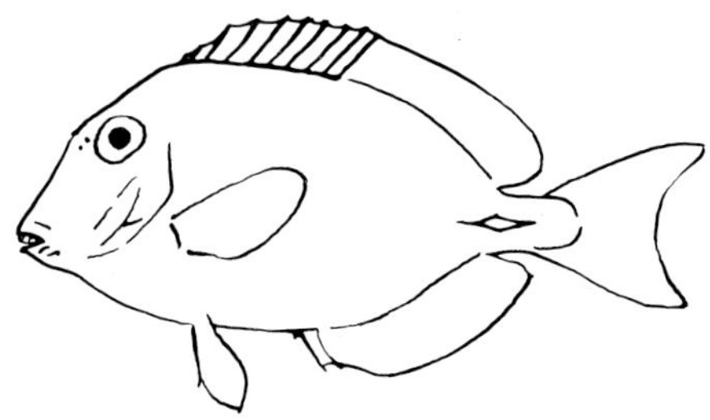

Acanthopterygii, Percomorpha, Perciformes, Acanthuroidei

Acanthuroids differ from the percoid teleosts principally by the anatomical relationship of the head to the rest of the skeleton. These fishes have compressed bodies and their life cycle includes a strange "acronurus" larval stage with vertical ridges on the body. These tropical, inshore fishes are herbivorous and diurnal.

The retina of *Xesurus scalprum* is well-developed. There are many single and double cones even in the nasal region, and the area retinae, which is not sharply demarcated. The area is found in the dorso-temporal region.

Reference: TAMURA (1957)

Visual Pigments: A_1 max 490–497; $A_2 > 10\%$ (ALI and HEUMANN, 1970, 1972; MUNZ and McFARLAND, 1973)

Siganidae (*Rabbitfishes*)

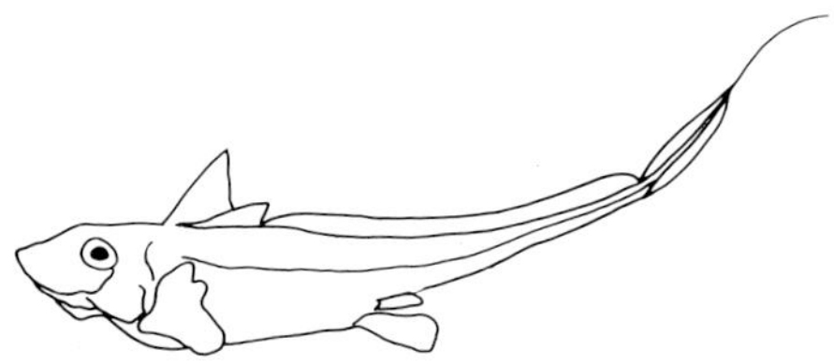

Acanthopterygii, Percomorpha, Perciformes, Acanthuroidei

Rabbitfishes are small, inshore teleosts related to the surgeonfishes, but differ from the latter by the presence of two spines on each pelvic fin. They are mostly herbivorous and are noted for rapid colour changes. The foxface (*L. vulpinus*) differs significantly in morphology from all the other siganids; it possesses a long, foxlike tubular snout. Siganids are confined to the tropical Indo-Pacific area.

The foxface (*L. vulpinus*) possesses a very well-developed retina that resembles those of other shorefishes such as the damselfishes and butterflyfishes. In the light-adapted state the pigment is confined to the epithelial cell processes (Plate 114.1). Both rods and cones are abundant, and the small single and equal double cones form regular square mosaics (Plate 114.3). This regular square arrangement can be followed in the horizontal cell layer (Plate 114.4). As seen in the ventral (Plate 114.1) and nasal (Plate 114.2) retinal regions, the bipolar and amacrine layers are almost equally well-developed.

Visual Pigments: A_1 max 495.5–495.6; $A_2 < 10\%$ (MUNZ and McFARLAND, 1973)

Plate 114.1-4. Transverse (1) and (2) and tangential (3) and (4) sections of the retina of the foxface, *L. vulpinus*. × 560 ▷

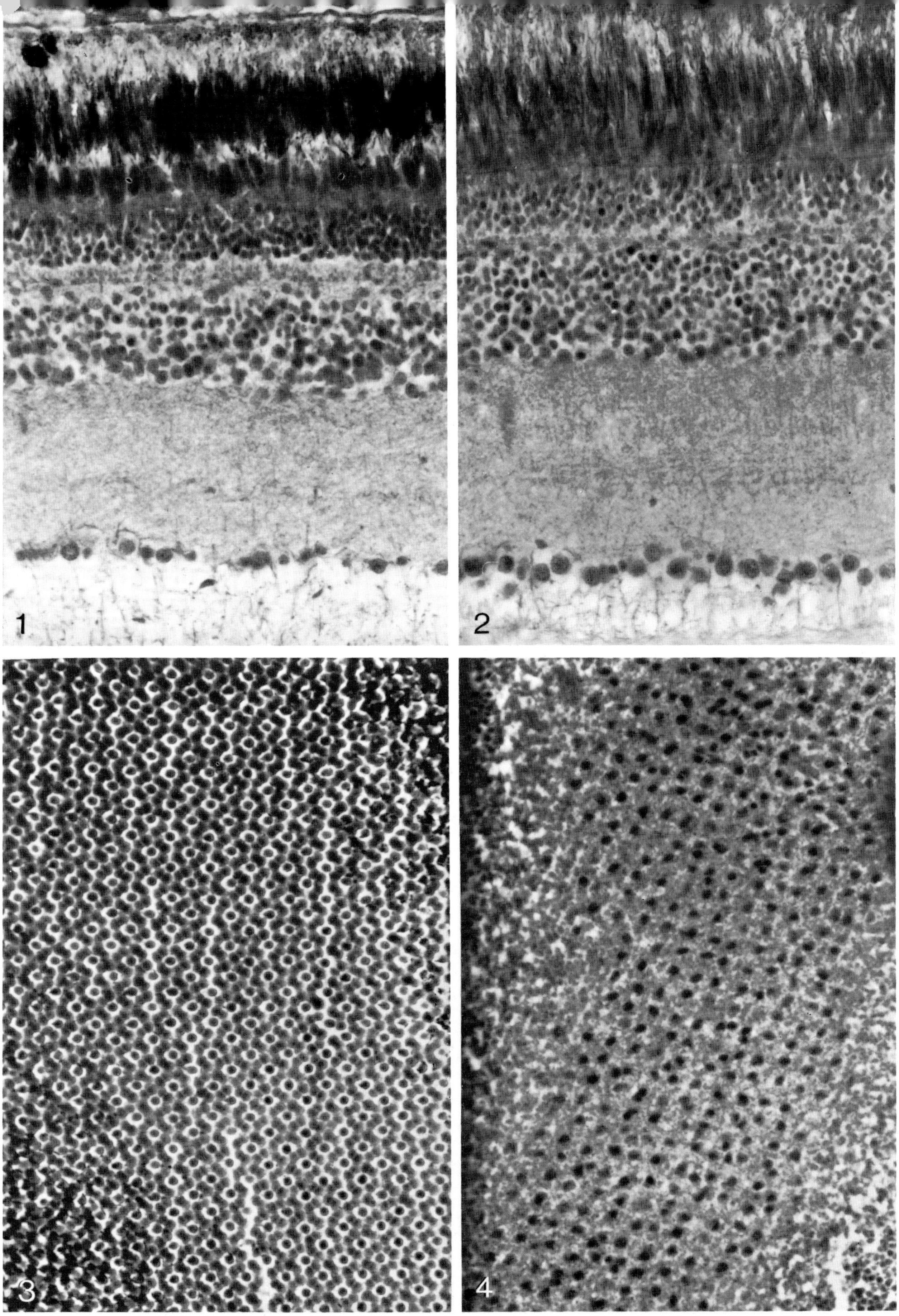
1
2
3
4

Trichiuridae (*Cutlassfishes*)

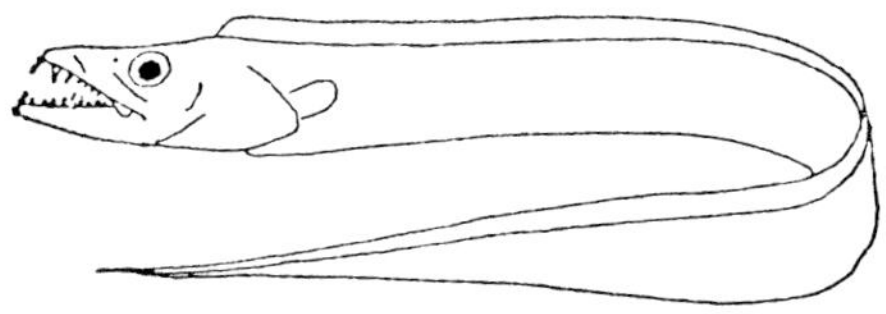

Acanthopterygii, Percomorpha, Perciformes, Scombroidei

Cutlassfishes are oceanic scombroid teleosts with narrow, compressed bodies, and snouts like a beak, armed with an impressive dentition. Most of them are pelagic, deep-water forms. They are good swimmers and active oceanic predators.

The retina of the Atlantic cutlass fish, *Trichiurus lepturus,* is characterised by the presence of epithelial cells with pigmented short processes, and of a thick visual cell layer composed of numerous slender rods and relatively large single cones (Plate 115.1). The rod ellipsoids are very tiny and distributed in several irregular banks. The single cones are rodlike in appearance, except for the tapering of the long outer segment. Three layers of horizontal cells are present, the median one being the largest (Plate 115.2). There are two or three layers of packed bipolars and amacrines in the thin internal nuclear layer.

Reference: BRAUER (1908), on deep-sea *Lepidopus tenuis,* a fish with a pure-rod retina

Plate 115.1, 2. Transverse sections of the retina of the Atlantic cutlassfish, *T. lepturus.* ×560. *H* horizontal cell layer

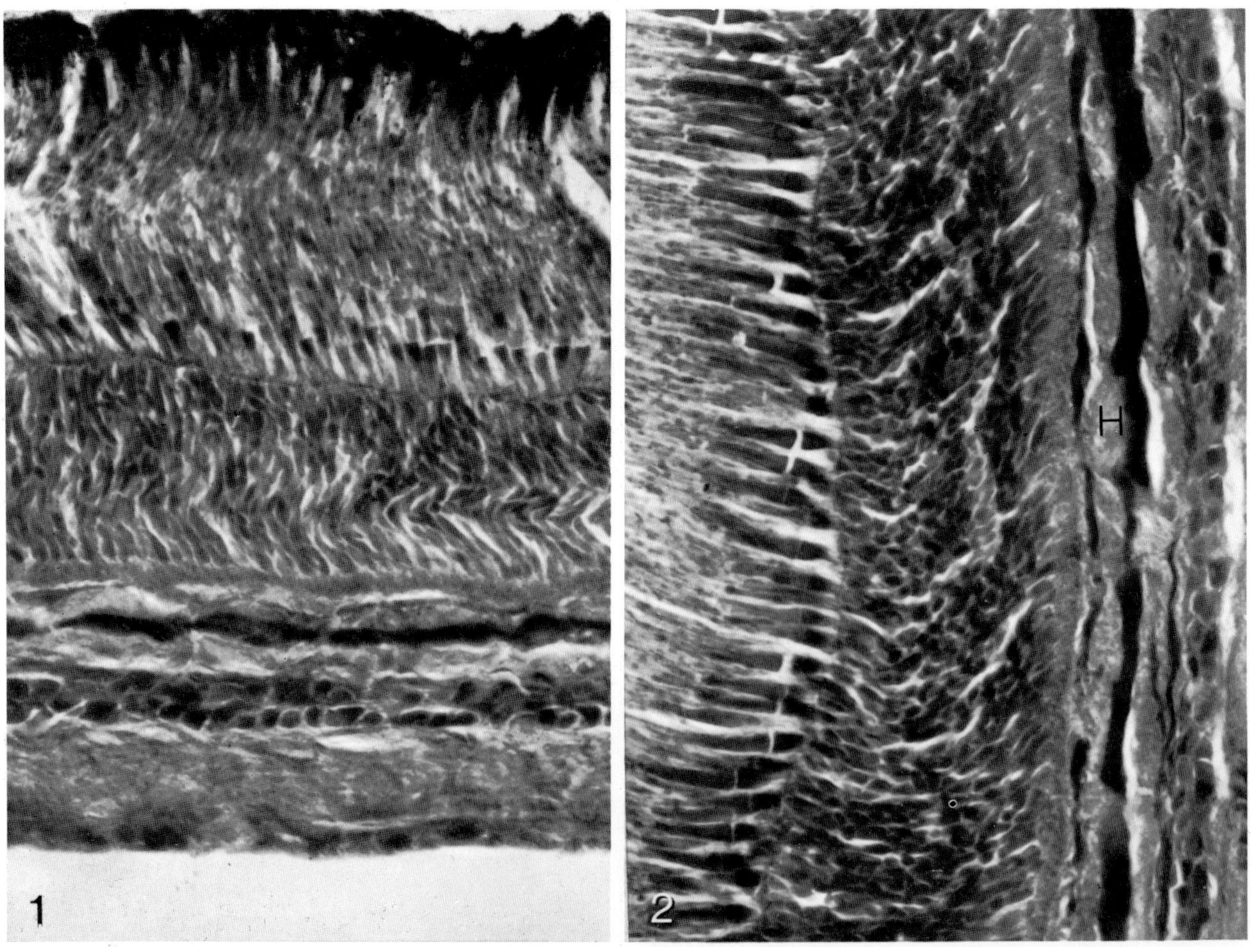

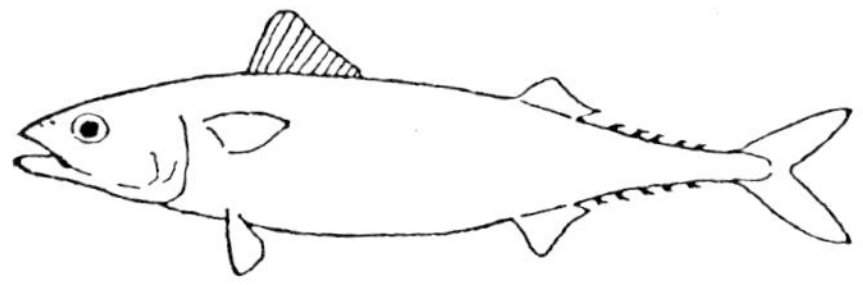

Scombridae (*Mackerels and Tunas*)

Acanthopterygii, Percomorpha, Perciformes, Scombroidei

This is a family of streamlined, fast swimmers living a pelagic life. A typical mackerel or tuna has a smooth, cigar-shaped body, a narrow caudal peduncle, and a large, strong tail divided into individual lobes. The second dorsal and the anal fin are followed by a series of small finlets.

The retina of *Scomber scombrus* possesses all the characters of a well-developed teleostean "duplex" retina. Its visual cell layer is composed of numerous rods, single and equal double cones. There are different retinomotor responses between the single and double cones, and in tangential sections the cone mosaic is made of alternate rows of single and double cones. Three horizontal cell layers are present. The area retinae of the scombrids is found temporally (*Scomber*) or ventro-temporally (*Acanthocybium, Euthynnus, Scomberomorous, Thunnus*).

References: Friis (1879); Verrier (1928a); Tretjakoff (1930); Tamura (1957); Engström (1963b); O'Connell (1963); Tamura and Wisby (1963); Wagner (1972); Kawamura and Tamura (1973); Ali and Wagner (1975a)

Visual Pigments: A_1 max 484; A_2 max 530 (Ali and Heumann, 1970; Munz and McFarland, 1973; Ali and Wagner, 1975b)

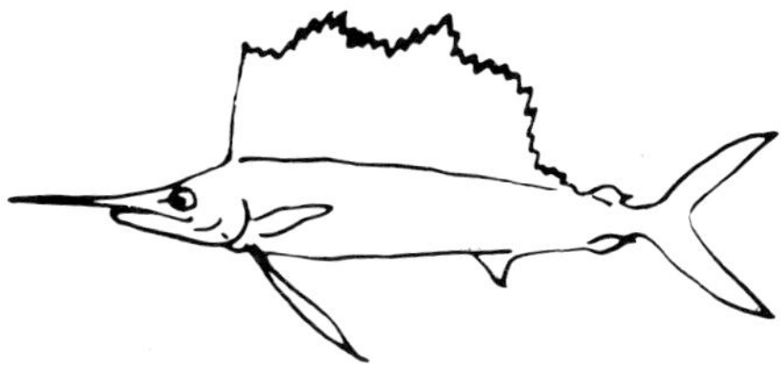

Istiophoridae (*Billfishes and Sailfishes*)

Acanthopterygii, Percomorpha, Perciformes, Scombroidei

Istiophorids are scombroid teleosts with long, rounded bills and two ridges on each side of the caudal peduncle. These large epipelagic fishes are active fish eaters.

The retinas of the sailfish (*Istiophorus albicans*) and white marlin (*Tetrapturus albidus*) are well-developed. The single and double cones are very numerous and there is a well-demarcated area retinae in the ventro-temporal region. In the tangential section cones are arranged in square mosaics.

Reference: Tamura and Wisby (1963)

Visual Pigments: A_1 max 484.4–486 (Munz and McFarland, 1973; Ali and Wagner, 1975b)

Scorpaenidae (*Scorpionfishes*)

Acanthopterygii, Percomorpha, Scorpaeniformes, Scorpaenoidei

Scorpionfishes (also known as rockfishes) are perchlike or basslike marine fishes, however they are related to the sculpins by having a bony stay stretching across the cheek area. These fishes are pelagic or benthic in the continental shelf waters. Their diet includes a variety of crustaceans and small fishes. *Sebastes marinus marinus* is usually confined to depths of 100–200 fathoms while *S. m. mentella* has been reported in waters 200–300 fathoms deep (a fathom is about two metres).

The retinal epithelial pigment is densely packed in the cell bodies sclerally. The rods are numerous and slender while the cones are comparatively few and bulky (Plate 116.1), almost exclusively represented by equal double cones (Plate 116.2). The internal nuclear layers (bipolars, ganglion cells) are poorly developed (Plate 116.1, 3). Note the difference in size and position of the cones between the retinas of *S. m. marinus* and *S. m. mentella* (Plate 116.1, 3).

References: BRAUER (1908); TAMURA (1957); WUNDER (1958); HANYU and ALI (1962); ALI and HANYU (1963); ALI and WAGNER (1975a)

Visual Pigments: A_1 max 491–500; $A_2 < 10\%$ (BEATTY, 1973; ALI and HEUMANN, 1970, 1972; MUNZ and MCFARLAND, 1973; ALI and WAGNER, 1975b)

Plate 116.1-3. Transverse (1) and (3) and tangential (2) sections of the retina of the redfish *S. marinus*. (1) and (3) ×440; (2) ×1,105

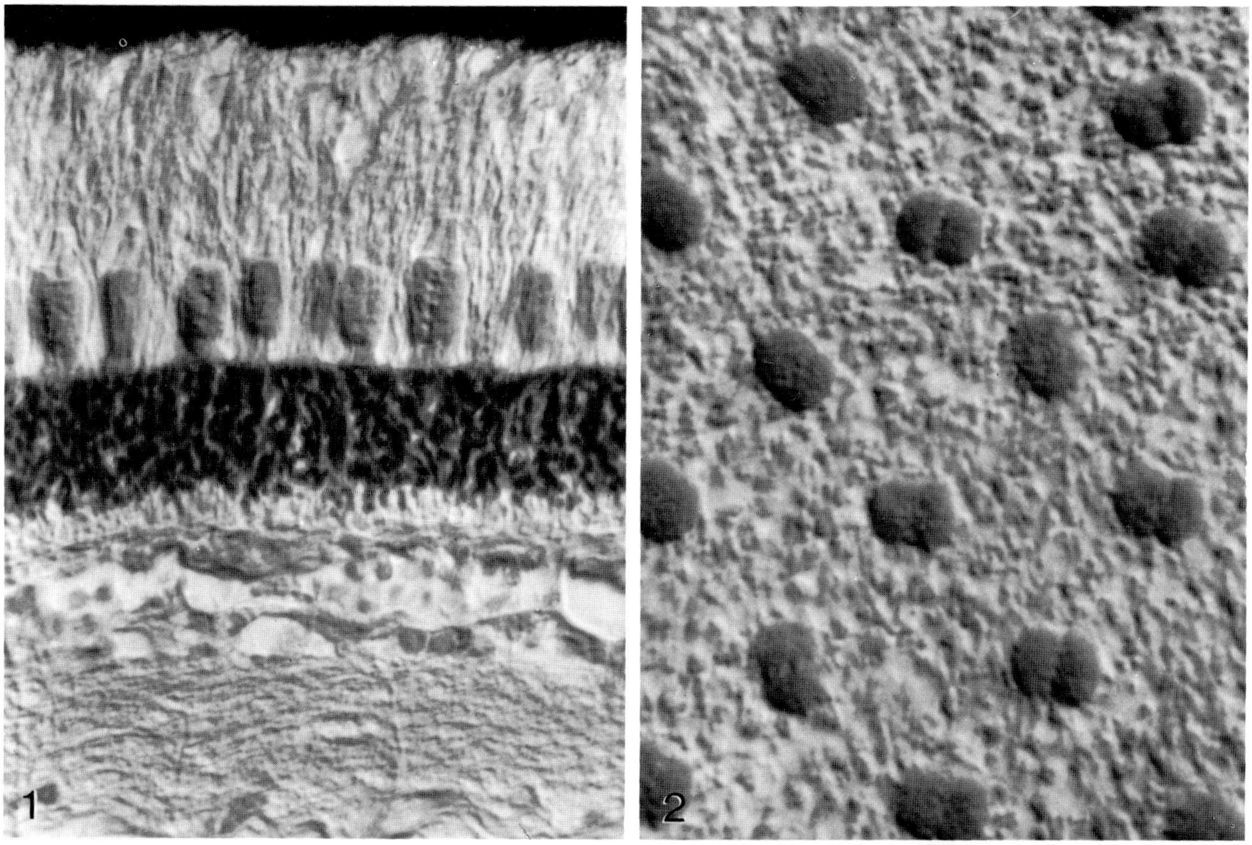

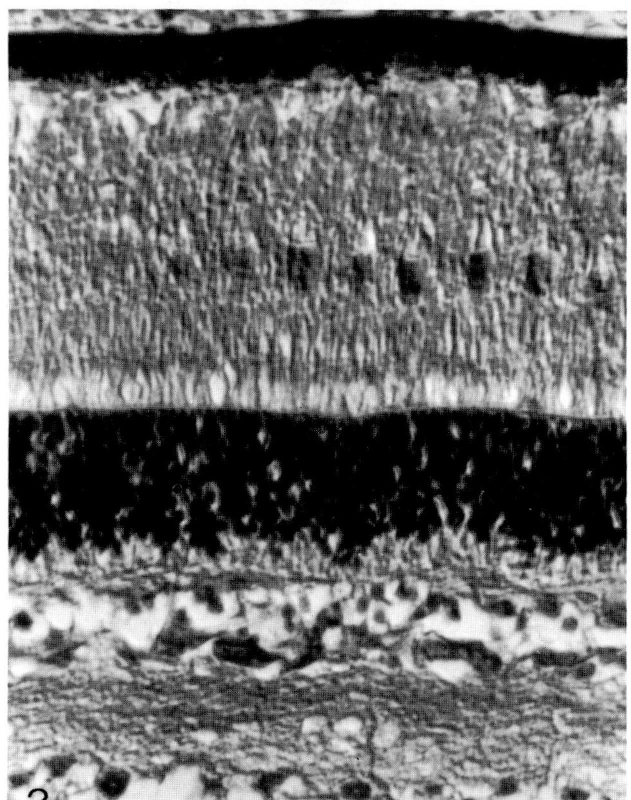

Plate 116.1-3. (Legend see opposite page)

Triglidae (*Sea Robins*)

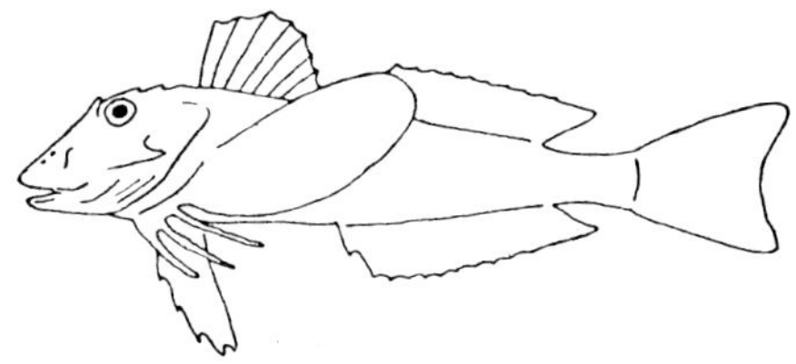

Acanthopterygii, Percomorpha, Scorpaeniformes, Scorpaenoidei

Sea robins are marine fishes from the continental shelf, and are characterised by bony heads armed with spines and pectoral fins, whose lower rays are modified into feelers. These feelers are used by the fish while "walking" on the bottom. These benthic fishes are found in shallow to moderately deep shelf waters, and they feed mainly on crustaceans and molluscs.

The retina is characterised by the presence of rods, single and equal double cones arranged in more or less regular square mosaics. Triple cones are also present. The area retinae is temporal (*Chelidonichthys kumu, Trigla gurnardus*). In *Peristedion rivers-andersoni,* the epithelial cell processes are few and short, and the epithelial pigment is packed sclerally in the cell bodies. The internal nuclear layers are poorly developed, except in the temporal region.

References: BRAUER (1908); VERRIER (1928a); TAMURA (1957); ENGSTRÖM (1963b)

Visual Pigments: A_1 max 491–511 (ALI and HEUMANN, 1970, 1972; ALI and WAGNER, 1975b)

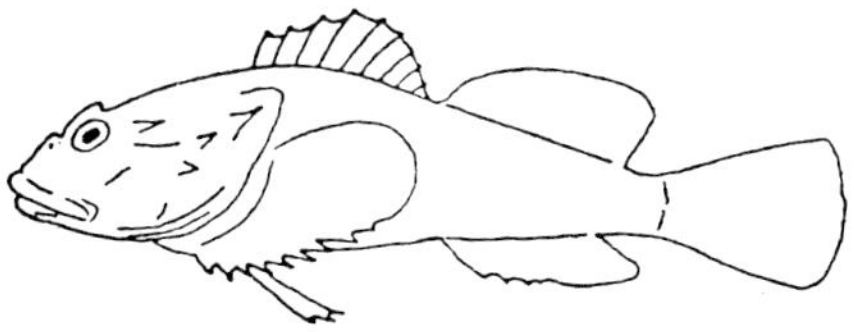

Cottidae (*Sculpins*)

Acanthopterygii, Percomorpha, Scorpaeniformes, Cottoidei

Sculpins are stout-bodied marine and freshwater fishes with large heads and prominent eyes, and with a preopercular bone often equipped with spines. They inhabit a variety of depths in the continental shelf area. The sea ravens (*Hemitripterus*) are found over rocky or hard sand bottoms to a depth of 200 m where they feed on whatever they find (invertebrates, fish).

The retina of the sea raven (*H. americanus*) is very much teleostean in appearance (Plate 117.1). The epithelial cell processes and pigment are well represented; rods are numerous, and long single cones as well as equal double cones are seen.

Mailed Sculpin

The mailed sculpin (*Triglops nybelini*) is a small bottom form, feeding on worms and various crustaceans.

Its retina resembles that of other sculpins (see the dorso-temporal retina near the ora terminalis, Plate 117.6) except that the area retinae is ventro-temporal and the cone densities are higher, while the internal nuclear layers are better organised.

Longhorn and Shorthorn Sculpins

These sculpins are omnivorous and voracious fishes, feeding chiefly on bottom invertebrates. The longhorn sculpin (*Myoxocephalus octodecemspinosus*) has a wider depth range than *M. scorpius* (Shorthorn sculpin).

The retina is essentially similar to that of the sea raven (cf. Plate 117.1). The epithelial processes are well-developed though the pigment is only moderately abundant (Plate 117.2, 3). Equal double cones and single cones form comparable square mosaics in both species (Plate 117.4, 5). The area retinae is temporal (Plate 117.5), and two horizontal layers are present (Plate 117.4).

References: (Cottids general): LYALL (1957); TAMURA (1957); ENGSTRÖM (1963b); GOODLAND (1966); ALI et al. (1968); ANCTIL (1969); BATHELT (1970); WAGNER (1972)

Visual Pigments: A_1 max 492–511 (BEATTY, 1973; ALI and WAGNER, 1975b)

Plate 117.1-6. Transverse (1), (2), (3) and (6) and tangential (4) and (5) sections of the retina of various cottid species (see text). (1, 2, 3) ×320; (4) and (5) ×560; (6) ×360

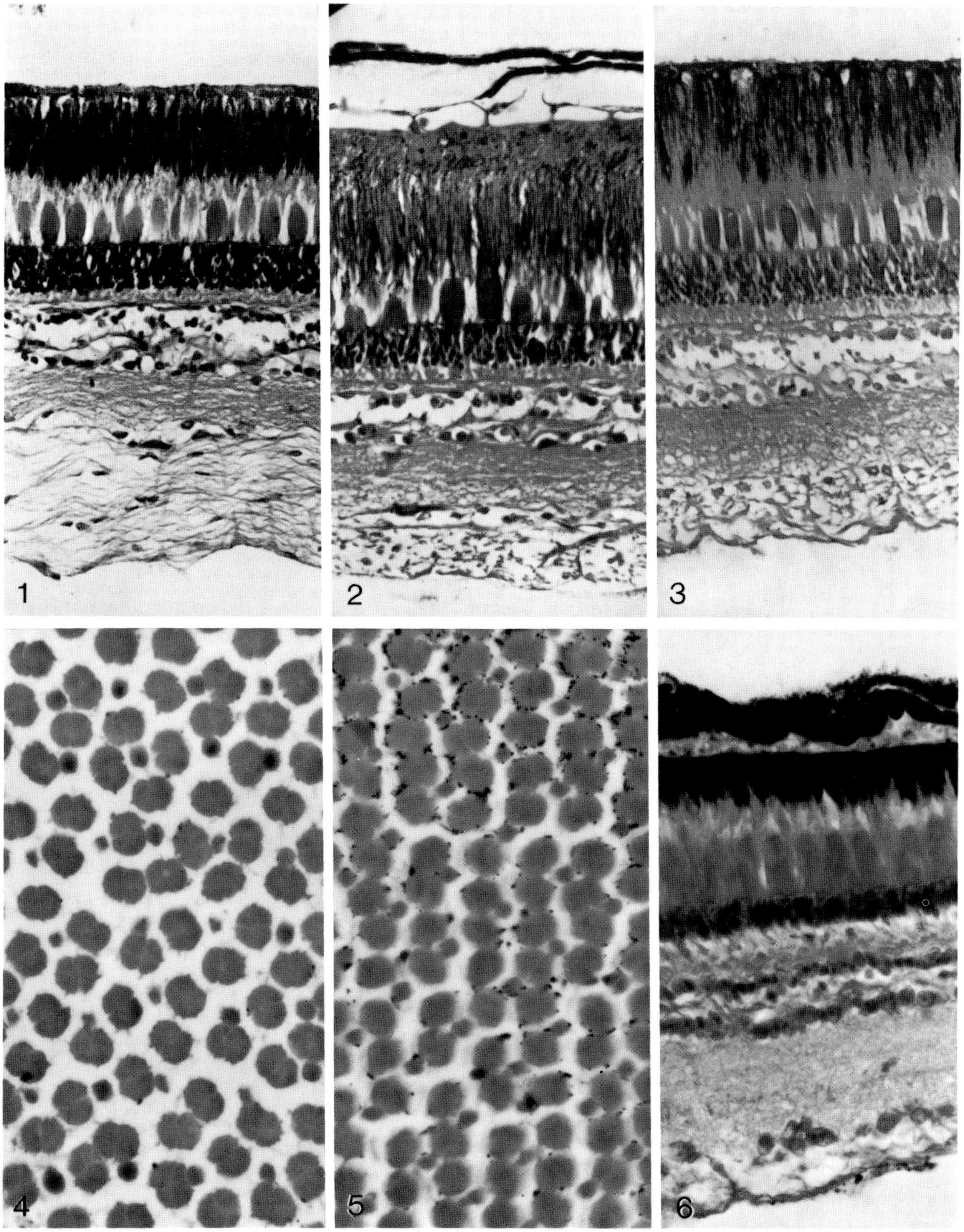

Plate 117.1-6. (Legend see opposite page)

Agonidae (*Poachers*)

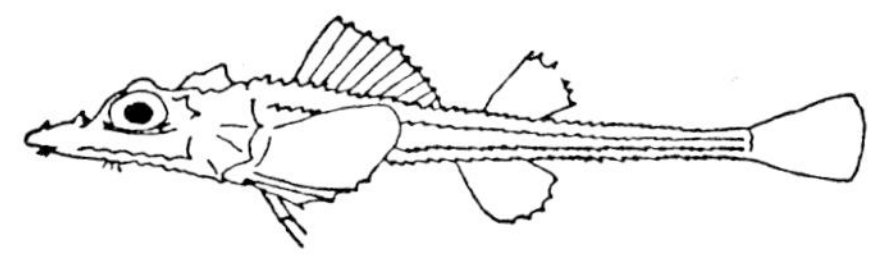

Acanthopterygii, Percomorpha, Scorpaeniformes, Cottoidei

Atlantic Poacher

Poachers are marine fishes, being related to the sculpins morphologically. Their bodies are armed with several rows of overlapping plates. These slender fishes inhabit the continental shelf on the bottom at various depths.

The retina of the Atlantic poacher is moderately developed; the epithelial cell processes are short and filled with pigment, the rods are relatively numerous (Plate 118.1), and the cones are few, mainly of the double cone type (Plate 118.2). Some triple cones are present.

References: ENGSTRÖM (1963b); ALI and HANYU (1963); ANCTIL (1969); ALI and WAGNER (1975a)

Alligatorfishes

The alligatorfish is also a benthic form, found over pebble, sand, or soft mud. As in the case of the Atlantic poacher, its depth range is about 20–200 m.

Plate 118.3 shows the dorsal retina of the alligatorfish with the epithelial cell layer removed. Rods are numerous and relatively short, while the cones (single and double) are fairly large. The bipolar and amacrine cells form two more or less defined layers. Note in Plate 118.4 the equal double cones, which tend to be organised in square mosaics.

Reference: ANCTIL (1969)

Visual Pigments: A_1 max 499 (BEATTY, 1973)

Plate 118.1-4. Transverse (1) and (3) and tangential (2) and (4) sections of the retinas of Atlantic poacher, *Agonus decagonus,* and the alligatorfish, *Aspidophoroides monopterygius.* (1) and (3) ×560; (2) ×630; (4) ×730 ▷

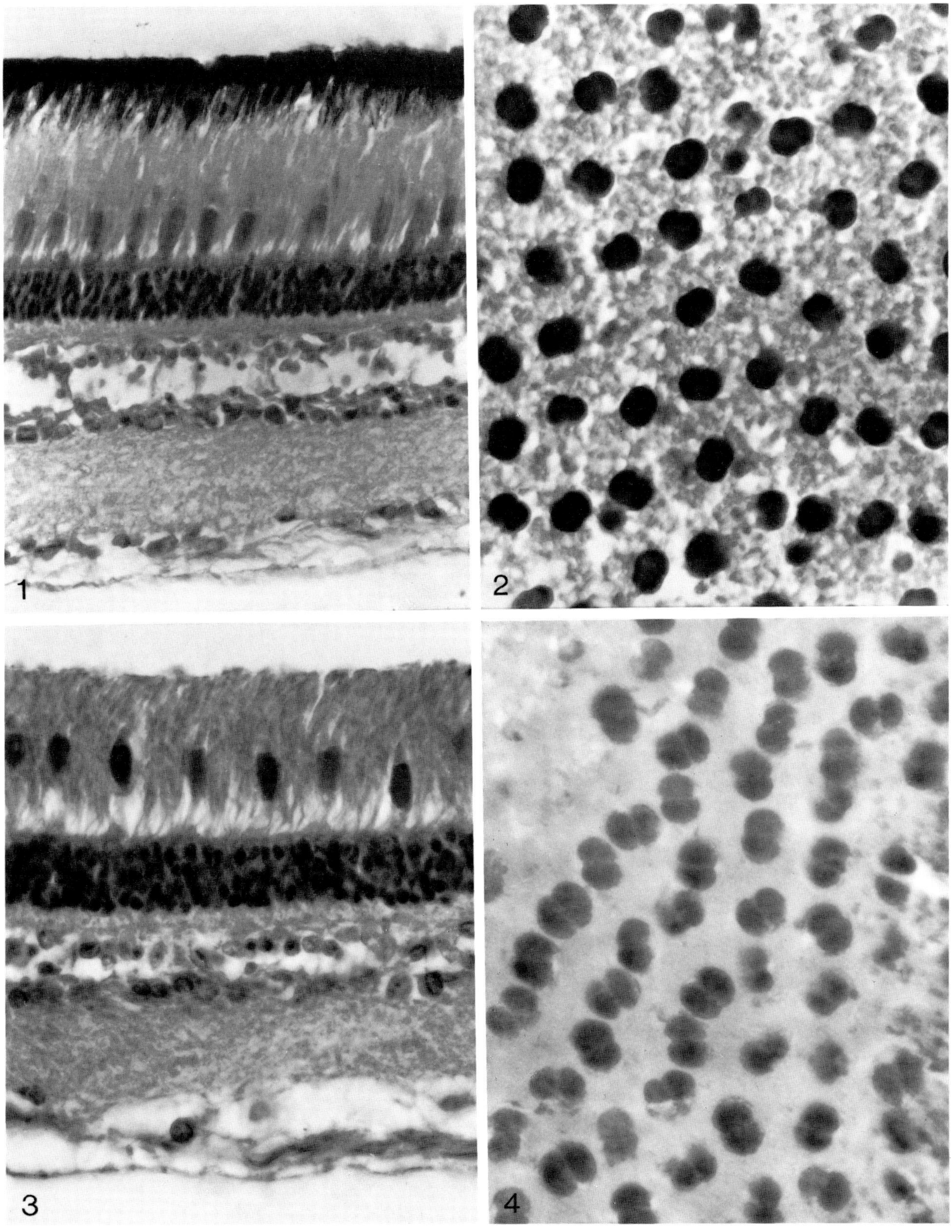

Plate 118.1-4. (Legend see opposite page)

Cyclopteridae (*Lumpfishes and Snailfishes*)

Acanthopterygii, Percomorpha, Scorpaeniformes, Cottoidei

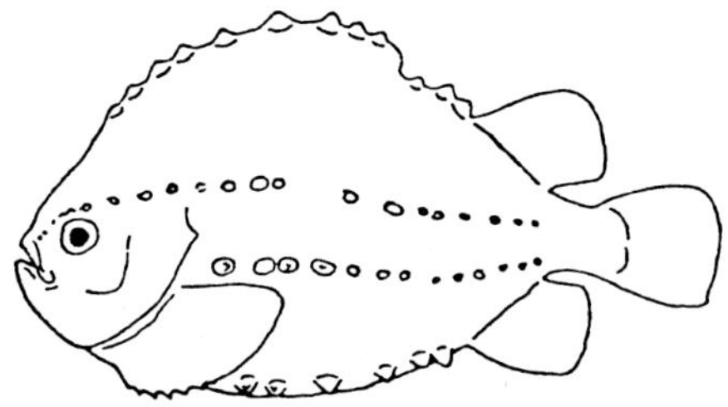

Lumpfish

Lumpfishes and snailfishes are stout-bodied marine fishes with pelvic fins modified to form an adhesive disc. The lumpfish is characterised by a short, thick, high-arched body, and a skin with tubercles. This is a bottom fish from shallow coastal waters; it is sometimes semipelagic, dwelling among floating masses of seaweed. Its diet includes euphausiid shrimps, jellyfish, amphipods and other invertebrates. The retina of the lumpfish (dorsal region, Plate 119.1) resembles that of the cottids except that rods are fewer and the internal nuclear layers (bipolar, ganglion cells) have few and scattered cells. Horizontal cells form 1–2 layers. The single and double cones (Plate 119.2) are arranged in square mosaics. The area retinae is ventral.

Reference: ANCTIL (1969)

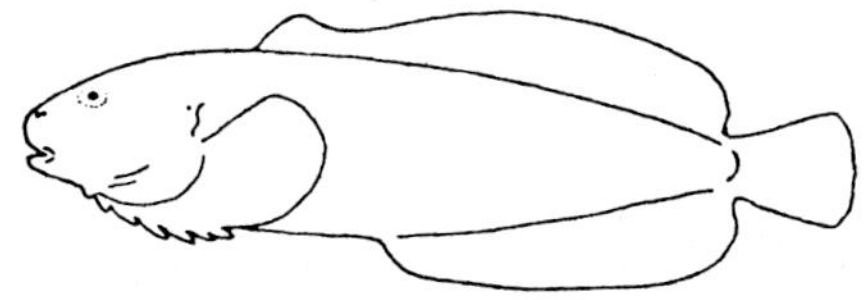

Snailfish

Snailfishes are shaped like tadpoles with soft bodies; they are distributed along the continental shelf from the intertidal zone to thousands of metres deep. They are benthic and sedentary in habits.

The retina of the snailfish is very poorly developed. This is a pure-rod retina; the rods are elongate and the rod nuclei form 6–8 layers. The heavily pigmented epithelial cell layer lacks processes (Plate 119.3). The internal nuclear layers are extremely poorly defined.

References: ENGSTRÖM (1963b, on *Liparis liparis,* whose retina possesses single, double, and triple cones as well as rods); ALI et al. (1968); ALI and WAGNER (1975a)

Plate 119.1-3. Transverse (1) and (3) and tangential (2) sections of the retinas of the lumpfish, *Cyclopterus lumpus,* and the snailfish, *Careproctus* sp. (1) ×260; (2) ×700; (3) ×320

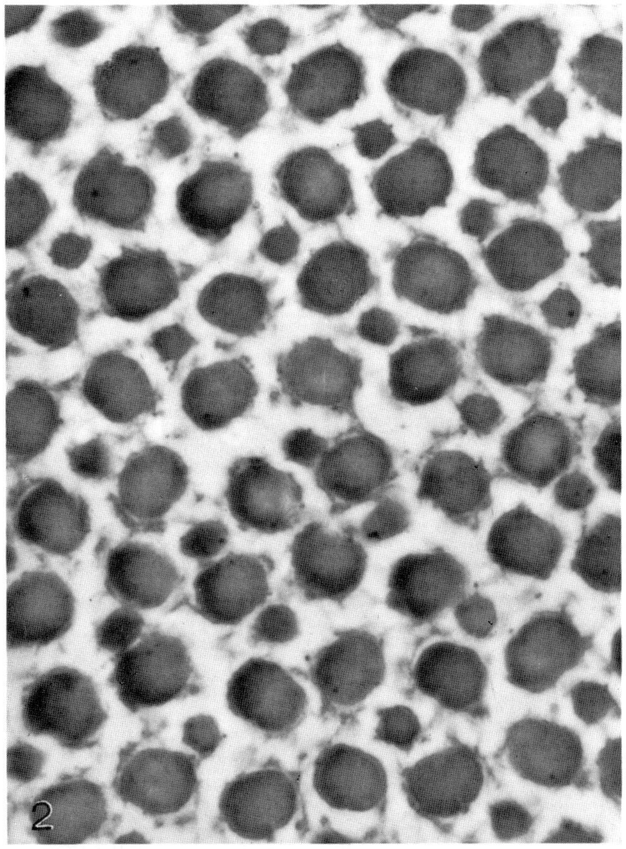

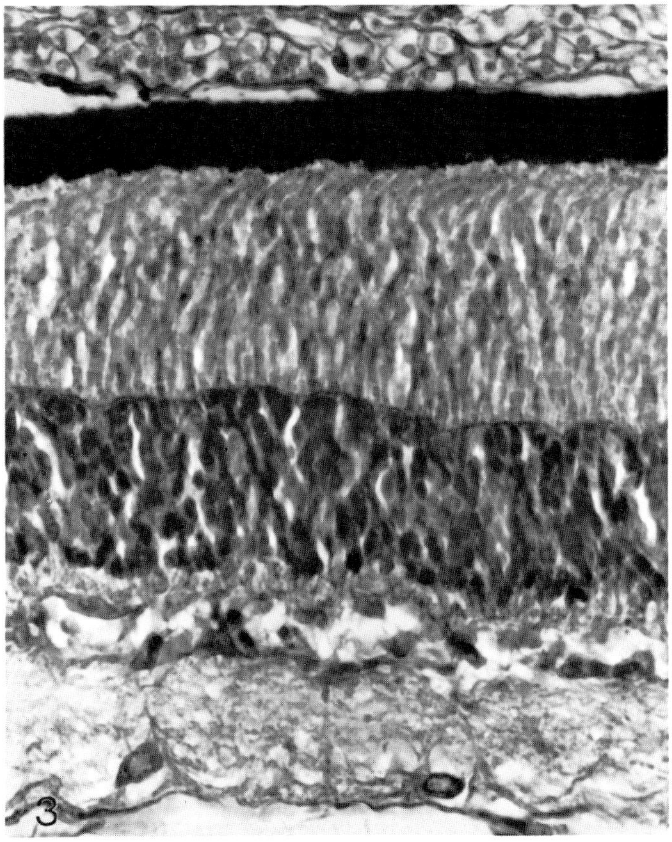

Plate 119.1-3. (Legend see opposite page)

Belontiidae (*Bettas*)

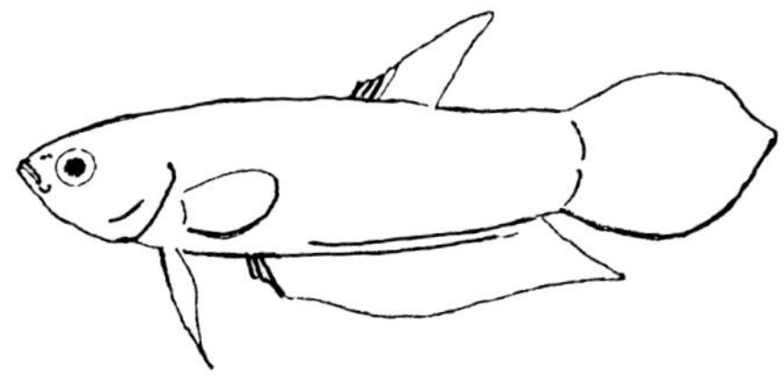

Acanthopterygii, Percomorpha, Perciformes, Anabantoidei

The anabantoid fish is a small, tropical freshwater fish distributed in Southeast Asia and Africa. Because of the labyrinthine breathing apparatus located in a cavity above each gill chamber and connected to it, the members of this specialised suborder are called labyrinth fishes. This accessory breathing apparatus enables the fish to breath air and to live in oxygen-deficient water. The Siamese fighting fish, a belontiid, is found in clear and weedy waters, as well as in irrigation ditches and dirty ponds.

Single and equal double cones are arranged in regular square mosaics in the retina of *Betta splendens*. The area retinae is located in the ventro-temporal region. Two horizontal cell layers are present.

References: ENGSTRÖM (1963b); WAGNER (1972)

Visual Pigments: A_1 max 500 (ALI and WAGNER, 1975b)

Osphronemidae (*Gouramies*)

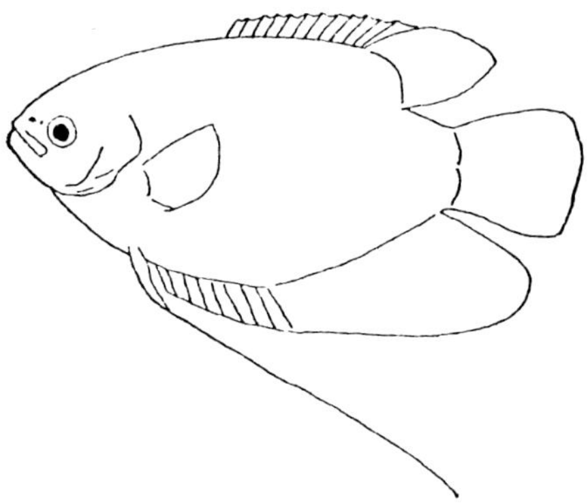

Acanthopterygii, Percomorpha, Perciformes, Anabantoidei

Osphronemids have essentially the same group characteristics as those of the bettas.

The retina of the dwarf gourami (*Colisa latia*) and the pearl gourami (*Trichogaster leeri*) is similar to that of the bettas. Both species possess single and equal double cones, and two layers of small horizontal cells.

Reference: ENGSTRÖM (1963b)

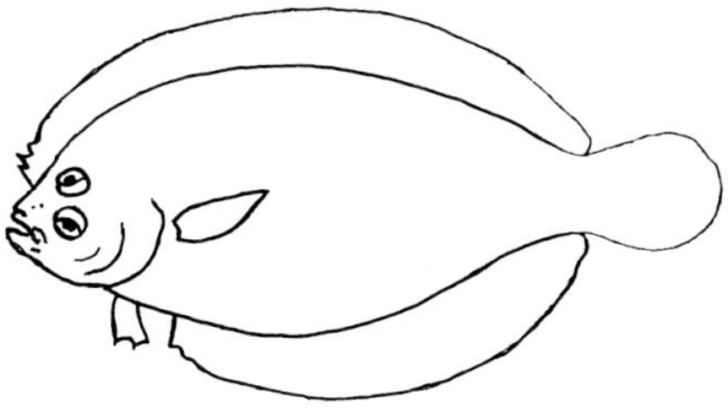

Bothidae (*Lefteye Flounders*)

Acanthopterygii, Percomorpha, Pleuronectiformes, Pleuronectoidei

The lefteye or bothid flounders are flatfishes with an oval, more or less compressed body, and a large, symmetrical mouth. The symmetrical mouth with uniformly distributed teeth indicates that the bothids, unlike the pleuronectids, catch their prey in midwater and not on the bottom.

Double cones (equal and unequal) are found in the retinas of *Bothus maximus* and *B. rhombus*.

Reference: Friis (1879)

Visual Pigments: A_1 max 486–517 (Ali and Heumann, 1972; Ali and Wagner, 1975b)

Pleuronectidae (*Righteye Flounders or Flatfishes*)

Acanthopterygii, Percomorpha, Pleuronectiformes, Pleuronectoidei

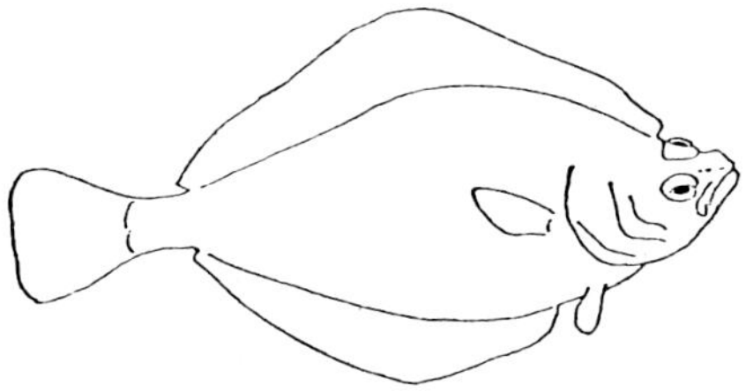

Flatfishes are specialised percomorph teleosts flattened laterally, and accompanied by a twist of the skull with the consequent migration of one eye to the upper side. These features are adaptations for a benthic life on sand bottoms of the continental shelf. The witch flounder, *Glyptocephalus cynoglossus,* inhabits deep shelf waters where it remains stationary in muddy sand and feeds on invertebrates.

Witch Flounder

The retina of the adult witch flounder (435 mm, Plate 120.1) is moderately developed. The epithelial cells are packed with pigment sclerally and have very short processes. The rods are extremely long and numerous, while the cones are small and relatively few except in the ventral area retinae and the temporal region (Plates 120.1,4). The single and equal double cones form regular square mosaics only in the area retinae (Plate 120.3), and the cones are irregularly arranged in other retinal regions (Plate 120.4). The internal nuclear layers are thin and poorly developed (Plate 120.1). As seen in Plate 120.2 the retina of the young witch flounder (ventral region) (130 mm), which probably has not yet settled on the bottom, is rich in cones and has relatively few rods; accordingly, the horizontal, bipolar, amacrine, and ganglion cells are very abundant. Note that the pigment epithelium has long processes.

Yellowtail

The yellowtail, *Limanda ferruginea,* is a shore water flatfish in comparison to the witch flounder. It lives on any type of sandy bottom and eats small crustaceans, small shellfish, and worms. It is considered to be a particularly sluggish fish.

The retina of the yellowtail is essentially similar to that of the adult witch flounder (Plate 121.1). The area retinae is dorsotemporal in this species. Single and double cones are arranged in square mosaics in the area retinae and other regions (ventral, Plate 121.2), while no regular patterns are present elsewhere in the retina (fundus, Plate 121.3). Triple cones are present in a confined spot near the area retinae.

Plate 120.1-4. Transverse (1) and (2) and tangential (3) and (4) sections of the retinas of the witch flounder, *G. cynoglossus.* (1) $\times 400$; (2) $\times 370$; (3) and (4) $\times 700$ ▷

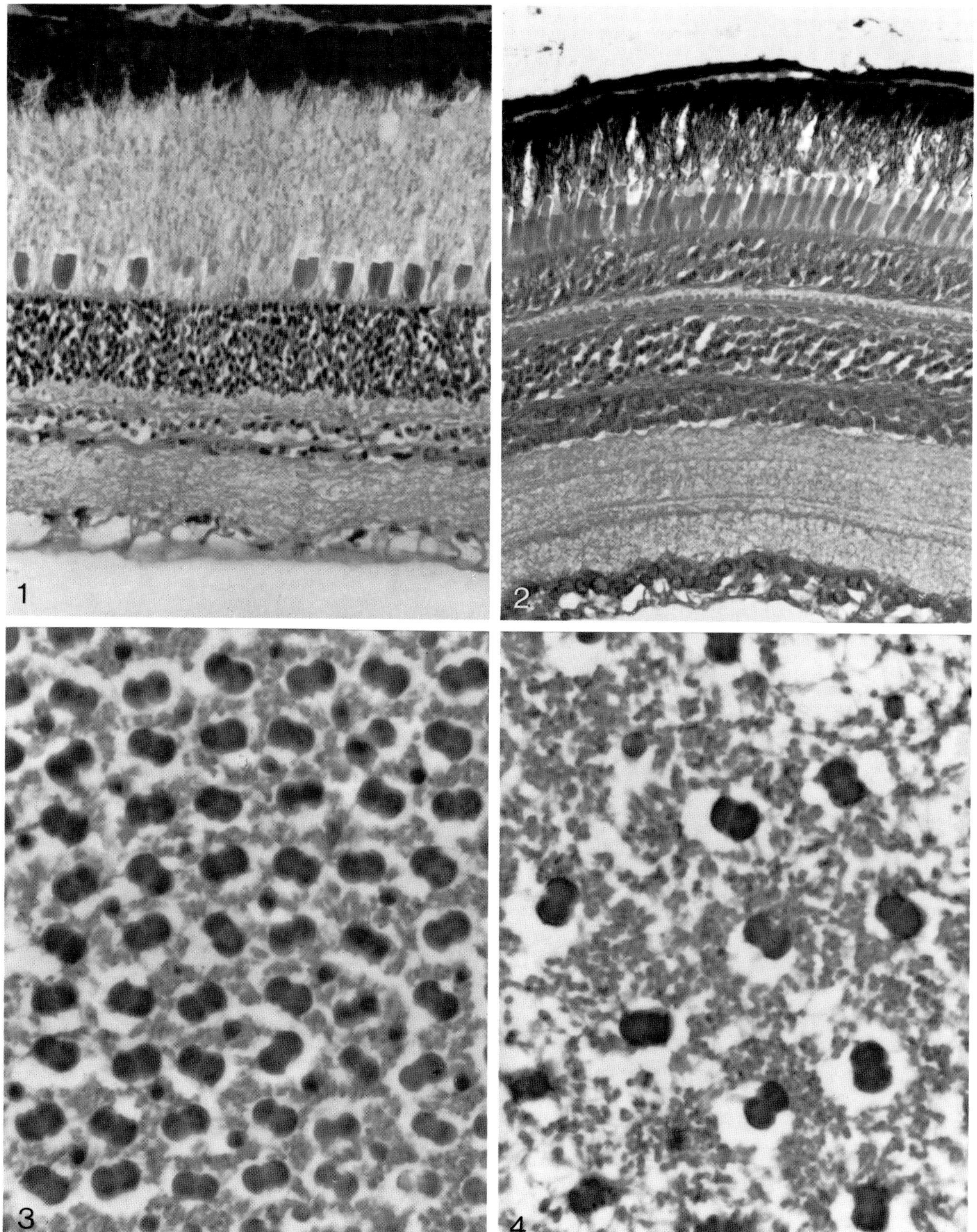

Plate 120.1-4. (Legend see opposite page)

Dabs

Dabs are flatfishes which inhabit a variety of depths, from tideline to 800 m. They prefer bottoms with a fine mixture of sand and mud. Their diet includes mainly echinoderms, various shrimps, and small fishes.

Contrary to the other flatfish mentioned, the retinal epithelium of the adult American plaice (dab) *Hippoglossoides platessoides,* possesses long, needlelike processes. The rods are numerous and moderately long, and the cones are relatively large. The internal nuclear layer and ganglion cell layer are very poorly developed (Plate 121.4).

References: HANNOVER (1843); FRIIS (1879); VERRIER (1928a); MENNER (1929, 1950); ENGSTRÖM and AHLBERT (1963); NICOL (1965); ALI et al. (1968); ANCTIL (1969); WAGNER (1972); ALI and WAGNER (1975a)

Visual Pigments: A_1 max 491–510 (BEATTY, 1973; ALI and WAGNER, 1975b)

Soleidae (*Soles*)

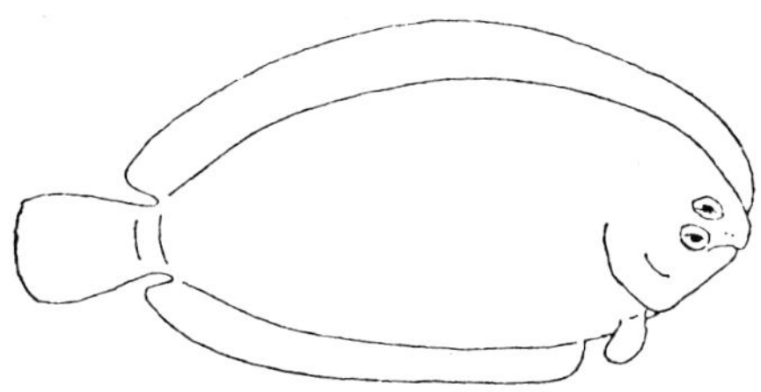

Percomorpha, Pleuronectiformes, Soleoidei

Soles are flatfishes in which both eyes are on the right side of the head. The migrated eye is very close to the fixed eye, unlike the pleuronectids. They have a small, asymmetrical mouth. Soles, like all the pleuronectids, are benthic.

The retina of the sole (*Solea solea*) is less developed than that of the pleuronectids. The epithelial pigment, rods, and cones undergo retinomotor changes. The single, equal double and triple cones do not form regular cone mosaics in tangential sections, except for more or less regular rows of double cones in some retinal regions.

References: CHABANAUD (1946); NICOL (1961b); ENGSTRÖM and AHLBERT (1963); ALI and WAGNER (1975a)

Visual Pigments: A_1 max 501 (ALI and HEUMANN, 1972; ALI and WAGNER, 1975b)

▷

Plate 121.1-4. Transverse (1) and (4) and tangential (2) and (3) sections of the retinas of the yellowtail, *L. ferruginea,* and the American dab, *H. platessoides.* (1) ×420; (2) and (3) ×700; (4) ×440

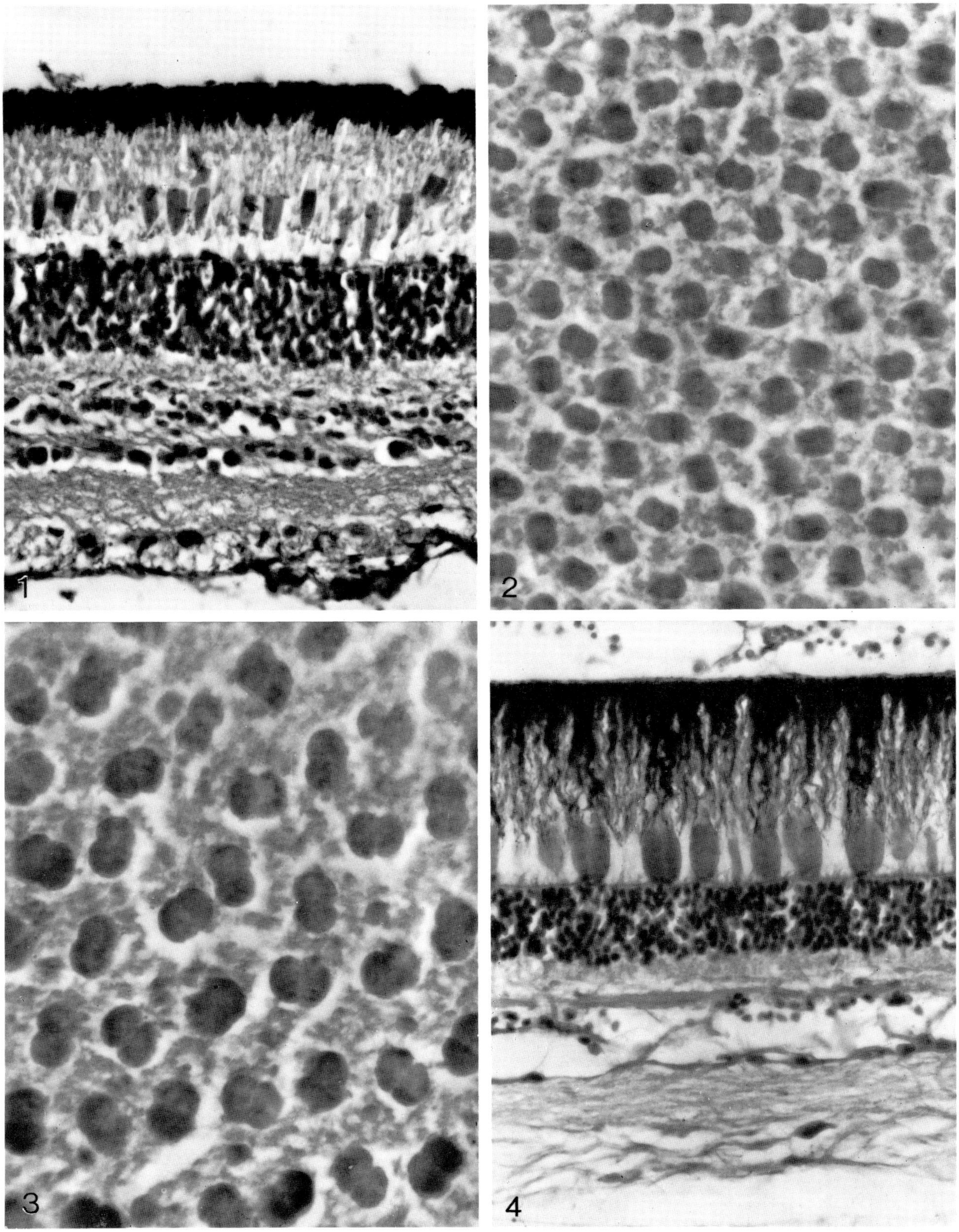

Plate 121.1-4. (Legend see opposite page)

Cynoglossidae (*Tonguefishes*)

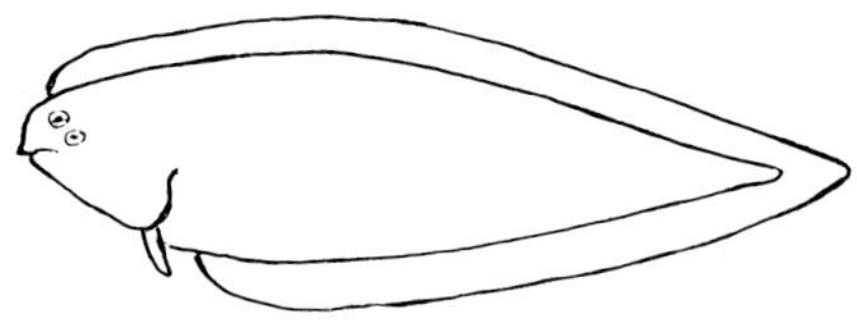

Acanthopterygii, Percomorpha, Pleuronectiformes, Soleioidei

Tongue soles are small flatfish characterised by the position of both eyes on the left side of the head, and the absence of ribs. Except for the position of the eyes and their more pointed tail they resemble the true soles.

The retinal structure of *Symphurus plagusia* is similar to that of other flatfish. The epithelial pigment is mostly retreated in the cell bodies (Plate 122). The rods are slender, densely packed, and numerous. The cones, of the equal double type are in majority, short, bulky, and few. The bipolar and amacrine cells form two or three layers.

Balistidae (*Triggerfishes*)

Acanthopterygii, Percomorpha, Tetraodontiformes, Balistoidei

Plectognaths or tetraodontiforms form a group of specialised teleosts restricted in distribution to shallow, inshore marine waters of the tropical regions. All of them have small gill openings and small mouths with strong teeth. The triggerfish have bony scales and lack pelvic fins. They are usually solitary and slow-moving in their habits, with small eyes.

The temporal area retinae is not sharply defined in *Cantherhines modestus*. Single and double cones are present in the retina of this species.

Reference: Tamura (1957)

Visual Pigments: A_1 max 491–502; $A_2 > 10\%$ (Ali and Heumann, 1972; Beatty, 1973; Munz and McFarland, 1973)

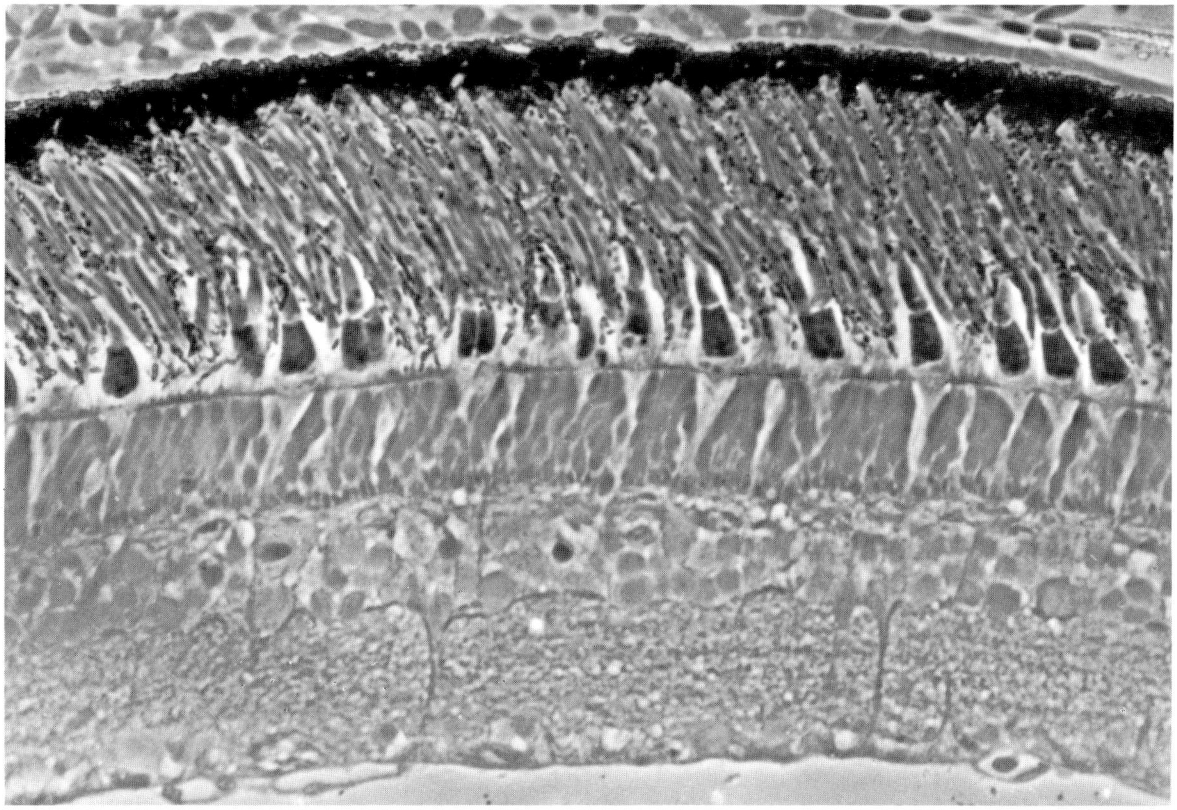

Plate 122. Transverse section of the retina of the blackcheek tonguefish, *S. plagiusa.* ×825

Tetraodontidae (*Puffers*)

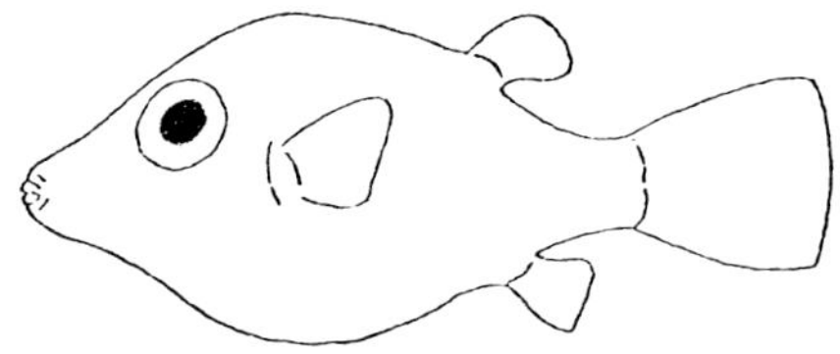

Acanthopterygii, Percomorpha, Tetraodontiformes, Tetraodontoidei

Puffer fishes are specialised teleosts that have their teeth fused together in each jaw. They possess the ability to inflate themselves with air or water when frightened. Puffers are shore fishes feeding mainly on crustaceans, molluscs, and echinoderms with the help of their powerful teeth.

In *Sphoeroides chrysops* and *S. nyphobles* single as well as double cones are present in the proportion 1:2. The area retinae is temporal in both species, but the cone density in this area differs greatly between *S. chrysops* (585 cones/0.01 mm^2) and *S. nyphobles* (107 cones/0.01 mm^2).

The retina of *S. spengleri* is very well-developed, with a large number of visual cells, principally cones. While the horizontal cells are arranged in one or two ill-defined layers, the bipolar and amacrine cells are packed in two easily distinguished layers (Plate 123.1), and the ganglion cells form a regular layer. The single and equal double cones are arranged in square patterns (Plate 123.3), and cone densities vary from one retinal region to another (Plate 123.3,4). The area retinae of these fishes is in the temporal region of the retina. Differences between dorsal and ventral regions are marked (Plates 124.1, 2, 3, 4)

References: TAMURA (1957); ENGSTRÖM (1963); WAGNER et al. (1976)

Visual Pigments: A_1 max 495–503; A_2 max 522–527 (ALI and HEUMANN, 1972; BEATTY, 1973; MUNZ and MCFARLAND, 1973; ALI and WAGNER, 1975b)

Plate 123.1-4. Transverse sections of the dorsal (1) and ventral (2) retinas of *S. spengleri*. (1) ×700; (2) ×880. Tangential sections (×1,380) of the dorsal (3) and ventral (4) retinas of the same fish ▷

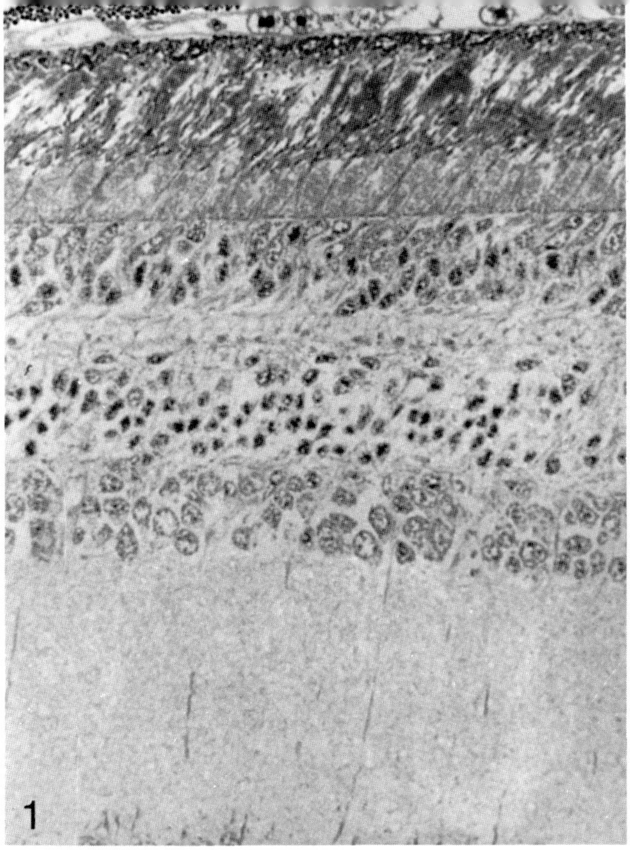
1

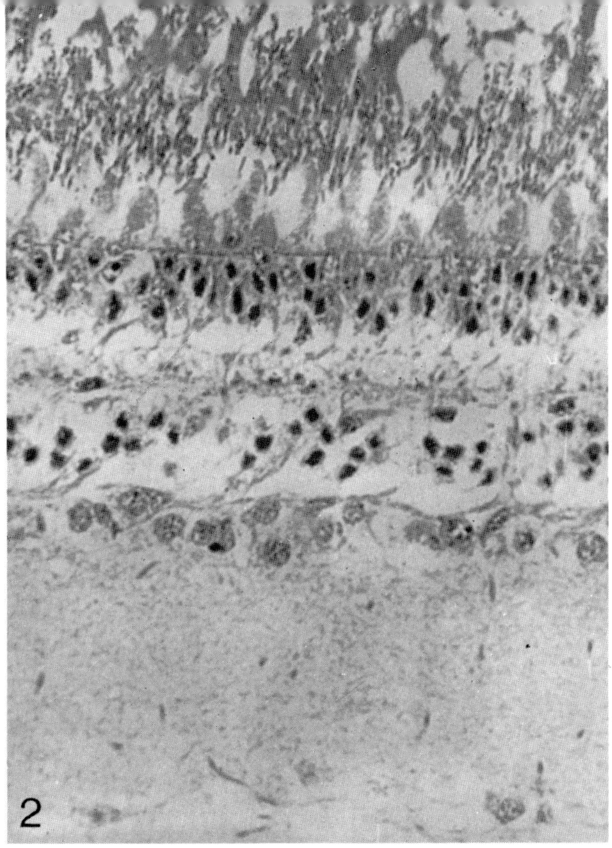
2

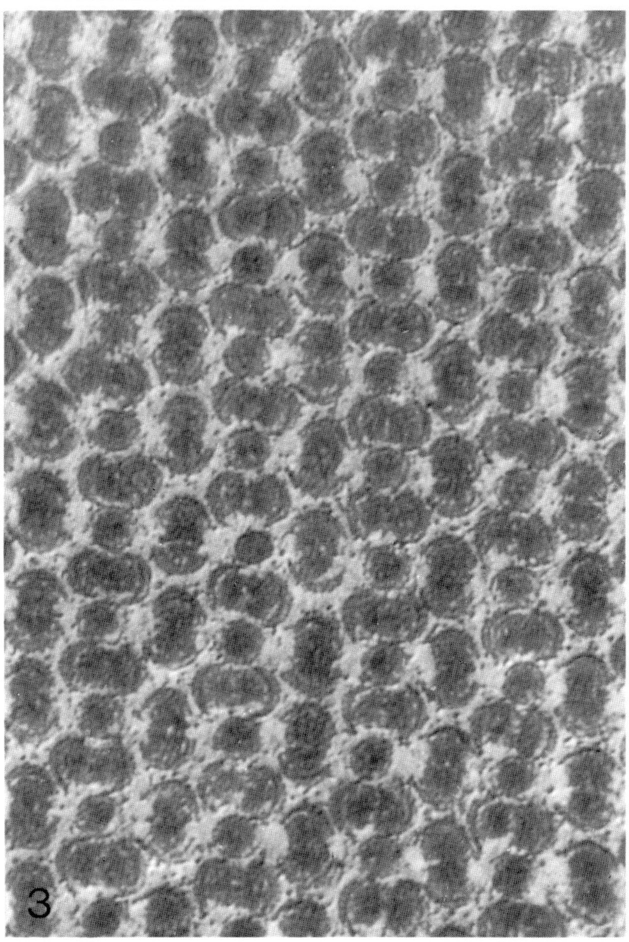
3

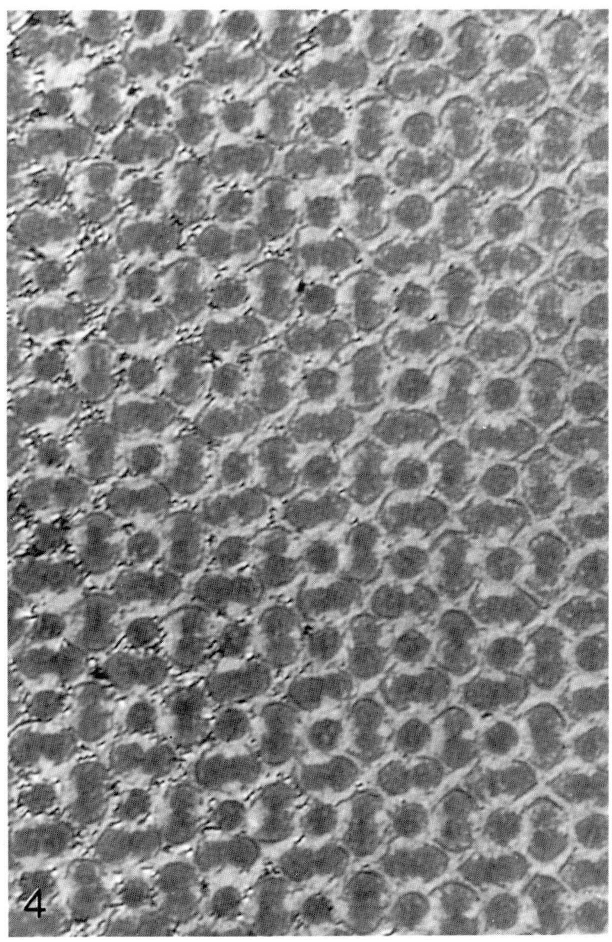
4

Lepidosirenidae (*South American Lungfishes*)

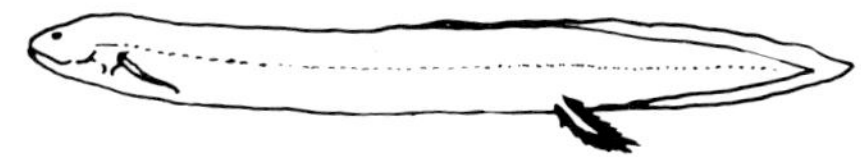

Teleostomi, Dipneusti, Ceratodi, Lepidosireniformes

The African (*Protopterus*) and South American (*Lepidosiren*) lungfishes differ in that the former has six gill arches and five gill clefts, whereas the latter has one less of each. *Lepidosiren* is more eel-like than *Protopterus* in appearance. On the whole they are much alike, not resembling, however, the Australian *Neoceratodus*. *Lepidosiren* is inactive and inhabits stagnant waters that become isolated and lacking of water during the dry seasons. The lungfish is predatory, feeding on bottom fishes, snails, mussels, etc.

The retina of *L. paradoxa* is essentially similar to that of *Protopterus*. There are large nuclei (Plate 124.1) throughout the retinal layers. Note the pigmented epithelial cell processes and the large number of bipolar and ganglion cells relative to the number of visual cell nuclei. Unlike in *Protopterus,* only single cones and rods are represented in the visual cell layer, with the characteristic oil droplet and paraboloid (Plate 124.2); note in (Plate 124.3) that the tangential section passed through the cone droplet and rod ellipsoid. Landolt's clubs (Plate 124.4) are also present.

References: KERR (1902); ROCHON-DUVIGNEAUD (1941, 1943, 1958); ALI and ANCTIL (1973); ALI and WAGNER (1975a)

Protopteridae (*African Lungfishes*)

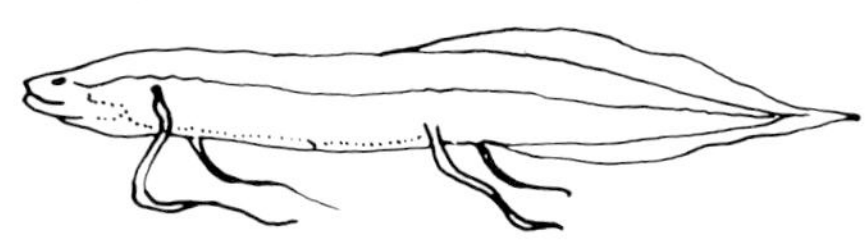

Teleostomi, Dipneusti, Ceratodi, Lepidosireniformes

Lungfishes are relics of a formerly abundant group that was plentiful from the Devonian to the Triassic. African lungfishes have small embedded scales, filamentous paired fins, and a long eel-like body. They possess two lungs. During the dry season *Protopterus* aestivates in mudballs and survives by breathing air with its lungs. *Protopterus* has very small eyes.

The whole retina of *P. aethiopicus* and *P. annectens* is characterised by the large size of its cells. The epithelial cells possess pigmented processes. Bulky rods, single, and unequal double cones are present. The rods have each a large paraboloid and a large oil droplet; the single cones also have a large oil droplet, with only a small, excentrically displaced paraboloid; the principal element of the dou-

---

Plate 124.1-4. Transverse (1), (2) and (4) and tangential (3) sections of the retina of the South American lungfish, *L. paradoxa*. A single one is outlined in (2). (1) and (3) ×560; (2) and (4) ×1,400. *LC* Landolt's club, *OD* oil droplet, *PA* paraboloid ▷

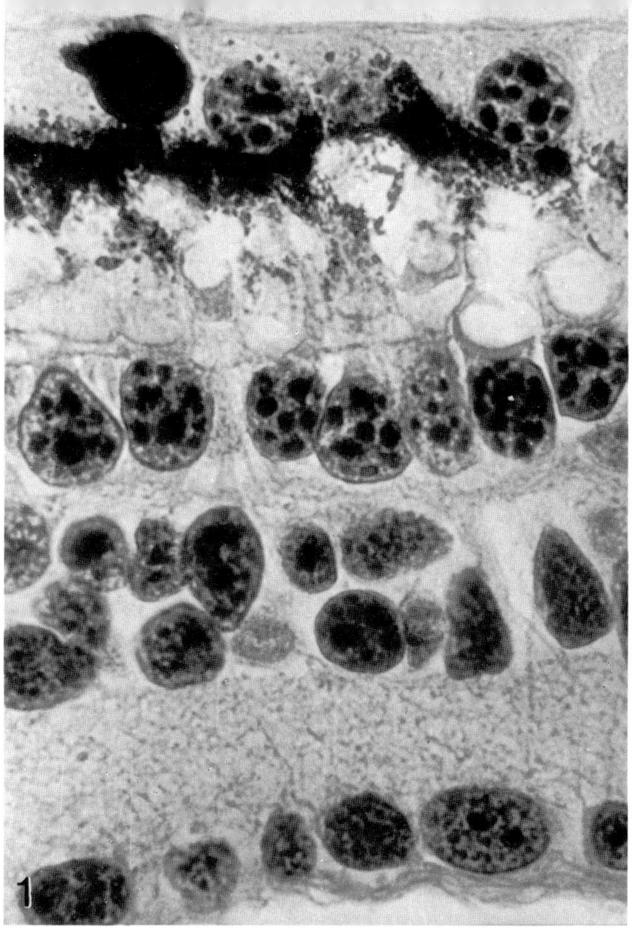

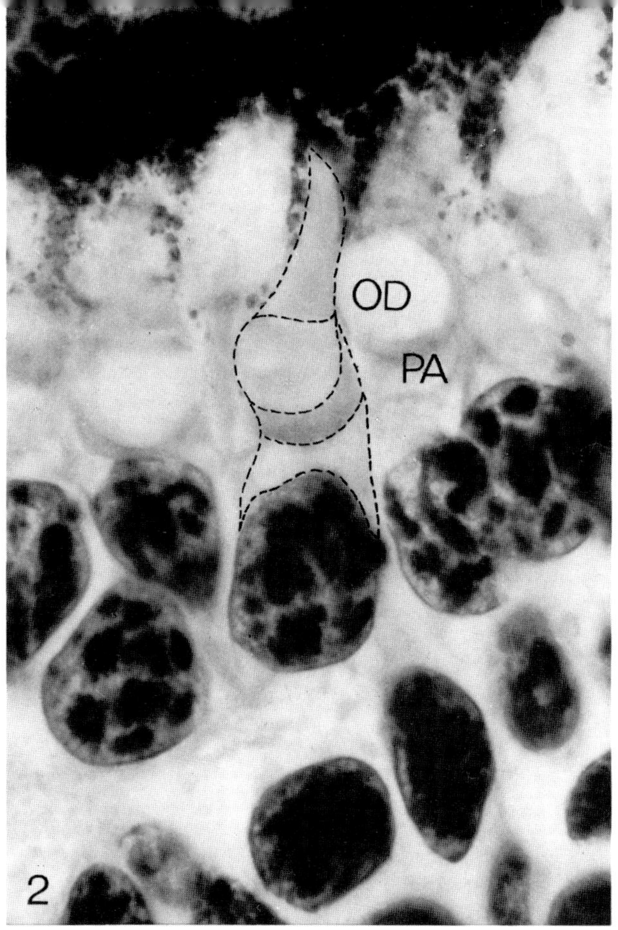
OD
PA

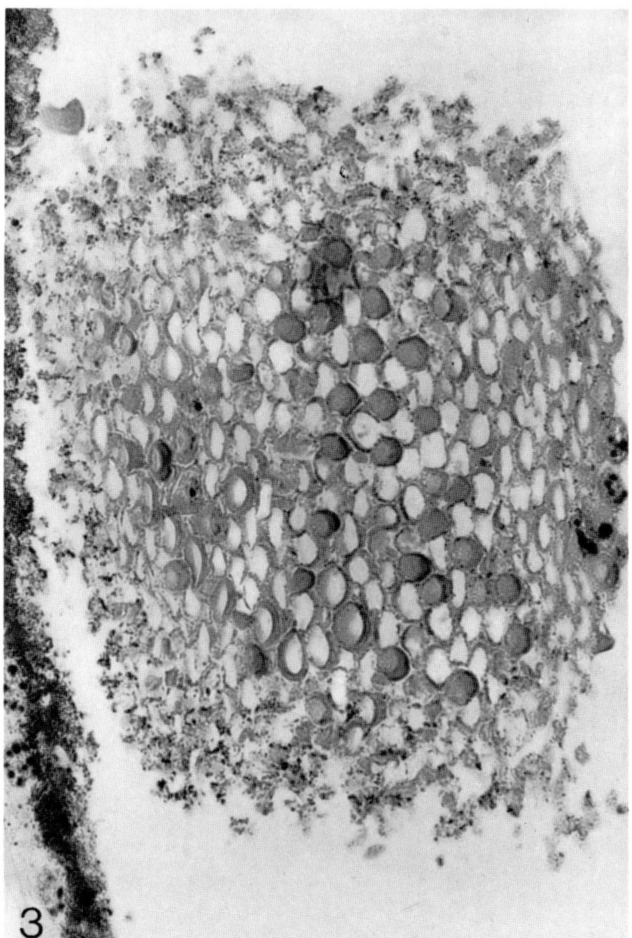

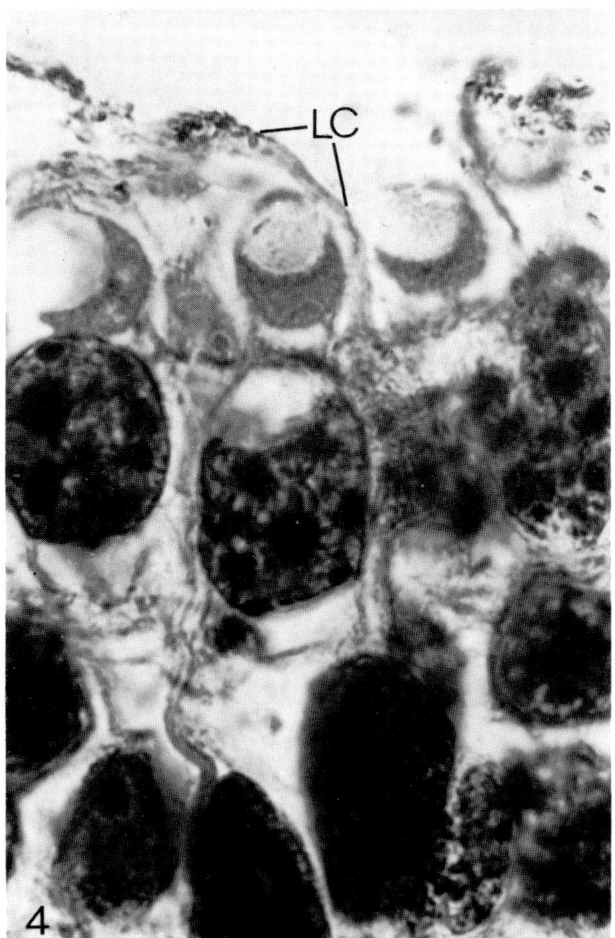
LC

ble cone has a large oil droplet, but the accessory element has only a paraboloid. Landolt's clubs derived from bipolar cells and projected in the sclerad direction are present; they have been described in urodele amphibians before. The internal nuclear and ganglion cell layers are well developed.

References: SCHIEFFERDECKER (1886); HOSCH (1904); GRYNFELTT (1911); WALLS (1942); MUNK (1964b); PFEIFFER (1968); LOCKET (1970a, b; 1975)

Latimeriidae (*Coelacanths*)

Crossopterygii, Coelacanthi, Coelacanthiformes

The coelacanth is a strong, heavy-bodied, carnivorous fish with pectoral and pelvic fins stalked with basal lobes. This fish, which lives in deep shelf waters off the southeastern African coast, is the only known living representative of the group that is ancestral to the land vertebrates. The coelacanth possesses a large lung, an almost linear heart, a sharklike intestine with a spiral valve, and an axial skeleton reduced to a hollow tube of cartilage. It is benthic in habit and negatively phototropic.

Latimeria chalumnae has an almost pure-rod retina (Plate 125). The epithelium is avascular and lacks melanin pigment entirely. The rods are numerous, densely packed, and possess long outer segments. Cones (single type) are very rare and possess a clear oil droplet. The internal nuclear and ganglion cell layers are made of sparsely distributed neuronal cells, whereas the Müller's supporting cells are very numerous.

References: MILLOT and CARASSO (1955); MILLOT and ANTHONY (1965); LOCKET (1972, 1973, 1975); GRIFFITH and THOMSON (1973); LAGIOS (1975)

Visual Pigments: A_1 max 473 (DARTNALL, 1972)

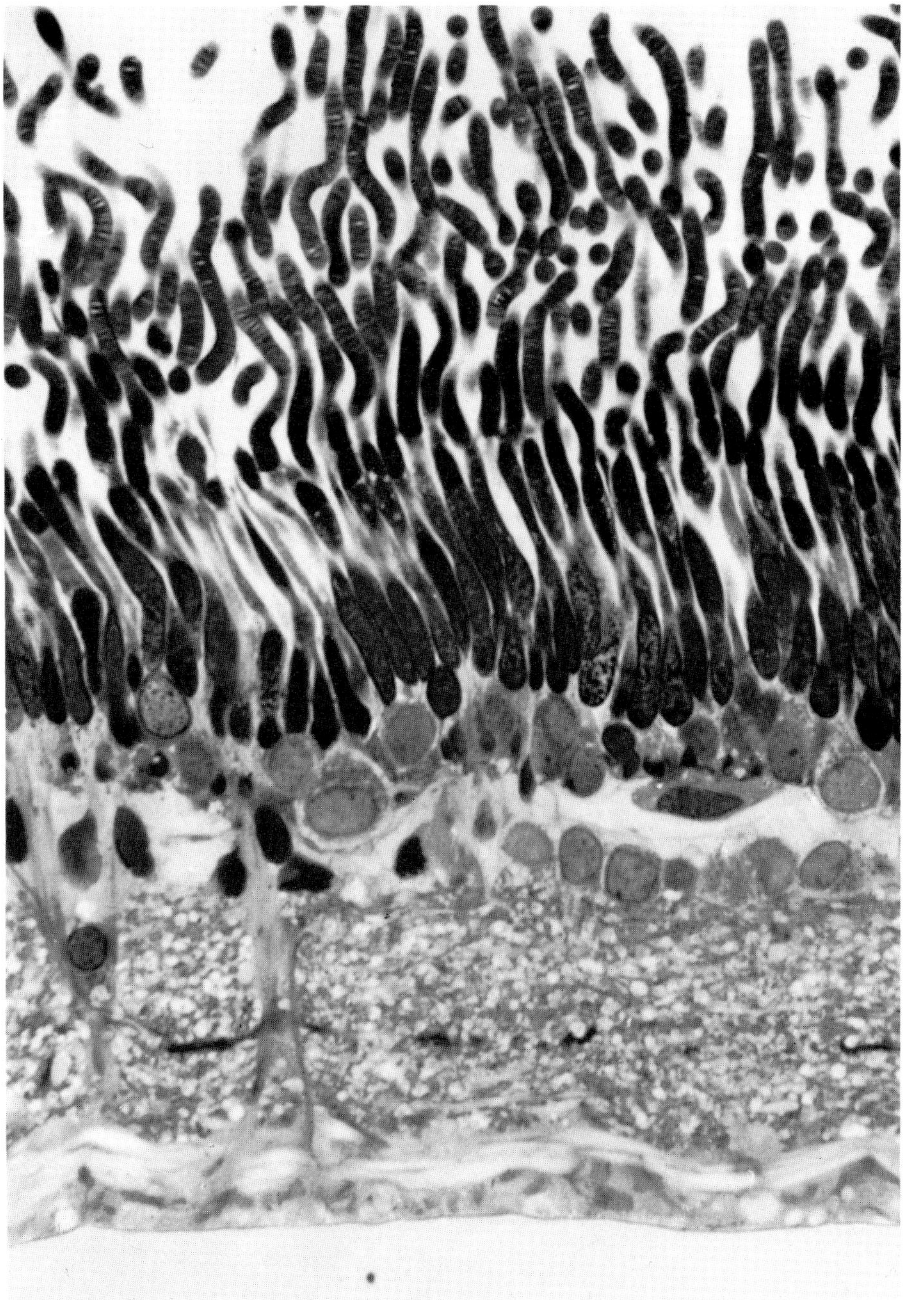

Plate 125. Photomicrograph of a transverse section of the retina of the coelacanth, *L. chalumnae*. × 850

5. BIBLIOGRAPHY

Ahlbert, I.B.: The organization of the cone cells in the retinae of four teleosts with different feeding habits (*Perca fluviatilis* L., *Lucioperca lucioperca* L., *Acerina carnua* L. and *Coregonus albula* L.) Ark. Zool. **22**, 445–481 (1969)

Ahlbert, I.-B.: Ontogeny of double cones in the retina of perch fry (*Perca fluviatilis,* Teleostei). Acta Zool. **54**, 241–254 (1973)

Ali, M.A.: The ocular structure, retinomotor and photobehavioural responses of juvenile Pacific salmon. Can. J. Zool. **37**, 965–996 (1959)

Ali, M.A.: Histophysiological studies on the juvenile Atlantic salmon (*Salmo salar*) retina. II. Responses to light intensities, wavelengths, temperatures and continuous light or dark. Can. J. Zool. **39**, 511–526 (1961)

Ali, M.A.: Retinal responses in enucleated eyes of Atlantic salmon (*Salmo salar*). Rev. Can. Biol. **21**, 7–15 (1962a)

Ali, M.A.: Influence of light intensity on retinal adaptation in Atlantic salmon (*Salmo salar*) yearlings. Can. J. Zool. **40**, 561–570 (1962b)

Ali, M.A.: Correlation of some retinal and morphological measurements from the Atlantic salmon (*Salmo salar*). Growth **27**, 57–76 (1963)

Ali, M.A.: Retinomotor responses of the goldfish (*Carassius auratus*) to unilateral photic stimulation. Rev. Can. Biol. **23**, 45–53 (1964a)

Ali, M.A.: Retinomotor responses in the enucleated eyes of the brown bullhead (*Ictalurus nebulosus*) and the goldfish (*Carassius auratus*). Rev. Can. Biol. **23**, 55–66 (1964b)

Ali, M.A.: Retina of the albino splake (*Salvelinus fontinalis* × *S. namaycush*). Can. J. Zool. **42**, 1158–1160 (1964c)

Ali, M.A.: Stretching of retina during growth of salmon (*Salmo salar*). Growth **28**, 83–89 (1964d)

Ali, M.A.: Retinal structure in the Arctic char (*Salvelinus alpinus* L.). J. Fish Res. Board Can. **22**, 221–223 (1965)

Ali, M.A.: Les réponses rétinomotrices: caractères et mécanismes. Vision Res. **11**, 1225–1288 (1971)

Ali, M.A. (Ed.): Vision in Fishes: New Approaches in Research. New York: Plenum Press 1975

Ali, M.A., Anctil, M.: Corrélation entre la structure rétinienne et l'habitat chez *Stizostedion vitreum vitreum* et *S. canadense*. J. Fish. Res. Board Can. **25**, 2001–2003 (1968)

Ali, M.A., Anctil, M.: Retina of the South American lungfish, *Lepidosiren paradoxa* Fitzinger. Can. J. Zool. **51**, 969–972 (1973)

Ali, M.A., Anctil, M.: Retinas of the electric ray, *Narcine brasiliensis* and the freshwater stingray *Paratrygon motoro*. Vision Res. **14**, 587–588 (1974a)

Ali, M.A., Anctil, M.: Retinomotor responses and isolated photoreceptor in *Amia calva* (Holostei, Amiidae). Copeia 379–395 (1974b)

Ali, M.A., Anctil, M., Mohideen, H.M.: Structure rétinienne et la vascularisation intraoculaire chez quelques poissons marins de la région de Gaspé. Can. J. Zool. **46**, 729–745 (1968)

Ali, M.A., Anctil, M., Raymond, N.: La rétine de quelques poissons marins du littoral brésilien. Rev. Biol. Lisb. **9**, 101–114 (1973)

Ali, M.A., Copes, P., Stevenson, W.R.: Correlation of morphological and intra-ocular measurements in the Atlantic salmon (*Salmo salar*) yearling. J. Fish. Res. Board Can. **18**, 259–272 (1961)

Ali, M.A., Hanyu, I.: A comparative study of retinal structure in some fishes from moderately deep waters of the Western North Atlantic. Can. J. Zool. **41**, 225–241 (1963)

Ali, M.A., Heumann, W.R.: Distribution of vitamins A_1 and A_2 in the retinas of some marine fishes from the Gulf of California. Vision Res. **10**, 1307–1310 (1970)

Ali, M.A., Heumann, W.R.: Distribution of vitamins A_1 and A_2 in the retinas of some marine fishes from the Gulf of California. II Vision Res. **12**, 2157–2159 (1972)

Ali, M.A., Kobayashi, H.: Electroretinogram—Flicker fusion frequency in albino trout. Experientia **24**, 454–455 (1968a)

Ali, M.A., Kobayashi, H.: Electroretinogram of albino and pigmented brook trout, *Salvelinus fontinalis* (Mitchill). Rev. Can. Biol. **27**, 145–161 (1968b)

Ali, M.A., Raymond, N.: La retine de piranha (*Serrasalmus marginatus*). Rev. Biol. Lisb. **8**, 27–32 (1972)

Ali, M.A., Stevenson, W.R., Press, J.S.: Histophysiological studies on the juvenile Atlantic salmon (*Salmo salar*) retina. I. Rates of light- and dark-adaptation. Can. J. Zool. **39**, 123–128 (1961)

Ali, M.A., Wagner, H.-J.: Distribution and development of retinomotor responses. In: Vision in Fishes: New Approaches in Research. Ali, M.A. (ed.). New York: Plenum Press, 1975a, pp. 369–396

Ali, M.A., Wagner, H.-J.: Visual pigments: Phylogeny and ecology. In: Vision in Fishes: New Approaches in Research. Ali, M.A. (ed.). New York: Plenum Press, 1975b, pp. 481–516

Allen, B.M.: The eye of *Bdellostoma stouti*. Anat. Anz. **26**, 208–211 (1905)

Anctil, M.: Intraocular vascular supply in some marine teleosts. Rev. Can. Biol. **27**, 347–355 (1968)

Anctil, M.: Structure de la rétine chez quelques télésteens marins du plateau continental. J. Fish. Res. Board Can. **26**, 597–628 (1969)

Anctil, M., Ali, M.A.: Retina of *Exocoetus volitans* and *Fodiator acutus* (Pisces: Exocoetidae). Copeia 43–48 (1970)

Anctil, M., Ali, M.A.: Giant ganglion cells in the retina of the Hammerhead shark (*Sphyrna lewini*). Vision Res. **14**, 903–905 (1974)

Anctil, M., Ali, M.A.: Cone droplets of mitochondrial origin in the retina of *Fundulus heteroclitus* (Pisces: Cyprinodontidae). Zoomorphol. **84**, 103–111 (1976)

Anctil, M., Ali, M.A., Couillard, P.: Isolated retinal cells of some lower vertebrates. Rev. Can. Biol. **32**, 107–119 (1973)

Arey, L.B.: The occurrence and significance of photomechanical changes in the vertebrate retina. An historical survey. J. Comp. Neurol. **25**, 535–554 (1915)

Arey, L.B.: The movements in the visual cells and retinal pigment of the lower vertebrates. J. Comp. Neurol. **26**, 121–201 (1916)

Arey, L.B.: Visual cells and retinal pigment. Section 25 in Special Cytology. Cowdry, E.V. (ed.). New York: Hoeber, 1928, Vol. 2, pp. 887–926

Arey, L.B., Mundt, G.H.: A persistent diurnal rhythm in visual cones. Anat. Rec. (Suppl.) **79**, 5 (1941)

Arnott, H.J., Maciolek, N.J., Nicol, J.A.C.: Retinal tapetum lucidum: A novel reflecting system in the eye of teleosts. Science **169**, 478–480 (1970)

Arnott, H.J., Nicol, J.A.C., Querfeld, C.W.: Tapeta lucida in the eyes of the seatrout (Sciaenidae). Proc. R. Soc. Lond. Ser. B **180**, 247–271 (1972)

Baburina, E.A.: The development of the eyes and vision in the black-back herring *Alosa kessleri kessleri* (Grimm). Vopr. Ikhtiol. **4**, 114–136 (1955)

Baburina, E.A.: The development of the eyes and their function in lithophilic fish of the carp family. Trud. Inst. Morfol. Zhiv. Akad. Nauk. SSSR **33**, 111–150 (1961a)

Baburina, E.A.: The development of the eyes and their function in embrys and larvae of the pike perch (*Lucioperca lucioperca* Linne). Trud. Inst. Morfol. Zhiv. Akad. Nauk. SSSR **33**, 151–172 (1961b)

Baburina, E.A.: Development and function of the eyes in the catfish *Silurus glanis* L. and *Parasilurus asotus* L. Trud. Inst. Morfol. Zhiv. Akad. Nauk. SSSR **40**, 138–156 (1962)

Baburina, E.A., Kovaleva, N.D.: Retinal structure in Caspian sprats. Dokl. Akad. Nauk. SSSR **125**, 1349–1352 (1959)

Bailey, R.M., Fitch, J.E., Herald, E.S., Lachner, E.A., Lindsey, C.C., Robins, C.R., Scott, W.B.: A List of Common and Scientific Names of Fishes from the United States and Canada. 3rd ed. Washington: American Fisheries Society

Bathelt, D.: Experimentelle und vergleichende morphologische Untersuchungen am visuellen System von Teleostiern. Zool. Jahrb. Abt. Anat. Ontog. **87**, 402–470 (1970)

Beatty, D.D.: Visual pigments of several species of teleost fishes. Vision Res. **13**, 989–992 (1973)

Beatty, D.D.: Visual pigments of the American eel *Anguilla rostrata*. Vision Res. **15**, 771–776 (1975)

Beer, T.: Die Accommodation des Fischauges. Pflügers Arch. ges. Physiol. **58**, 623–650 (1894)

Berger, E.R.: On the mitochondrial origin of oil drops in the retinal double cone inner segments. J. Ultrastruct. Res. **14**, 143–157 (1966)

Bernstein, M.H., Dietrich, T.S.: Electron microscopic studies on the reflecting structures of elasmobranch and teleost eyes. Biol. Bull. Woods Hole **119**, 303–304 (1960)

Bertelsen, E., Theisen, B., Munk, O.: On a postlarval specimen, anal light organ, and tubular eyes of the argentinoid fish *Rhynchohyalus natalensis* (Gilchrist and von Bonde). Vidensk. Medd. Dansk. Naturh. Foren. **128**, 357–371 (1965)

Bigelow, H.B., Schroeder, W.C.: Fishes of the Western North Atlantic. Memoir No. I, Part I: Cyclostomes and sharks; Part 2: Sawfishes, guitarfishes, skates, rays and chimaeroids. New Haven, Conn. Sears Foundation for Marine Research, Yale Univ. 1948

Bimes, G., Guilhem, A., Peyraud, C., Serfaty, A.: Mise en évidence chez un poisson téléostéen: la carpe (*Cyprinus carpio* L.) de l'influence d'un broyat de rétine adaptée à l'obscurité sur la dynamique in vitro, des photorecepteurs d'un oeil exposé à la lumière. C.R. Séances Acad. Sci. Ser. D. **263**, 668–670 (1966a)

Bimes, C., Guilhem, A., Peyraud, C., Serfaty, A.: Mise en évidence d'un mécanisme humoral dans la dynamique rétinienne d'un poisson téléostéen: la carpe commune (*Cyprinus carpio* L.). C.R. Séances Soc. Biol. **160**, 2470–2473 (1966b)

Blaxter, J.H.S.: The eyes of larval fish. In: Vision in Fishes: New Approaches in Research. Ali, M.A. (ed.). New York: Plenum Press 1975, pp. 427–443

Blaxter, J.H.S., Jones, M.P.: The development of the retina and retinomotor responses in the herring. J. Mar. Biol. Assoc. U.K. **47**, 677–697 (1967)

Blaxter, J.H.S., Staines, M.: Pure-cone retinae and retinomotor responses in larval teleosts. J. Mar. Biol. Assoc. U.K. **50**, 449–460 (1970)

Borwein, B., Hollenberg, M.J.: The photoreceptors of the four-eyed fish, *Anableps anableps* L. J. Morphol. **140**, 405–442 (1973)

Braemer, W.: Verhaltensphysiologische Untersuchungen am optischen Apparat bei Fischen. Z. vergl. Physiol. **39**, 374–398 (1957)

Brauer, A.: Die Tiefseefische. II. Anatomische Teil. B. Augen, 266p. Wissenschaftliche Ergebnisse der deutschen Tiefsee-Expedition auf dem Dampfer „Valdivia“ 1898–1899, Bd. 15. Jena: Gustav-Fischer 1908

Brett, J.R., Ali, M.A.: Some observations on the structure and photomechanical responses of the Pacific salmon retina. J. Fish. Res. Board Can. **15**, 815–829 (1958)

Butcher, E.O.: The structure of the retina of *Fundulus heteroclitus* and the regions of the retina associated with the different chromatophoric responses. J. Exp. Biol. **79**, 275–297 (1938)

Byzov, A.L., Trifonov, Y.A., Chailakhyan, L.M.: Effect of polarization of the pike retina horizontal cells on their potential propagation. Neirofiziologiya **4**, 90–96 (1972)

Cahn, P.: Comparative optic development in *Astyanax mexicanus* and in two of its blind derivatives. Bull. Am. Mus. Nat. Hist. **115**, 70–112 (1958)

Carrière, J.: Die Sehorgane der Thiere, vergleichend-anatomisch dargestellt. Munich: R. Othenburg 1885

Chabanaud, P.: Soléidés spécifiquement affectés d'une atrophie totale de l'oeil migrateur. C.R. Séances Acad. Sci. **223**, 486–487 (1946)

Chirichigno, F., Norma: Clave para identificar los peces marinos del Peru. Instituto del Mar del Peru, Informe No. 44, 1974

Cobbold, T.S.: Histological observations on the eye of the codfish (*Morrhus vulgaris*) with special reference to the choroid gland and the cones of the retina. J. Linn. Soc. Lond. Zool. **6**, 145–152 (1862)

Cohen, A.I.: Rods and cones and the problem of visual excitation. In: The Retina; Morphology, function and clinical characteristics. UCLA Forum in Medical Sciences No. 8. Bradley, R.S., Hall, M.O., Allen, R.A. and Crescitelli, F. (eds.). Berkeley, University of California Press, 1969, pp. 31–62.

Cohen, A.I.: Rods and cones. In: Physiology of Photoreceptor Organs, Fuortes, M.G.F. (ed.). Berlin: Springer-Verlag, 1972, pp. 63–110

Contino, F.: Das Auge des *Argyropelecus hemigymnus*. Morphologie Bau, Entwicklung und Refraktion. v. Graefes Arch. Ophthalmol. **140**, 390–441 (1939)

Dartnall, H.J.A. (ed.): Photochemistry of vision. In: Handbook of Sensory Physiology. Berlin: Springer-Verlag, 1972, Vol. VII/I

Dartnall, H.J.A.: Visual pigment of the coelacanth. Nature **239**, 341–342 (1972)

Dartnall, H.J.A., Lythgoe, J.N.: The clustering of fish visual pigments around discrete spectral positions and its bearing on chemical structure. In: Colour Vision: Physiology and Experimental Psychology. De Reuck, A.V.S., Knight, J. (eds.). London: Churchill, 1965

Dathe, H.H.: Vergleichende Untersuchungen an der Retina mitteleuropäischer Süßwasser-Fische. Z. Mikr.-Anat. Forsch. **80**, 269–319 (1969)

Davson, H., Graham, L.T.: The Eye; Comparative Physiology. New York: Academic Press, Vol. 6, 1974

Denissenko, G.: Einiges über den Bau der Netzhaut des Aales. Arch. Mikrosk. Anat. **21**, 1–20 (1882)

Detwiler, S.R.: Vertebrate Photoreceptors. New York: Macmillan, 1943

Dowling, J.E., Ripps, H.: Visual adaptation in the retina of the skate. J. Gen. Physiol. **56**, 491–520 (1970)

Dücker, M.: Über die Augen der Zyklostomen. Jena Z. Naturwiss. **60**, 471–528 (1924)

Duke-Elder, S.: System of Ophthalmology. Vol. I: The Eye in Evolution. London: Henry Kimpton, 1958

Eggert, B.: Der Bau des Auges und der Hautsinnesorgane bei den Gobiiformes *Amblyopus brachygaster* Gthr. und *Trypauchen vagina* Bl. Schn. Z. wiss. Zool. **138**, 68–87 (1931)

Eigenmann, C.H.: The eyes of the blind vertebrates of North America. I. The eyes of the amblyopsidae. Arch. Entwicklungs-Mech. **8**, 545–617 (1899)

Eigenmann, C.H.: Cave vertebrates of America. A study in degenerative evolution. Carnegie Inst. Washington Publ. No. **104**, 1–241 (1909)

Eigenmann, C.H., Shafer, G.D.: The mosaic of single and twin cones in the retina of fishes. Am. Nat. **34**, 109–118 (1900)

Engström, K.: Cone types and cone arrangement in the retina of some cyprinids. Acta Zool. **41**, 277–295 (1960)

Engström, K.: Cone types and cone mosaic in the retina of some gadids. Acta Zool. **42**, 227–243 (1961)

Engström, K.: Structure, organization and ultrastructure of the visual cells in the teleost family labridae. Acta Zool. **44**, 1–41 (1963a)

Engström, K.: Cone types and cone arrangements in teleost retinae. Acta Zool. **44**, 179–243 (1963b)

Engström, K., Ahlbert, I.B.: Cone types and cone arrangement in the retina of some flatfishes. Acta Zool. **44**, 119–129 (1963)

Engström, K., Rosstorp, E.: Photomechanical responses in different cone types of *Leuciscus rutilus*. Acta Zool. **44**, 145–160 (1963)

Fernholm, B., Holmberg, K.: The eyes in three genera of hagfish (*Eptatretus, Paramyxine* and *Myxine*)—A case of degenerative evolution. Vision Res. **15**, 253–259 (1975)

Feuerwerker, E., Ali, M.A.: La vision chez les poissons; historique et bibliographie analytique. Rev. Can. Biol. **34**, 221–285 (1975)

Frank, S.: The Pictorial Encyclopedia of Fishes. London: Hamlyn, 1969

Franz, V.: Zur Anatomie, Histologie und funktionellen Gestaltung des Selachierauges. Jena Z. Naturwiss. **40**, 697–840 (1905)

Franz, V.: Die japanischen Knochenfische der Sammlungen Haberer und Doflein. Abhandl. Bayer. Akad. Wiss. Suppl. **4**, 1–132 (1910)

Franz, V.: Sehorgan. In: Lehrbuch der vergleichenden mikroskopischen Anatomie der Wirbeltiere. Oppel, A. (ed.). Jena: Gustav-Fischer, 1913, Vol. 7

Franz, V.: Zur mikroskopischen Anatomie der Mormyridae. Zool. Jahrb. Abt. Anat. **42**, 91–148 (1920)

Franz, V.: Die Akkommodation des Selachieraugers und seine Abblendungsapparate, nebst Befunden an der Retina. Zool. Jahrb. Abt. Zool. Physiol. **49**, 323–461 (1931)

Franz, V.: Vergleichende Anatomie des Wirbeltierauges. In: Handbuch der vergleichenden Anatomie der Wirbeltiere. Bolk, L., Goppert, E., Kallius, E., Lubosch, W. (eds.). Berlin: Urban und Schwarzenberg, 1934, Vol. 2

Fries, B., Ekström, C.V., Sunnevall, C.: A History of Scandinavian Fishes. 2nd ed. revised and completed by A. Smitt. Stockholm and London, 1893

Friis, G.: Fiskeiet. Köpenhaven, 1879

Fürst, C.M.: Zur Kenntnis der Histogenese und des Wachstums der Retina. Acta Univ. Lund. **40**, 1–45 (1904)

Gabe, M.: Techniques histologiques. Paris: Masson, 1968

Gallego, A.: Nota sobre las celulas horizontales de la retina de tiburon, *Ginglymostoma cirratum*. Madrid: Arch. Fac. Med. **21**, 237–245 (1972)

Garten, S.: Die Veränderungen der Netzhaut durch Licht. In: Graefe-Saemisch Handbuch der gesam. Augenheilkunde. Leipzig: Vol. 2, 1907, pp. 1–30

Gentle, M.J.: The eye and colour change in the minnow *Phoxinus phoxinus* L. Exp. Biol. **57**, 701–707 (1972)

GILBERT, P.W.: The visual apparatus of sharks and its probable role in predation. Pacific Sci. Congr. Abstr. Symp. Paper **10**, 176–177 (1961)

GILBERT, P.W.: The visual apparatus of sharks. In: Sharks and Survival. Chapt. 9, GILBERT, P.W. (ed.). Boston: Heath, D.C., 1965

GOODLAND, H.: The ultrastructure of the inner plexiform layer of the retina of *Cottus bubalis*. Exp. Eye Res. **5**, 198–200 (1966)

GOVARDOVSKY, V.I., TSIRULIS, T.P.: Mitochondrial genesis in photoreceptor ellipsoids of the lamprey. Citologia **11**, 499–502 (1969)

GRAMONI, R., ALI, M.A.: L'electrorétinogramme et sa fréquence de fusion chez *Amia calva* (LINNÉ). Rev. Can. Biol. **29**, 353–363 (1970)

GRANIT, R.: Sensory Mechanisms of the Retina with an Appendix on Electroretinography. New York: Hafner, 1963

GREEFF, R.: Cited in Vertebrate Photoreceptors, by S.R. DETWILER. New York: Macmillan, 1899, pp. 39, 62

GREENWOOD, P.H., ROSEN, D.E., WEITZMANN, S.A., MYERS, G.S.: Phyletic studies of teleostean fishes, with a provisional classification of living forms. Am. Mus. Nat. Hist. Bull. **131**, 341–455 (1966)

GRIFFITH, R.W., THOMSON, K.S.: *Latimeria chalumnae*: Reproduction and conservation. Nature **242**, 617–618 (1973)

GRUBER, S.H.: Duplex vision in the elasmobranchs: histological, electrophysiological and psychophysical evidence. In: Vision in Fishes: New Approaches in Research. ALI, M.A. (ed.). New York: Plenum Press, 1975, pp. 525–540

GRUBER, S.H., HAMASAKI, D.I., BRIDGES, C.D.B.: Cones in the retina of lemon shark (*Negaprion brevirostris*). Vision Res. **3**, 397–399 (1963)

GRYNFELTT, E.: Etudes anatomiques et histologiques sur l'oeil du *Protopterus annectens*. Bull. Mens. Acad. Sci. Lett. Montpellier 3, 210–232 (1911)

GRZIMEK, H.C.B.: GRZIMEK's Animal Life Encyclopedia. Vol. 3–5, Mollusks and Echinoderms, Fishes and Amphibia. Translated from the German. New York: Van Nostrand, 1970

HAFFNER, R.: The zoogeography, biology and systematics of the chauliodontidae. Ph. D. Thesis, Yale University New Haven, Conn., 1952

HAMASAKI, D.I., GRUBER, S.H.: The photoreceptors of the nurse shark *Ginglymostoma cirratum* and the sting ray, *Dasyatis sayi*. Bull. Mar. Sci. **15**, 1051–1059 (1965)

Handbook of Sensory Physiology: Vol. I: Principles of Receptor Physiology, Vol. VII/1: Photochemistry of Vision, Vol. VII/2: Physiology of Sense Organs, Vol. VII/3A: Central Visual Information, Vol. VII/3B: Central Visual Information. Berlin: Springer-Verlag, 1971

HANNOVER, A.: Mikroskopiske Undersögelser af Nervesystemet. Vid. I Sel. Naturvid. Mathem. Afh. **10**, 9–112 (1843)

HANYU, I.: On the falciform process vitreal vessels and other related structures of the teleost eye. I. Various types and types and their interrelationships. Bull. Jap. Soc. Sci. Fish. **25**, 595–613 (1959)

HANYU, I.: Intraocular vascularization in some fishes. Can. J. Zool. **40**, 87–106 (1962)

HANYU, I., ALI, M.A.: Intra-sub-specific variation in retinal structure in *Sebastes marinus mentella*. Nature **196**, 554–556 (1962)

HART, J.L.: Pacific Fishes of Canada. Fish. Res. Board Can. Bull. **180** (1973)

HERALD, E.S.: Living Fishes of the World. New York: Doubleday, 1961

HIBBARD, E.: Grid patterns in the retinal organization of the cichlid fish *Astronotus ocellatus*. Exp. Eye Res. **12**, 175–180 (1971)

HOLMBERG, K.: Hagfish eye: ultrastructure of retinal cells. Acta Zool. **50**, 179–183 (1969)

HOLMBERG, K.: The hagfish retina: fine structure of retinal cells in *Myxine glutinosa* L. with special reference to receptor and epithelial cells. Z. Zellforsch. **111**, 517–538 (1970)

HOLMBERG, K.: The hagfish retina: electron microscopic study comparing receptor and epithelial cells in the Pacific hagfish, *Polistotrema stouti,* with those in the Atlantic hagfish, *Myxine glutinosa*. Z. Zellforsch. **121**, 249–269 (1971)

HOSCH, F.: Das Sehorgan von *Protopterus annectens*. Arch. Mikrosk. Anat. **64**, 99–110 (1904)

JOHN, K.R.: Further studies on retinomotor rhythms in the teleost *Astyanax mexicanus*. Physiol. Zool. **42**, 60–70 (1969)

JOHN, K.R., GRING, D.M.: Retinomotor rhythms in the bluegill, *Lepomis* macrochirus. J. Fish. Res. Board Can. **25**, 373–381 (1968)

JOHN, K.R., HAUT, M.: Retinomotor cycles and correlated behavior in the teleost *Astyanax mexicanus* (FILLIPI). J. Fish. Res. Board Can. **21**, 591–595 (1964)

JOHN, K.R., KAMINESTER, L.H.: Further studies on retinomotor rhythms in the teleost *Astyanax mexicanus*. Physiol. Zool. **42**, 60–70 (1969)

JOHN, K.R., SEGALL, M., ZAWATZKY, L.: Retinomotor rhythms in the goldfish, *Carassius auratus*. Biol. Bull. Woods Hole **132**, 200–210 (1967)

JOLLIE, M.T.: The general anatomy of *Lampanyctus leucopsarus* (EIGENMANN and EIGENMANN). Ph. D. Thesis, Stanford University Cal., 1954

KAHMANN, H.: Über das Vorkommen einer Fovea centralis im Knochenfischauger. Zool. Anz. **106**, 49–55 (1934)

KAHMANN, H.: Über das foveale Sehen der Wirbeltiere. I. Über die Fovea centralis und die Fovea lateralis bei einigen Wirbeltieren. v. Graefes Arch. Ophthalmol. **135**, 265–276 (1936)

KARSTEN, H.: Das Auge von *Periophthalmus* KÖHLREUTERI. Jena Z. Naturwiss. **59**, 115–154 (1923)

KATO, S.: Histology of the retinas of the Pacific sharks *Carcharhinus melanopterus* and *Trisenodon obsesus*. M. Sc. Thesis, University of Hawaii, 1962

KAWAMURA, G., TAMURA, T.: Morphological studies on the retina of two teleosts *Scomber tapeinocephalus* and *Halichoeres poecilopterus*. Bull. Jap. Soc. Sci. Fish. **39**, 715–726 (1973)

KERR, J.G.: The development of *Lepidosiren paradoxa*. III. Development of the skin and its derivatives. Quart. J. Micros. Sci. N.S. **46**, 417–460 (1902)

KLEEREKOPER, H.: The sense organs. In: The Biology of Lampreys. HARDISTY, M.W., POTTER, I.C. (eds.). London: Academic Press, 1972

KOBAYASHI, H.: Notes on retinomotor phenomena in some fishes under the various light conditions. J. Shimonoseki Coll. Fish. **7**, 169–177 (1957)

KOBAYASHI, H.: On the photoperceptive function in the eye of the hagfish, *Myxine garmani* JORDAN and SNYDER. J. Shimonoseki Univ. Fish. **13**, 141–157 (1964)

KOBAYASHI, H., ALI, M.A.: Electroretinographic determination of spectral sensitivity in albino and pigmented trout (*Salvelinus fontinalis,* MITCHILL). Can. J. Physiol. Pharmacol. **49**, 1030–1037 (1971)

KOHL, C.: Rudimentäre Wirbelthieraugen. Bibliotheca Zool. **13**, 193–204 (1892)

KOLMER, W.: Die Netzhaut. In: Handbuch der mikroskopischen Anatomie des Menschen. VON MOLLENDORFF, W. (ed.). Berlin: Springer, 1936

KUENZER, P., WAGNER, H.J.: Bau und Anordnung der Sehzellen und Horizontalen in der Retina von *Nannacara anomala* (Cichlidae, Teleostei). Z. Morphol. Tiere **65**, 209–224 (1969)

KUNZ, Y.W., KELLY, D.: The histology of the retina of *Pomatoschistus (Gobius) microps* (KRØYER). Experientia **29**, 837–839 (1973)

LAGIOS, M.D.: The pituitary gland of the coelacanth *Latimeria chalumnae* SMITH. Gen and Com Endoc. **25**, 126–146 (1975)

LANG, H.-J.: Eine lichtmikroskopische Untersuchung der Guppynetzhaut. Zeiss-Mitteilungen **3**, 415–438 (1965)

LAUFER, M., VANEGAS, H.: The optic tectum of a perciform teleost. II. Fine structure. J. Comp. Neurol. **154**, 61–96 (1974)

LEGENDRE, V.: Les Poissons d'Eau Douce; Clef des Poissons de Pêche, Sportive et Commerciale de la Province de Québec. 2e ed. française. Montréal: Société canadienne d'Ecologie, 1954 (Tome I.)

LEIM, A.H., SCOTT, W.B.: Fishes of the Atlantic Coast of Canada. Bull. Fish. Res. Board Can. **155**, (1966)

LOCKET, N.A.: The retina of *Poromitra nigrofulvus* (GARMAN). An optical and electron microscope study. Exp. Eye Res. **8**, 265–275 (1969)

LOCKET, N.A.: Retinal structure in a deep-sea fish, *Sternoptyx diaphana,* HERMANN. Exp. Eye Res. **9**, 22–27 (1970a)

LOCKET, N.A.: Landolt's club in the retina of the African lungfish, *Protopterus aethiopicus,* HECKEL. Vision Res. **10**, 299–306 (1970b)

LOCKET, N.A.: Retinal structure in *Platytroctes apus* a deep-sea fish with a pure rod fovea. J. Mar. Biol. Assoc. U.K. **51**, 79–91 (1971a)

LOCKET, N.A.: Retinal anatomy in some scopelarchid deep-sea fishes. Proc. R. soc. Lond. Ser. B **178**, 161–184 (1971b)

LOCKET, N.A.: The reflecting structure in the iridescent cornea of the serranid teleost *Nemanthias carberryi*. Proc. R. Soc. Lond. Ser. B. **182**, 249–254 (1972)

LOCKET, N.A.: Retinal structure in *Latimeria chalumnae*. Phil. Trans. R. Soc. Lond. Ser. B. **266**, 493–521 (1973)

LOCKET, N.A.: Possible discontinuous retinal rod outer segment formation in *Latimeria chalumnae*. Nature **244**, 308–309 (1973)

LOCKET, N.A.: Landolt's club in some primitive fishes. In: Vision in Fishes: New Approaches in Research. ALI, M.A. (ed.). New York: Plenum Press, 1975, pp. 471–480

LÜLING, K.H.: Zur Augenreduktion des aus mexikanischen Höhlen stammenden blinden Salmlers *Anoptichthys jordani* (HUBBS and INNES). Photographie und Forschung **6**, 138–143 (1955a)

LÜLING, K.H.: Untersuchungen am Blindfisch *Anoptichthys jordani* HUBBS und INNES (Characidae). III. Vergleichend anatomisch-histologische Studien an den Augen des *Anoptichthys jordani*. Zool. Jahrb. Abt. Anat. Ontol. **74**, 339–400 (1955b)

LÜLING, K.H.: Morphologisch-anatomische und histologische Untersuchungen am Auge des Schützenfisches *Toxotes jaculatrix* (Pallas 1766) (toxotidae), nebst Bemerkungen zum Spuckgehaben. Z. Morphol. Ökol. Tiere **47**, 529–610 (1958)

LYALL, A.H.: The growth of the trout retina. Quart. J. Micros. Sci. **98**, 101–110 (1957a)

LYALL, A.H.: Cone arrangements in teleost retinae. Quart. J. Micros. Sci. **98**, 189–201 (1957b)

LYTHGOE, J.N.: List of vertebrate pigments. In: Handbook of Sensory Physiology, Vol. VII/I. DARTNALL, H.J.A. (ed.). Berlin: Springer-Verlag, 1972, pp. 604–624

LYTHGOE, J., LYTHGOE, G.: Fishes of the Sea; the Coastal Waters of the British Isles, Northern Europe and the Mediterranean. A Photographic Guide in Colour. London: Blandford Press 1971

MARSHALL, N.B., THINÈS, G.: Studies on the brain, sense organs and light sensitivity of a blind cave fish *(Typhlogaria widdowsoni)* from Irak. Proc. Zool. Soc. Lond. **131**, 441–456 (1958)

McAllister, D.E.: The evolution of branchiostegals and the classification of teleostome fishes, living and fossil. Bull. Nat. Mus. Can. **221**, 1–239 (1968)

McEwan, M.R.: A comparison of the retina of the mormyrids with that of various other teleosts. Acta Zool. **19**, 427–465 (1938)

Menner, E.: Untersuchungen über die Retina mit besonderer Berücksichtigung der Äußeren Körnerschicht. Z. vergl. Physiol. **8**, 761–826 (1929)

Menner, E.: Über den Feinbau der Außenglieder der Sehzellen bei Wirbeltieren. Verhandl. Deut. Zool. Ges. **15**, 124–139 (1950)

Meyer-Rochow, V.B.: The larval eye of the deep-sea fish *Cataetyx memorabilis* (Teleostei, Ophidiidae) Z. Morphol. Tiere **72**, 331–340 (1972)

Millot, J., Anthony, J.: Anatomie de *Latimeria chalumnae*. II. Système nerveux et organes des sens. Paris: Cent. Nat. Rech. Sci., 1965, p. 130

Millot, J., Carasso, N.: Note préliminaire sur l'œil de *Latimeria chalumnae* (Crossoptérygien Coelacanthidé) C.R. Acad. Sci. **241**, 576–577 (1955)

Mirzaliev, V., Koloss, E.: Characteristics of certain eye structure in river eel. Uch. Zap. Anat. Gistol. Embriol. Republ. Shredn. Asii. Kazakhstana **1**, 118–122 (1964)

Moore, G.A.: The retinae of two North American teleosts with special references to their tapeta lucida. J. Comp. Neurol. **80**, 369–379 (1944)

Moore, G.A., McDougal, R.C.: Similarity in the retinae of *Amphiodon alosoides* and *Hiodon tergisus*. Copeia, 298 (1949)

Moore, G.A., Pollock, H.R., Lima, D.: The visual cells of *Ericymba buccata* (Cope). J. Comp. Neurol. **93**, 289–295 (1950)

Moskalkova, K.I.: Adaptative changes of the retina during the development of the goby, *Gobius melanostomus* (Pall). Dokl. Akad. Nauk. SSSR Ser. Biol. **198**, 1225–1227 (1971)

Müller, H.: Bau und Wachstum der Netzhaut des Guppy *(Lebistes reticulatus)*. Zool. Jahrb. Physiol. **63**, 276–324 (1952)

Munk, O.: The eyes of *Ipnops murrayi* Günther, 1878. Galathea Rep. **3**, 79–87 (1959)

Munk, O.: The eye of *Stomias boa ferox* Reinhardt. Vidensk. Medd. Dansk. Naturh. Foren. **125**, 353–359 (1963)

Munk, O.: The eyes of three benthic deep-sea fishes caught at great depths. Galathea Rep. **7**, 137–149 (1964a)

Munk, O.: The eyes of some ceratioid fishes. Dana Rep. **61**, 1–15, (1964b)

Munk, O.: The eye of *Calamoichthys calabaricus* Smith, 1865 (Polypteridae, Pisces) compared with the eye of other fishes. Vidensk. Medd. Dansk. Naturh. Foren. **127**, 113–126 (1964c)

Munk, O.: Ocular degeneration in deep-sea fishes. Galathea Rep. **8**, 21–31 (1965a)

Munk, O.: *Omosudis lowei* Günther, 1887, a bathypelagic deep-sea fish with an almost pure-cone retina. Vidensk. Medd. Dansk. Naturh. Foren. **128**, 341–355 (1965b)

Munk, O.: Ocular anatomy of some deep-sea teleosts. Dana Rep. **70**, 1–62 (1966a)

Munk, O.: On the retina of *Diretmus argenteus* Johnson, 1863 (Diretmidae, Pisces). Vidensk. Medd. Dansk. Naturh. Foren. **129**, 73–80 (1966b)

Munk, O.: On the eye and the so-called preorbital light organ of the isospondylous deep-sea fish *Bathylaco nigricans* Goode and Bean, 1896. Galathea Rep. **9**, 211–218 (1968a)

MUNK, O.: The eyes of *Amia* and *Lepisosteus* (Pisces, Holostei) compared with the brachiopterygian and teleostean eyes. Vidensk. Medd. Dansk. Naturh. Foren. **131**, 109–127 (1968b)

MUNK, O.: The eye of the "four-eyed" fish *Dialommus fuscus*. (Pisces, Blennioidei, Clinidae). Vidensk. Medd. Dansk. Naturh. Foren. **132**, 7–24 (1969a)

MUNK, O.: On the visual cells of some primitive fishes with particular regard to the classification of rods and cones. Vidensk. Medd. Dansk. Foren. **132**, 25–30 (1969b)

MUNK, O.: On the occurrence and significance of horizontal band-shaped retinal areae in teleosts. Vidensk. Medd. Dansk. Naturh. Foren. **133**, 85–120 (1970)

MUNTZ, W.R.A., CHURCH, E., DARTNALL, J.H.A.: Visual pigment of the freshwater stingray, *Paratrygon motoro*. Nature **246**, 517 (1973)

MUNZ, F.W., MCFARLAND, W.N.: The significance of spectral position in the rhodopsins of tropical marine fishes. Vision Res. **13**, 1829–1874 (1973)

NEUMAYER, L.: Der feinere Bau der Selachier-Retina. Arch. Mikrosk. Anat. **48**, 83–111 (1897)

NICOL, J.A.C.: The tapetum in *Scyliorhinus canicula*. J. Mar. Biol. Ass. U.K. **41**, 271–277 (1961a)

NICOL, J.A.C.: Photomechanical changes in the eyes of fishes. I. Retinomotor changes in *Solea solea*. J. Mar. Biol. Ass. U.K. **41**, 695–698 (1961b)

NICOL, J.A.C.: Some aspects of photoreception and vision in fishes. Adv. Mar. Biol. **1**, 171–208 (1963)

NICOL, J.A.C.: Retinomotor changes in flatfishes. J. Fish. Res. Board Can. **22**, 513–520 (1965)

NICOL, J.A.C.: The tapetum lucidum of the sturgeon. Contrib. Mar. Sci. **14**, 5–18 (1969)

NICOL, J.A.C., ARNOTT, H.J., BEST, A.C.G.: Tapeta lucida in bony fishes *(Actinopterygii)*. Can. J. Zool. **51**, 69–81 (1973)

NICOL, J.A.C., ZYZNAR, E.S., THURSTON, E.L., WANG, R.T.: The tapetum lucidum in the eyes of cusk-eels (Ophidiidae). Can. J. Zool. **53**, 1063–1079 (1975)

NICOL, J.A.C., ZYZNAR, E.S.: The tapetum lucidum in the eye of the big-eye *Priacanthus arenatus* Cuvier. J. Fish. Biol. **5**, 519–522 (1973)

NIKOLSKY, G.V.: The Ecology of Fishes. Translated from the Russian by L. BIRKETT. New York: Academic Press, 1963

O'CONNELL, O.P.: The structure of the eye of *Sardinops caerulea, Engraulis mordax,* and four other pelagic marine teleosts. J. Morphol. **113**, 287–329 (1963)

O'DALY, J.A.: Ultrastructura y citoquímica de la retina de los teleósteos. Tesis doctoral, Universidad Central de Venezuela, Caracas, 1967

ÖHMAN, P.: A fluorescence microscopic study of the retina of the river lamprey *(Lampreta fluviatilis)* and the Atlantic hagfish *(Myxine glutinosa)*. Acta Zool. **51**, 179–181 (1970)

ÖHMAN, P.: The photoreceptor outer segments of the river lamprey *(Lampreta fluviatilis)*: an electron-fluorescence- and light microscopic study. Acta Zool. **52**, 287–297 (1971)

OLLA, B.L., MARCHIONI, W.W.: Rhythmic movements of cones in the retina of blue-fish, *Pomatomus saltatrix,* held in constant darkness. Biol. Bull. Mar. Biol. Lab. Woods Hole **135**, 530–536 (1968)

PAPPENHEIM, S.: Die specielle Gewebelehre des Auges mit Rücksicht auf Entwicklungsgeschichte und Augenpraxis. Breslau: ADERHOLZ, G.P., 1842

PARTHE, V.: Celulas horizontales y amacrinas de la retina. Acta Cient. Venezol. Suppl. **33**, 240–249 (1967)

PARTHE, V.: Horizontal, bipolar and oligopolar cells in the teleost retina. Vision Res. **12**, 395–406 (1972)

PAVAN, C.: Observations and experiments on the cave fish *Pimelodella kronei* and its relatives. Am. Nat. **80**, 343–361 (1946)

PELLEGRIN, J.: Contribution à l'étude anatomique, biologique et taxonomique des poissons de la famille des cichlidés. Mem. Soc. Zool. France, 41–399 (1903)

PERCY, R., LEATHERLAND, J.F., BEAMISH, F.W.H.: Structure and ultrastructure of the pituitary gland in the sea lamprey, *Petromyzon marinus* at different stages in its life cycle. Cell. Tiss. Res. **157**, 141–164 (1975)

PETERS, N., PETERS, G.: Die Degeneration der Augen von *Typhlogarra widdowsoni* (Cyprinidae, Pisces). Roux Arch. Entwicklungsmech. **156**, 344–362 (1965)

PETERS, N., PETERS, G.: Das Auge zweier Hohlenformen von *Astyanax mexicanus* (Philippi). (Characinidae, Pisces). Roux Arch. Entwicklungsmech. **157**, 393–414 (1966)

PEYRAUD, C.M.: Contribution à l'étude de la dynamique des pigments et des photorécepteurs rétiniens chez la carpe, *(Cyprinus carpio* L.*)* C. R. Séances Acad. Sci. **263**, 65–67 (1966)

PFEIFFER, W.: Retina und Retinomotorik der Dipnoi und Brachiopterygii. Z. Zellforsch. **89**, 62–72 (1968)

POLYAK, S.: The Vertebrate Visual System. Chicago, Ill.: University of Chicago Press, 1957

PORA, E.A., WITTENBERGER, G., RUSDEA, D., DRAGOS, M.: Observations on retinomotor phenomena among some fishes of the Black Sea. Stud. Cercet. Biol. **14**, 299–304 (1963)

PROTASOV, V.R.: Some functional peculiarities of the retina in nine fish species of the Barents Sea. Vopr. Ikhtiol. **14**, 139–155 (1960)

PROTASOV, V.R., ALTUKHOV, Y.P., KOVALEVA, N.D.: Morphological and functional features of the transition from day vision to twilight vision in some Black Sea fish. Dokl. Akad. Nauk. SSSR. **134**, 195–198 (1960)

QUAGUEBEUR, M.: Onderzoekingen over blinde grotvissen verband met de reductie in de ontwikkeling van hun ogen. Thesis, University of Louvain, Belgium, 1955

RAMON Y CAJAL, S.: Histologie du Système Nerveux de l'Homme et des Vertébres. Tome II. Traduite de l'Espagnol par L. Azoulay. Madrid: Consejo Superior de Invest. Cient., 1955

REICH, M.: Zur Histologie der Hechtretina. Arch. Ophthalmol. **20**, 1–14 (1874)

RETZIUS, G.: Zur Kenntnis der Retina der Selachier. In: Zoologiska Studier: Festskrift for WILHELM LILLJEBORG. Uppsala: Almgrist & Wiksells Boktryckeri-Aktiebolag. 1896, p. 3–12

RETZIUS, G.: Zür Kenntnes vom Bau der Selachierretina. Biol. Untersuch. N.F. **12**, 55 (1905)

RITTER, W.E.: On the eyes, the integumentary sense papillae and its integument of the San Diego blind fish (*Typhlogobius californiensis* Steindachner). Bull. Mus. Comp. Zool. Harvard **24**, 51–102 (1893)

ROBERTS, T.R.: Ecology of fishes in the Amazon and Congo basins. Bull. Mus. Comp. Zool. **143**, 117–147 (1972)

ROCHON-DUVIGNEAUD, A.: L'œil de *Lepidosiren paradoxa*. C.R. Séances Acad. Sci. **212**, 307–309 (1941)

ROCHON-DUVIGNEAUD, A.: Les Yeux et la Vision des Vértébrés. Paris: Masson, 1943

ROCHON-DUVIGNEAUD, A.: L'œil et la vision. Tome 13 (2) in: Traité de Zoologie. GRASSE, P.P. (ed.). Paris: Masson, 1958, p. 1099–1142

ROCHON-DUVIGNEAUD, A., ROULE, L.: Observation sur le comportement visuel et la structure de l'œil chez *Blennius basilicus* C.V. Bull. Mus. Natl. Hist. Nat. **2**, 139–145 (1927)

Rosen, D.E., Patterson, C.: The structure and relationships of the paracanthopterygian fishes. Bull. Am. Mus. Nat. Hist. **141**, 359–474 (1969)

Ryder, J.A.: An arrangement of the retinal cells in the eyes of fishes partially simulating compound eyes. Proc. Acad. Nat. Sci. Philadelphia, 1895, pp. 161–166

Schiefferdecker, P.: Studien zur vergleichenden Histologie der Retina. Arch. Mikrosk. Anat. **28**, 305–396 (1886)

Schneider-Orelli, M. von: Untersuchungen über das Auge von *Anableps tetrophthalmas.* Mitt. Naturforsch. Gesellsch., Bern 1907, pp. 87–113

Schoebitz, K., Echandia, R.E.L., Campos, H.: Complex mitochondria in the retinal cones of the teleost *Galaxia platei.* J. Microsc. **18**, 109–114 (1973)

Scholes, J.H.: Colour receptors and their synaptic connexions, in the retina of a cyprinid fish. Phil. Trans. R. Soc. Lond. Ser. B. **270**, 61–118 (1975)

Schroeder, D.M., Ebbesson, S.O.E.: Cytoarchitecture of the optic tectum in the nurse shark. J. Comp. Neurol. **160**, 443–461 (1975)

Schwassmann, H.O.: Visual projection upon the optic tectum in foveate marine teleosts. Vision. Res. **8**, 1337–1348 (1968)

Schwassmann, H.O., Kruger, L.: Experimental analysis of the visual system of the four-eyed fish *Anableps microlepis.* Vision Res. **5**, 269–281 (1965)

Scott, W.B., Crossman, E.J.: Freshwater Fishes of Canada. Bull. Fish. Res. Board Can. **184**, (1973)

Selvin de Testa, A.: Morphological studies on the horizontal and amacrine cells of the teleost retina. Vision Res. **6**, 51–59 (1966)

Shafer, G.D.: The mosaic of the single and twin cones in the retina of *Micropterus salmoides.* Arch. Entwicklungsmech. Organismen **10**, 685–691 (1900)

Somiya, H., Tamura, T.: On the eye of "yellow lens" fish *Chlorophthalmus albatrossis.* Bull. Jap. Soc. Sci. Fish. **37**, 840–845 (1971)

Stell, W.K.: Correlation of retinal cytoarchitecture and ultrastructure in Golgi preparation. Anat. Rec. **153**, 389–398 (1965)

Stell, W.K.: The structure and relationships of horizontal cells and photoreceptor bipolar synaptic complexes in the goldfish retina. Am. J. Anat. **120**, 401–424 (1967)

Stell, W.K.: The structure and morphologic relations of rods and cones in the retina of the spiny dogfish *Squalus.* Comp. Biochem. Physiol. **42A**, 141–151 (1972)

Stell, W.K.: Horizontal cell axons and axon terminals in goldfish retina. J. Comp. Neurol. **159**, 503–520 (1975)

Stell, W.K., Detwiler, P.B., Wagner, H.G., Wolbarsht, M.L.: Giant retinal ganglion cells in dogfish (*Mustelus*): Electrophysiology of single on-center units. In: Vision in Fishes: New Approaches in Research. Ali, M.A. (ed.). New York: Plenum Press, 1975, pp. 99–112

Stell, W.K., Hárosi, F.J.: Cone structure and visual pigment content in the retina of the goldfish. Vision Res. **16**, 647–657 (1976)

Stell, W.K., Lightfoot, D.O.: Color-specific interconnections of cones and horizontal cells in the retina of the goldfish. J. Comp. Neurol. **159**, 473–502 (1975)

Stell, W.K., Witkovsky, P.: Retinal structure in the smooth dogfish, *Mustelus canis:* general description and light microscopy of giant ganglion cells. J. Comp. Neurol. **148**, 1–32 (1973a)

Stell, W.K., Witkovsky, P.: Retinal structure in the smooth dogfish, *Mustelus canis:* Light microscopy of photoreceptor and horizontal cells. J. Comp. Neurol. **148**, 33–46 (1973b)

Sterba, G.: Freshwater Fishes of the World. London: Vista Books, 1962

Studnitz, G. von: Physiologie des Sehens. Leipzig: Akademische Verlagsgesellschaft, 1952

Sverdlick, J.: Conos y bastoncitos. Tesis doctoral, Facultad de Ciencias Médicas de Buenos Aires, 109 p., 1940

Sverdlick, J.: Significación histobiologica de los conos dobles y gemelos en la retina de ciertos peces oseos (teleosteos). Arch. Histol. Normal Patol. **6**, 117–122 (1955)

Tamura, T.: A study of visual perception in fish, especially on resolving power and accommodation. Bull. Jap. Soc. Sci. Fish. **22**, 536–557 (1957)

Tamura, T., Wisby, W.J.: The visual sense of pelagic fishes, especially the visual axis and accommodation. Bull. Mar. Sci. Gulf Caribbean. **13**, 433–448 (1963)

Tansley, K.: Vision in Vertebrates. London: Chapman and Hall, 1965

Thinès, G.: Sensory degeneration and survival in cave fishes. Symp. Zool. Soc. Lond. **3**, 39–52 (1960)

Tretjakoff, D.: Das Auge und die Nasenhöhle der Makrele. Z. wiss. Zool. **137**, 550–577 (1930)

Vanegas, H., Ebbesson, S.O.: Retinal projections in the perch-like teleost *Eugerres plumieri*. J. Comp. Neurol. **151**, 331–357 (1973)

Verheijen, F.J.: A peculiar nystagmus and a corresponding foveal structure in the eye of the herring (*Clupea harengus* L.). Experientia **15**, 443–447 (1959)

Verrier, M.L.: Sur la structure de l'oeil de *Ameiurus nebulosus* Lesueur et de *Clariss batrachius* L., ses rapports avec l'habitat et le comportement biologique de ces deux Siluridés. Bull. Soc. Zool. Fr. **52**, 581–588 (1927)

Verrier, M.L.: Recherches sur les yeux et la vision des poissons. Bull. Biol. Fr. Belge Suppl. **11**, 1–222 (1928a)

Verrier, M.L.: Sur la presence et la structure d'une fovéa rétinienne chez un Percidé: *Serranus cabrilla* L. C.R. Seances Acad. Sci. **186**, 457 (1928b)

Verrier, M.L.: Contribution à l'étude de la vision chez les Sélaciens. Annal Sci. Nat. Zool. IOE Series **13**, 5–61 (1930a)

Verrier, M.L.: Contribution à l'étude des yeux des poissons de grands fonds. Comparison avec les yeux du carassin doré macrophtalme. Arch. Zool. Exp. Gén. **70**, 58–69 (1930b)

Verrier, M.L.: Les yeux de *Pholis gunnellus*. Bull. Bull. Sta. Oceanogr. Salammbo **20** (1931)

Verrier, M.L.: Recherches sur les foveae des poissons. Etude des yeux de *Julis giofredi* Risso. C.R. Séances Soc. Biol. **113**, 134–135 (1933)

Villegas, G.M.: Electron microscopic study of the vertebrate retina. J. Gen. Physiol. **43**, Part 2, 15–43 (1960)

Villegas, G.M.: Comparative ultrastructure of the retina in fish, monkey and man. In: The Visual System: Neurophysiology and Psychophysics. Jung, R., Kornhuber, H. (eds.). Berlin: Springer-Verlag, 1961, pp. 1–13

Villegas, G.M., Villegas, R.: Neuron-glia relationship in the bipolar cell layer of the fish retina. J. Ultrastruc. Res. **8**, 84–106 (1963)

Vilter, V.: Dissociation spatiale des champs photo-sensoriels à cônes et à bâtonnets chez un poisson marin, le *Callionymus lyra*. C.R. Séances Soc. Biol. **141**, 344–346 (1947a)

Vilter, V.: Dissociation spatiale des cônes et des bâtonnets dans la retina du Callionyme et des relations avec l'architectonique neuronale de l'appareil visuel. C.R. Séances Soc. Biol. **141**, 346–348 (1947b)

VILTER, V.: Existence d'un double area dans la retine du *Lebistes reticulatus* et signification fonctionnelle des structures aréales. C.R. Séances Soc. Biol. **142**, 292–294 (1948)

VILTER, V.: Adaptation biologique de l'appareil visuel et les structures rétiniennes de la sardine. C.R. Séances Soc. Biol. **144**, 200–203 (1950)

VILTER, V.: Bases cyto-architectoniques de l'acuité visuelle chez un poisson abyssal, le *Lampanyctus crocodilus*. C.R. Séances Soc. Biol. **145**, 52–54 (1951 a)

VILTER, V.: Intervention probable de la lumière dans la naissance des structures rétiniennes révélée par l'étude comparée de la rétine chez les anguilles normales et cavernicoles. C.R. Séances Soc. Biol. **145**, 54–56 (1951 b)

VILTER, V.: Existence d'une rétine à plusieurs mosaiques photoréceptrices chez un poisson abyssal bathypélagique, *Bathylagus benedicti*. C.R. Séances Soc. Biol. **147**, 1937–1939 (1953)

VILTER, V.: Differenciation fovéale dans l'appareil visuel d'un poisson abyssal le *Bathylagus benedicti*. C.R. Séances Soc. Biol. **148**, 59–63 (1954 a)

VILTER, V.: Interpretation biologique des trames photoréceptrices superposées de la rétine du *Bathylagus benedicti*. C.R. Séances Soc. Biol. **148**, 327–330 (1954 b)

VILTER, V.: Relations neuronales dans la fovea à bâtonnets du *Bathylagus benedicti*. C.R. Séances Soc. Biol. **148**, 466–469 (1954 c)

VILTER, V., THIBAULT, C.: Etude du mécanisme photoadaptif du pigment rétinien mélanique chez la carpe par les éclairements unilateraux. C.R. Séances Soc. Biol. **142**, 290–292 (1948)

VRABEC, F.: Sur les cônes triples de la rétine des téléostéens. Acta Soc. Zool. Bohemoslov. **19**, 183–186 (1955)

VRABEC, F.: A new finding in the retina of a marine teleost *Callionymus lyra* L. Folia Morphol. **14**, 143–147 (1966)

VRABEC, F.: Sur la présence d'une fovéa dans la rétine du *Lepadogaster*. Vie et Milieu, Sèr. A. **20**, 245–250 (1969)

WAGNER, H.-J.: Der Bau der Retina und der multiplen optischen Papille bei zwei *Synodontis*-Arten (Teleostei, Siluroidea). Z. Morphol. Tiere **68**, 69–82 (1970)

WAGNER, H.-J.: Vergleichende Untersuchungen über das Muster der Sehzellen und Horizontalen in der Teleostier-Retina (Pisces). Z. Morphol. Tiere **72**, 77–130 (1972)

WAGNER, H.-J.: Darkness-induced reduction of the number of synaptic ribbons in fish retina. Nature **246**, 53–55 (1973)

WAGNER, H.-J.: Die nervösen Netzhautelemente von *Nannacara anomala* (Cichlidae Teleostei). I. Darstellung durch Silberimprägnation. Z. Zellforsch. **137**, 63–86 (1973 a)

WAGNER, H.-J.: Die nervösen Netzhautelemente von *Nannacara anomalia* (Cichlidae, Teleostei). II. Quantitative Verteilung. Z. Zellforsch. **137**, 87–95 (1973 b)

WAGNER, H.-J.: Die Entwicklung der Netzhaut von *Nannacara anomalia* (Regan) (Cichlidae, Teleostei) mit besonderer Berücksichtigung regionaler Differenzierungsunterschiede. Z. Morphol. Tiere **79**, 113–131 (1974)

WAGNER, H.-J., MENEZES, N.A., ALI, M.A.: Retinal adaptations in some Brazilian tide pool fishes. Zoomorphol. **83**, 209–226 (1976)

WALLS, G.L.: An experimental study of the retina of the brook lamprey, *Entosphenus appendix* (Dekay). J. Comp. Neurol. **46**, 465–473 (1928)

WALLS, G.L.: The visual cells of lampreys. Brit. J. Ophthalmol. **19**, 129–148 (1935)

WALLS, G.L.: The Vertebrate Eye and its Adaptive Radiation. New York: Hafner, 1942

WANG, C.S.J.: The eye of fishes with special reference to pigment migration. Ph. D. Thesis, Cornell University. 291 p., 1968

WARD, J.W., GUNTER, G.: Description of the eyes of a blind specimen of the Red Drum, *Sciaenops ocellate*. Copeia 440–442 (1962)

WELSH, J.H., OSBORN, O.M.: Diurnal changes in the retina of the catfish *Ameiurus nebulosus*. J. Comp. Neurol. **66**, 349–359 (1937)

WITKOVSKY, P., SHAKIB, M., RIPPS, H.: Interreceptoral junctions in the teleost retina. Invest. Ophthalmol. **13**, 996–1009 (1974)

WITKOVSKY, P., STELL, W.K.: Retinal structure in the smooth dogfish, *Mustelus canis:* Light microscopy of bipolar cells. J. Comp. Neurol. **148**, 47–60 (1973)

WITTENBERG, J.B., WITTENBERG, B.A.: Active secretion of oxygen into the eye of a fish. Nature **194**, 106–107 (1962)

WUNDER, W.: Physiologische und vergleichend-anatomische Untersuchungen an der Knochenfischnetzhaut. Z. vergl. Physiol. **3**, 1–63 (1925)

WUNDER, W.: Über den Bau der Netzhaut bei Süßwasserfischen, die in großer Tiefe leben (Coregonen, Tiefseesaibling). Z. vergl. Physiol. **4**, 22–36 (1926)

WUNDER, W.: Bau und Funktion der Netzhaut beim Zander (*Lucioperca sandra*) und einigen anderen im Balatonsee häufigen Fischarten. Z. vergl. Physiol. **11**, 749–766 (1930)

WUNDER, W.: Über den Bau der Netzhaut bei drei verschiedenen Formen der kleinen Maraene (*Coregonus albula* L.). Z. Morphol. Ökol. Tiere **27**, 684–690 (1933)

WUNDER, W.: Physiologie der Süßwasserfische Mitteleuropas Band II, B. In: Handbuch der Binnenfischerei Mitteleuropas. DEMOLL, R., MAIER, H.N.E. (eds.). Stuttgart: Schweigerbarth'sche Verlagsbuchhandlung, 1936

WUNDER, W.: Biologie und Bau der Netzhaut beim Rotbarsch (*Sebastes marinus* L.). Zool. Anz. **160**, 94–105 (1958)

YAMADA, E., ISHIKAWA, T.: The fine structure of the horizontal cells in some vertebrate retinae. Cold Spring Harbor Symp. Quant. Biol. **30**, 383–392 (1965)

YAMADA, E., ISHIKAWA, T.: The so-called "synaptic ribbon" in the inner segment of the lamprey retina. Arch. Histol. Jap. **28**, 411–417 (1967)

YAMAMOTO, TOKI-O.: Medaka (Killifish) Biology and Strains. Tokyo: Keigaku Publishing Co., 1975

YAMANOUCHI, T.: The visual acuity of the coral fish *Microcanthus strigatus* (Cuvier Valenciennes). Publ. Seto Mar. Biol. Lab. **5**, 133–156 (1956)

YOUNG, R.W.: A difference between rods and cones in the renewal of outer segment protein. Invest. Ophthalmol. **8**, 222–231 (1969)

ZYZNAR, E.S., ALI, M.A.: An interpretative study of the organization of the visual cells and tapetum lucidum of *Stizostedion*. Can. J. Zool. **53**, 180–196 (1975)

ZYZNAR, E.S., NICOL, J.A.C.: Reflecting materials in the eyes of three teleosts, *Orthopristes chrysopterus, Dorosoma cepedianum* and *Anchoa mitchilli*. Proc. R. Soc. Lond. Series B. **184**, 15–27 (1973)

6. SYNOPSIS OF FAMILIES

Subject Index

Numbers in *italics* refer to pages including photographic plates for that particular species.

N

O

N. J. Strausfeld

Atlas of an Insect Brain

81 figures, partly coloured, 71 plates
XII, 214 pages. 1976

Contents: A Historical Commentary.–The Structure of Neuropil.–The Primary Compartments of the Brain.–The Coordinate System.–Some Quantitative Aspects of the Fly's Brain.–The Atlas: Sections through the Brain.–The Forms and Dispositions of Neurons in the Brain.–Appendix 1: Histological Methods.–Appendix 2: Dictionary of Terms.

This atlas is the first presentation of the main regions and pathways of an arthropod brain to combine both mass-staining of fibres and selective impregnation of neurons. It displays in detail the basic structures of the neuropils and schematizes them into a comprehensive and simple plan of sensory compartments and core neuropil. The main section of the book illustrates serial sections through the brain of the fly *Musca domestica* with reference to a coordinate system that relates covert structures to the head capsule. There follow detailed drawings of the forms and locations of Golgi-stained elements in the brain. The introductory chapters summarize the history of insect neuroanatomy, sketch the cellular constituents of neuropils, and outline the neuropil's basic organization. There is an appendix on histological methods applicable to insects and a multilingual glossary of terms relating to brain structures. The atlas is richly illustrated with 160 carefully prepared photographs and many beautiful drawings.

Springer-Verlag
Berlin Heidelberg New York

R.G. Kessel · C.Y. Shih

Scanning Electron Microscopy in Biology

A Students' Atlas on Biological Organization

22 figures, 132 plates, XI, 345 pages
Corrected Reprint. 1976

Contents: One-Celled Organisms.–Cells in Culture.–Prokaryotes.–Fungi and Algae.–Multicellular Plants.–Organ Systems of Angiosperms.–Multicellular Animals.–Tissue and Organ Systems of Animals.–Development.

The recent development of the scanning electron microscope has resulted in increasingly wider use of this instrument as a research and teaching tool in the biological and natural sciences. This altlas covers theory and techniques, protozoa, cells in culture, prokaryotes, myxomycetes, multicellular plants, organ systems of angiosperms, multicellular animals, tissue and organ systems of animals and development. In each case, the emphasis is on the view obtained with the scanning electron microscope, since this information sometimes represents new knowledge or expands and clarifies previously existing information.

The major significance of this well-accepted book is that it gives coverage in a phylogenetic and systematic way to both the animal and plant kingdoms. It also attempts to disseminate knowledge of recent acquisition to students beginning their study of biology and medicine.

Because of the broad information coverage, in terms both of illustrations and text this atlas has proven to be particularly useful to students and teachers of general biology, including cell biologists, anatomists, zoologists, botanists, entomologists, and electron microscopists.

Springer-Verlag
Berlin Heidelberg New York